Building the Most Complex Structure on Earth

Building the Most Complex Structure on Earth

An Epigenetic Narrative of Development and Evolution of Animals

Nelson R. Cabej

Department of Biology

University of Tirana

Tirana, Albania

AMSTERDAM • BOSTON • HEIDELBERG • LONDON • NEW YORK • OXFORD
PARIS • SAN DIEGO • SAN FRANCISCO • SINGAPORE • SYDNEY • TOKYO

Elsevier
32 Jamestown Road, London NW1 7BY, UK
225 Wyman Street, Waltham, MA 02451, USA

First edition 2013

British Library Cataloguing-in-Publication Data
A catalogue record for this book is available from the British Library

Library of Congress Cataloging-in-Publication Data
A catalog record for this book is available from the Library of Congress

ISBN: 978-0-12-401667-5

For information on all Elsevier publications
visit our website at store.elsevier.com

This book has been manufactured using Print On Demand technology. Each copy is produced to order and is limited to black ink. The online version of this book will show color figures where appropriate.

To my wife, Nikoleta

Contents

Preface

In this book, I intent to succinctly present my theory of the epigenetic mechanisms of evolution and development. By making its presentation accessible to a wide range of readership, I hope not to have sacrificed scientific rigor.

An earlier extensive substantiation of this theory and its biological ramifications in *Epigenetic Principles of Evolution* (2008, 2012) allows me to focus on the principal ideas—and present only the most representative empirical evidence in support of the theory—in this work. I have also expanded my epigenetic view of development and evolution to the world of unicellulars and marginally to the plant kingdom.

Herein, epigenetics is dealt with as a biological discipline on its own rather than as a branch, or frontier area, of genetics or any other discipline. I extend the epigenetics' object beyond the classical areas of the DNA methylation, histone acetylation, and chromatin remodeling, which in this book are considered downstream elements of signal cascades at the systemic level.

I include in epigenetics the vast areas of the nongenetic mechanisms of reproduction, growth, cell differentiation, development, and evolution. It is in this broader context that epigenetics promises to be the genetics of the twenty-first century.

As it occurs often in the study of biological phenomena, an overlap of epigenetics with the objects of the study of other disciplines is unavoidable. So, for example in studying epigenetic mechanisms of homeostasis, its object overlaps with that of physiology, and in studying mechanisms of reproduction and development, it overlaps with the disciplines of genetics and endocrinology.

At the core of my comprehensive vision of epigenetics is the concept of the epigenetic control and regulation of gene expression to include changes in the developmental mechanisms that produce evolutionary novelties without changes in genes. In this vision, epigenetics is function of the integrated control system, which, I think, represents the core of the epigenetics as the new northstar of the biological research.

The predominant genetic approach to the study of inheritance has naturally led to the use of the adjective *genetic* synonymously to *inherited*. By expanding the field of study of the phenomenon of inheritance to nongenetic mechanisms, epigenetics imposes a reconsideration of the synonymous use of these words. In our time, biological inheritance is a shared object of study where epigenetics and genetics come into a complex, but clear relationship, characterized by epigenetic control and regulation of genetic functions and structures.

My son's, Redon, help in preparing the manuscript of this book can hardly be overestimated.

1 Control Systems in the Living World

The Nature of Living Systems

Although biologists are still arguing about the nature and definition of life, humans have always been able to distinguish intuitively between living and nonliving things. It is obvious that the Matterhorn mountain is inanimate, while the tree alongside it is alive (Figure 1.1).

We viscerally distinguish between a mountain as a nonliving entity and a tree as a living thing. The key difference is the biological pattern of the tree, which is repeated regularly. Matterhorn is unique, naturally unrepeatable, and there is no Matterhorn pattern. This unmistakable biopattern is unique to living things.

But, ill defined as it is, the "biopattern" may not always be a reliable indication of life, and most will agree that the pictures below may defy our perception of biopattern and animal pattern. At first glance, one may not notice the difference between a stick insect and a dead twig (Figure 1.2) or a living one (Figure 1.3), but our doubts disappear as soon as we see them move or react to our touch. Our gut instinct, again, is that only living things are irritable and display motile avoidance behavior. Besides, humans also always knew that living things grow and reproduce.

Aristotle ('Αριστοτέλης, 384–322 BC) used a similar empirical approach 23 centuries ago to develop his classification of the living world, which was based on visual perceptions of living things and phenomena. He believed both living and nonliving things *exist* and are distinguished by the *soul*, which is found only in living organisms. Aristotle characterized the plants as living entities with souls, which enables them to *grow* and *reproduce*; in addition, an evolved type of soul allows animals to *perceive* the external world, *move*, and react to it instinctively, while the "human soul" enables us to do everything animals do, plus *use logic* and *think*.

Since Aristotle, biology has made considerable progress in attempting to know the structure and function of living systems and understand their essential properties behind the visual perception. Two centuries after the discovery of the cell by the English scientist Robert Hooke (1635–1703), zoologist Theodor Schwann (1810–1882) and biologist Matthias Jakob Schleiden (1804–1881) suggested that cells were the basic units of all living beings, unicellulars and multicellulars, animals and plants. This great generalization had a profound heuristic effect on biological studies and represents a landmark in the development of biology and in its gradual transformation from a descriptive into a causal science.

The progress in the study of the cell and living systems in general has created a detailed picture of high organization and functional complexity. Life is a process that biological systems have to perform in order to maintain and to perpetuate, via reproduction, their highly improbable structures. Life is an inseparable manifestation of

Building the Most Complex Structure on Earth. DOI: http://dx.doi.org/10.1016/B978-0-12-401667-5.00001-8

Figure 1.1 The Matterhorn in the Swiss Alps and a tree.

Figure 1.2 A stick insect on a dead branch.

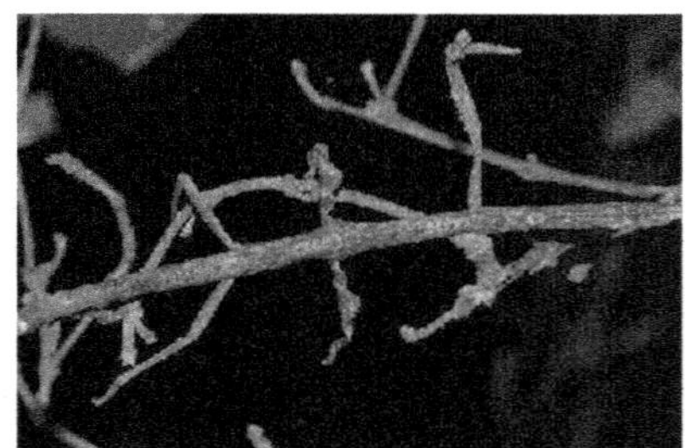

Figure 1.3 A stick insect on a living twig.

the existence of living systems, and all living systems, from a unicellular prokaryote to a human being, have in common several essential properties.

Living Systems Have Clear-Cut Boundaries

Living systems build clear boundaries that separate them from the environment, thus determining the range of action or "territoriality" of the organism's homeostatic mechanisms. On the inner side of the boundary is the system; on the outer is its surrounding. In unicellulars, this boundary is represented by the cell membrane. In multicellulars, it is represented by skin/integument (animals) or bark (plants). Living cells build boundaries to control the flow of matter and energy rather than isolate themselves from the environment. The cell membrane is an integral part of the living system. It is designed to allow for the controlled intake of nutrients and the excretion of waste and nonusable energy (heat), while preventing the free diffusion of solutes that is necessary to maintain differences in concentration between the cell and its environment. It represents the front line for the antientropic drama of the living system, to vanquish thermodynamic forces of disorder and to build, maintain, and perpetuate its physically improbable structure.

Metabolism

In performing their vital functions, living systems obtain energy by breaking down nutrients, depleting their reserves of matter and free energy (catabolism). In order to maintain their structure and function, living systems have to compensate for the loss by synthesizing the lost components through the nutrients they take in with food (anabolism). The equilibrium between the catabolism and anabolism in living systems represents their normal metabolism, which enables them to maintain a state of dynamic material and energetic equilibrium.

Since the seventeenth century, metabolism has been considered a defining property of living systems (viruses are metabolic parasites, and it is the host cell that goes astray to produce its own killers). The maintenance of the structure and functions of the cell require spatiotemporal coordination of a multitude of anabolic and catabolic reactions occurring in the living cell. But the maintenance of the naturally eroding cell structure requires that the cell somehow "knows" or has information on the structure to be maintained and does the species-specific work at the right places and at the right times within the cell's nanospace.

How does the cell accomplish this? If metabolism is understood to be the work that the cell does to retain its structural identity, then how does the cell get the information on the changes occurring in the system, how does it detect the deviations from the norm, and, finally, how does it generate instructions to restore the normal structure and send them to the changed structures? But if the living cell is not controlled by external forces, as is clearly the case, metabolism implies the presence of a built-in control system.

Control Systems Are Prerequisites of Living Systems

A cell is a supercomplex microscopic structure that performs thousands of reactions coordinated perfectly in space and time every moment, and from the perspective of physics, it is clearly an improbable construction. Nevertheless, it survives and perpetuates its structure via reproduction. Theoretically, there are two alternative ways that the living cell might accomplish this marvel of nature's biotechnology; all these reactions are spontaneously coordinated, or, alternatively, the cell has evolved a control system to coordinate that myriad of chemical reactions within the cell.

The first possibility, that thousands of spatiotemporally precisely coordinated biochemical reactions within the cell can spontaneously occur, seems next to impossible. *Emergence* is a descriptive term that does not explain or help us explain anything. The same can be said of self-regulation. Loose as they are, these terms only avoid the questions on how cell structures arise and are maintained.

The alternative explanation of the phenomenon is that the wonderful spatiotemporal coordination of many thousands of chemical reactions occurring in a cell, and many more in a multicellular organism, are controlled by a specialized system. Human experience shows that even the simplest artificial devices or machines cannot function without control systems involving continued human supervision and regulation of material and energetic supply or built-in control systems.

An example of a simple, one-variable control system with a built-in controller is a thermostat used for regulating a room's temperature. The control system consists of a thermostat and a furnace. The thermostat is the controller of the system; it receives information on the temperature of the room via a sensor (thermometer) and compares it to the desired temperature, a selected set point. When the temperature is lower than the set point, the thermostat switches on the circuit, which causes the furnace to produce heat. When the temperature exceeds the set point, the circuit opens, and the furnace switches off until the temperature falls below the set point again, and the cycle repeats. But if regulation of a single variable, room temperature, cannot be achieved without a control system, what should one think of incomparably complex systems such as living cells or multicellular organisms, which have to both control and regulate thousands of different variables, in differential patterns in tens or hundreds of different cell types? The control system is a *sine qua non* of the existence of all living organisms. The emergence and evolution of living systems are inseparable from the evolution of the control system; the evolution of complex animal structures and functions is associated by a parallel increase in the complexity of the control systems.

If a control system with a controller is necessary for regulation of a single variable such as the temperature of a room, it is absolutely necessary to regulate hundreds and thousands of variables coordinated in time and the nanospaces of a cell. In multicellular animals, the development and maintenance of normal structure are a function of an integrated control system (Figure 1.4). It is a hierarchical system of controls on several levels of organization, in which higher levels of control impose restrictions on lower levels to minimize the noise in the transmission of information downward to the cell level, where gene expression is regulated and patterns of gene expression are determined.

The continued evolution of control systems increased the independence of living systems from their environment, and as a rule, the degree of complexity of a living system parallels the degree of the complexity and sophistication of the control system. More complex systems require more complex and sophisticated control systems.

Recognition of the presence of a control system that is capable of maintaining the normal structure of the organism implies that it "knows" what the normal structure is. But if it has information about its own structure, there is no reason to doubt that it is capable of transmitting it to its offspring.

Biological Reproduction

However successful they are in maintaining their normal structure, living systems have to succumb to the thermodynamic forces of disintegration and decomposition sooner or later; their life expectancy is temporally limited, varying from minutes in unicellulars to thousands of years in trees such as olives. Yet life on Earth has been prospering and evolving for more than 3 billion years because living systems invented a special "trick" of circumventing the second law of thermodynamics. In order to avoid their unavoidable demise, they live or subsist long enough to reproduce themselves before dying. The progeny will also be engaged in the same

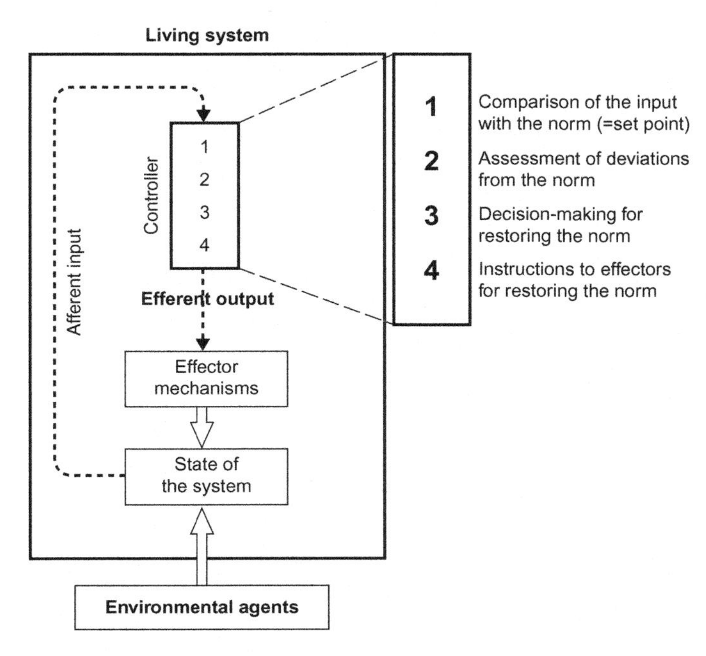

Figure 1.4 A generalized and simplified diagram of the integrated control system in metazoans, with a central nervous system (CNS) acting as controller of the system. Metazoan structure degrades continually due to intrinsic, thermodynamically determined causes, as well as a result of adverse influences of the environment. Changes in the structure and function of the organism and environmental changes are monitored by a pervasive network of interoceptors and exteroceptors and communicated to the CNS. In the CNS, the afferent input is compared to the neurally determined set points (1). Deviations from the norm are identified (2), and pathways for restoring the norm are determined (3). "Instructions," or commands for restoring the norm (4), are sent to effectors (pituitary, target endocrine glands, or cells in target tissues) through signal cascades. Via the molecular and afferent feedback input, the controller receives continual information on the restored/degraded state of the system (Cabej, 2012).

struggle against the hostile thermodynamic forces, but ultimately it will give in to these forces after producing progeny and so forth, in recurring cycles of reproduction that enable the species to survive at the expense of the individual. Thus, a species' existence is perpetuated by sacrifices of individual lives, or the species owes its existence to the reproduction of mortal individuals.

Given the exceptionally high degree of the functional and structural complexity of living systems, unrivaled by anything existing in nature or ever created by humans, their capability to reproduce is far from self-explanatory. The genome or genetic information, as an answer, is out of the question for obvious qualitative and quantitative reasons (see later in Chapter 2, section "Is there any program in the genome?"). Erecting animal structure requires more than the production of proteins, which are the only products genes are known to account for. An animal organism is more complex than a bag of proteins. Cells, not proteins, are the basic unit of life. Under no circumstances can a protein, gene, or genome, or any combination of them, function as a living system; only a cell can (or at least a cell of a unicellular organism can).

As pointed out earlier, the ability of the control system of an organism to monitor the structure, to identify deviations from the norm and send instructions for restoring the norm, clearly shows that it has information on the structure and is capable of transmitting that information to the offspring.

Unicellulars reproduce via binary fission or sexual reproduction. In strict terms, their reproduction does not fit well into the conventional mother-to-daughter reproduction scheme. Cell division does not produce two "daughter" cells; the result of the division is only two cells, rather than three. It would not be correct to say that it produces one daughter cell, because the semiconservative mechanism of cell duplication, from both the genetic and epigenetic viewpoint, makes it impossible to determine which of the cells resulting from cell division is the mother or daughter. Hence, each of the resulting cells is the twin of the other. Cell division, thus, leads to production of two "twin cells." Each of the twin cells is equally ancestral to, and descendant of, the other. The distinction can be clear only at the generational level, where we can speak of successive generations or ancestral and descendant generations.

From this viewpoint, unicellular forms also defy our traditional dichotomic concept of death and life—unicellular forms of life virtually are potentially immortal—the dividing cell does not die, but half-lives in the structure and functions of the two cells of the next generation. This semantic aspect aside, what is essential is the fact that in unicellulars, the two sister cells after division are fully capable of independent life and reproduction, which is in marked contrast with what is observed in multicellular organisms.

The epigenesis-preformation dichotomy also seems hardly applicable in the case of the reproduction of unicellular organisms. Transformation of an original organism into two implies quantitative change; that is, duplication of existing cell mass, including duplication of the genetic information and epigenetic information contained in epigenetic structures of the unicellular organism.

At the multicellular level of organization, biological systems shifted from the reproduction mode of unicellulars. They do not reproduce semiconservatively, producing a twin of themselves within the existing system and sharing its original structure with it, as unicellular organisms do. The basic mode of reproduction in multicelllulars—sexually and nonsexually reproducing multicellulars alike—is via gametes. Theirs is an epigenetic mode of reproduction that comprises both qualitative (biological development) and quantitative (growth) changes via the sequential steps of cell differentiation, organogenesis, and morphogenesis of the egg/zygote into an adult organism.

The idea that multicellulars produce copies of themselves is also controversial. What we actually observe are gametes, eggs, and sperm cells (eggs only in parthenogenetic organisms), or newborns that follow an *independent complex development* until they become "copies" of their parents. Parents provide gametes with the epigenetic information necessary to develop to an early embryonic stage, the phylotypic stage, when only one organ system, the nervous system, is operational. At this early embryonic stage, when the maternal epigenetic information provided to the embryo via gametes is exhausted, the CNS is already capable of stepwise computation of the epigenetic information necessary for the development of the adult metazoan supracellular structures.

The capability of living systems to reproduce their kind leads to two other properties of the living systems: evolvability and growth.

Evolvability

This is a relatively new biological concept, and its definition depends on the question addressed (Pigliucci, 2008). In this context, evolvability is the ability of living organisms to adapt their phenotype by changes in developmental pathways. Since evolutionary changes occur in the process of development, the evolution and evolvability of living systems is related to, and is enabled by, the phenomenon of biological reproduction. Evolvability is thought to evolve (Kirschner and Gerhart, 1998), as is clearly indicated by evidence of the acceleration of the rate of evolution.

Genetic changes are too rare and overwhelmingly deleterious to account for the huge diversity of forms in the living world. The prevailing idea that changes in genes are necessary for the evolution of living systems is challenged by numerous biological phenomena. The concept of the phenotype as a result of the interaction of genes with the environment fails to explain how, in concrete terms, a change in a gene or DNA can produce an *adaptive morphological change*. I emphasize the word *adaptive* because it is well known that mutations in genes can lead to phenotypic changes at the molecular level; that is, deleterious changes *sensu* Archibald Garrod's (1857–1936) "inborn errors of metabolism." Genome sequencing of various species of unicellular and multicellular organisms, conservation of the genetic toolkit, biological phenomena such as developmental plasticity (intragenerational developmental plasticity and especially transgenerational plasticity), reversion of ancestral morphological characters, metamorphosis in invertebrates and vertebrates, cell differentiation, loss of morphological characters, etc., suggest that it is not changes in genes or DNA, but epigenetically determined changes in patterns of gene expression in the process of individual development that may be responsible for evolution of structure and morphology.

Growth

In unicellular organisms, growth is a stage in the process of their reproduction. It consists of a stepwise and ordered increase in the size of the cytoplasm, including the increase in the number (e.g., ribosomes mitochondria) or duplication of organelles, (chromosomes, centrosomes, cell nuclei, etc.). In multicellular organisms, as the founders of the cell theory determined almost two centuries ago, growth consists of the growth of the number of cells in the process of development, comprising prephylotypic development, histogenesis, and organogenesis.

Homeostasis and Adaptability

Both these properties of living systems are regulated by the control system. In order to function normally, unicellular and multicellular organisms must maintain a relatively constant internal environment under considerably varying conditions in the external environment.

Homeostasis and mechanisms of its regulation are better known in mammals, especially in humans. These organisms maintain the level of many physicochemical parameters in their fluids constant (see later section "Homeostasis - The maintenance of Constant Internal Environment is Prerequisite of Living Systems", in this chapter). Among the most important homeostatic parameters they regulate are body temperature, pH, levels of electrolytes, hormones, growth factors, secreted proteins, etc. So, for example, in winter when the external temperature falls, in warm-blooded animals, a specialized part of the controller of their control system, the hypothalamus, activates neuroendocrine mechanisms to increase heat generation. On the other hand, in the summer when the environmental temperature rises above the normal body temperature, it activates mechanisms that reduce heat production and increase heat loss. Living organisms can respond adaptively to changes in the environment with changes in their structure, function, behavior, and life history that tend to neutralize the harmful effects of environmental changes or agents and maintain homeostasis. Numerous described cases of developmental plasticity, both intragenerational and transgenerational (see Chapter 4), illustrate the high capability of living systems to adapt and survive even under unfavorable environmental conditions. The adaptability increased the independence of living systems from their environment.

The above properties are unique to living systems, and the control system is fundamental to all of them: reproduction, growth, evolution, adaptability, and evolution. An organism can live without reproducing, evolving, or growing for considerable spans of time, from a few minutes for a bacterium, to one century for a human being, to thousands of years for some trees. But any human being would almost instantly perish in the absence of the control system. From this view, all the essential properties and functions of living systems are subordinate to the control system.

The Principle of Entropy and Erosion of Material Structures

From experience, anyone knows that inanimate objects always tend to reach stabler states. We consider quite natural to see a book fall from a shelf, a brick from a wall, or a shingle to the ground, but we do not expect the fallen book, brick, or shingle to go back to their original positions on the shelf, wall, or roof, respectively. Such miracles could happen in the world of myth and fiction but not in the real world ruled by the rigorous laws of nature. We consider the above objects to have found their stabler or more probable, although less-ordered, states.

All the above-mentioned examples are unidirectional and less-ordered states, and the spontaneous return to the original ordered state is impossible. But exceptions exist. Nests of birds and beaver dams are ordered structures, artifacts that cannot

Figure 1.5 A complex termite mound in Cape York, Australia.

arise naturally. Termites build mounds, which clearly are structures that cannot occur spontaneously (Figure 1.5). They build their mounds above their subterranean multichambered nests, and the mounds may be of different sizes, shapes, and heights, with complex mazes of tunnels and shafts used for ventilation. But whether a bird's nest, a beaver dam, or a termite mound, left under natural conditions, sooner or later all of them are doomed to lose their order and break down.

Over centuries, human civilizations around the globe have added order and created naturally highly ordered structures by investing work and information. Yet observe what remains of Athens's ancient Acropolis, or of thousands of remnants of ancient and prehistoric works of art, inhabited centers, fortifications, and castles. No one could expect that under natural conditions, the heads of the US presidents carved in granite on Mount Rushmore will remain as originally sculpted by Gutzon and Lincoln Borglum (Figure 1.6) 60 years ago. In fact, fractures in the granite have already occurred. No spontaneous process or event can improve the structure or function of a Porsche; only the opposite is possible. Since the probability of occurrence of less-ordered states is infinitely greater than the ordered state, less-ordered states are more probable, and hence statistically more stable. This explains the observation that all objects in nature tend to reach stabler states: stabler states are statistically more likely to occur.

Such observations of the natural trend of the loss of order in nature, which goes as far back as the origins of humanity, found a theoretical explanation only around the second half of the nineteenth century with the discovery of the second law of

Figure 1.6 Mount Rushmore National Memorial.

thermodynamics, which, in the opinion of many scientists, may be the most universal law governing the universe. Central to the second law of thermodynamics is the principle of entropy. For the purpose of this discussion, a simple and classical definition of the law is:

> *In an isolated system, only processes that are associated with increase of entropy can occur.*

The movement of molecules of a gas under moderate temperatures may be described as disordered, but molecules of water are less so, while nitrogen bases in DNA are highly ordered. Working on a mechanical theory of gases, Rudolf Clausius (1822–1888) coined the term "transformation content" (*Verwandlungsinhalt*) or entropy (from the Greek *en+trope*—"toward"+"turn") for describing the direction of the flow of heat from a hotter compartment of a gas system to a colder one. This leads to a state of equilibrium in the system where the temperature (the average speed of gas molecules in the whole system) equalizes. Reversion to the original state (i.e., segregation of high-speed gas molecules from low-speed ones in the system) cannot occur. Extrapolating this to the universe, Clausius concluded that the entropy of the universe tends to be at a maximum. Later, Ludwig Boltzmann (1844–1906) popularized the idea of entropy as a measure of disorder in a system. He tried to explain why material systems tend to increase their entropy or states of disorder; this is because the number of *disordered states* in a gas system is infinitely larger than the *ordered state*. Accordingly, the stabler state is the one that the system has the highest probability to reach.

Boltzmann tried to explain why the entropy increases in one direction only: past→present→future. To account for this time asymmetry of entropy, he suggested that in the endless universe, isolated spaces still exist, which have not reached the equilibrium state (maximal entropy) and are still increasing their entropy. Boltzmann reasoned that since the probability of disordered states in these "isolated spaces" is incomparably higher than the ordered one, it follows that these spaces will continue to increase their entropy as time progresses. He argued that the direction of the increase in entropy in the part of the universe we live in might determine our perception of the movement of time from the present to the future, not the present back to the past. Accordingly, in regions of the universe where the entropy is maximal, time may be symmetrical, or there should be no time direction.

First, let us elucidate the meaning of *entropy* from Boltzmann's physical view as a measure of atomic disorder. Under moderate temperatures, water molecules are more disordered than in lower temperatures when they transform into ice, where atoms are arranged in highly determined spatial patterns in three-dimensional crystal lattices. But if conditions change (e.g., if the temperature rises above the melting point or if the ice crystal plunges into water), the crystalline structures vanish; atoms dissociate from the lattice and begin moving randomly. In the liquid state, the order of the crystalline structure is lost, and thus, the system increased its molecular disorder or entropy. In both cases, the temperature rise increases the entropy or the disorder of the system. The opposite will occur when the water temperature falls, and it can be imagined that by lowering the temperature to absolute zero (i.e., −273.15°C (−459.67°F or 0 K)), the entropy of the system will decline to zero (when virtually no atomic movements occur) and order will be maximal.

The second law of thermodynamics concerns isolated systems that exchange neither matter nor energy with the external world. Since such systems exist in human imagination rather than in nature, it follows that the second law does not forbid the increase of order or decrease of entropy in open systems. Living systems, by definition, are open systems (i.e., out of the realm (range of action) of the second law), and they do not mind if open systems find ways to decrease their entropy. This is what living systems actually do, although not very honorably (fairly)—they do it by stealing the neighborhood's existing order.

It is this restricted application of the second law that gives rise to the increase of order in nature; the growth of vegetation, animals, and even human population and growth of ecosystems in general. Universal as it is, the second law has its loopholes. More soberly, living systems abide by, rather than defy, the second law, which is valid only for isolated systems, the category to which only the universe may belong.

Living systems are highly ordered at all levels of organization: molecular, cellular, and supracellular (of tissues and organs), as well as at the organismic level. In view of the fact that the ordered state of a complex system is statistically highly improbable compared to the infinite states of disorder, from a physicist's view, living systems represent highly improbable structures. But the improbable nature of living systems implies a low thermodynamic stability, which is the cause of the continuous erosion of their structure. Regardless, living systems fare well and have inundated the Earth, both at sea and on land. They survived, flourished, and evolved to rule the inorganic world.

Can material systems overcome the thermodynamic barrier of the second law? In 1871, Maxwell designed a famous thought experiment to investigate (explore) how entropy could be decreased in a system: two chambers of a container filled with gas are connected by a door. A demon standing there opens the door so that only faster-moving gas molecules may pass into one of the chambers. The temperature will increase in this chamber but will fall in the other. The demon, thus, ideally, would violate the second law by displacing the system from the initial state of equilibrium to a state of nonequilibrium, with different temperatures of gas in both chambers. This is not done free of charge; the price the demon has to pay is to invest work and information into the system (for now, we ignore the fact that the information used and work done also increase the entropy of the system!). Thus, Maxwell's thought experiment suggests that the second law can locally be avoided by a demon willing to spend information and energy to do the work.

Living systems' ability to maintain their structure, in apparent defiance of the second law of thermodynamics, has long amazed and perplexed scientists. What makes living systems so extraordinary and unique? How do they succeed in maintaining their improbable structures for considerable periods of time? Where in the living system does Maxwell's demon reside?

In *What Is Life?* the Austrian physicist Erwin R.J.A. Schrödinger (1887–1961) (Schrödinger, 1944) dealt with this issue and popularized the concept of entropy in living systems. Echoing the idea of his compatriot Boltzmann, he introduced the concept of *negative entropy*, shorterend to *negentropy* by Louis Brillouin (1854–1948), which is the opposite of the entropy in the system. He attempted to resolve the enigma of the existence and maintenance of the thermodynamically improbable structure of living organisms by assuming that they "suck" order or negentropy from the environment. In other words, they use external sources of order to compensate for the order they inescapably lose, thus avoiding structural degradation. Living systems solved life's dilemma of "to be or not to be" by acquiring order from sources in their environment. Plants, both unicellular and multicellular, succeeded in evolving a mechanism of photosynthesis for generating their order by using sun radiation and CO_2 to synthesize polysaccharides, whereas animals choose to use amino acids, carbohydrates, lipids, vitamins, and other elements of the plant kingdom. Accordingly, the ultimate source of information and order in the biosphere is the difference of temperature between the sun and Earth (it is no wonder that most people in antiquity worshipped the sun as their most important deity).

With the benefit of hindsight, we know that Schrödinger's pioneering idea is incongruous with some basic physiological knowledge. Most nutrients contain no *utilizable* order of the specific type of metazoan needed to build their structures. What an animal utilizes from the ingested food for generating its species-specific molecular and cytological building blocks is the *raw material* and the *free energy* it contains. The organism is a generator of its own order rather than a consumer of external order. The first step that an animal takes after eating food is to break most of the ingested organic compounds down to simpler, low-molecular organic compounds (amino acids, monosaccharides, fatty acids, glycerol, nitrogen bases, pentoses, etc.). Destroying the original order, it creates its own order by synthesizing *de novo* species-specific proteins, carbohydrates, lipids, and other compounds resulting from the activity of its proteins. Living systems are not stores of extrinsic order and information; they transform the order they get from environmental sources into their own species-specific order. As mentioned earlier, this is done at the price of spending free energy to do the necessary work to synthesize polysaccharides and protein molecules. For living systems, the acquisition of the external order would be worthless if they did not "know" how to convert it into usable form. As an old Albanian adage says: Knowing trumps having.

What Does It Take to Build Improbable Structures?

Improbable structures are complex, but as Orgel (1973) argued, although necessary, complexity is not a sufficient condition for biological improbable structures. Unlike inorganic structures, biological structures are specified structures that cannot arise by chance. Although it is very difficult to define what an improbable structure is, human beings have some idea about it, and an intuitive definition of an improbable structure would be "a complex specified structure that cannot arise spontaneously, without some kind of biological, unconscious, or conscious, effort." The more ordered a structure becomes, the more improbable it becomes.

The production of improbable structures of high structural order involves antientropic processes. For such processes to occur, some kind of work needs to be done, implying that free energy must be used. The relatively abundant sources of free energy are constantly doing work on the Earth's surface and interior. But despite the huge amount of energy the nature discharges on Earth, there is no obvious increase in the inanimate world's order, although quasi-improbable ephemeral (transient) structures arise in nature. And the reason is simple: the work done by natural forces is undirected.

Nature's Artworks

Ordered structures sometimes arise in nature because of the actions of the various sources of energy, such as sea waves, winds, and rivers. Spontaneous forces of nature, such as orogenic activity (formation of mountains) and erosion, may also "create" different kinds of order, such as the undulating traces created by sea waves on beach sand, oval pebbles on riverbeds, or the Balanced Rock in Arches National Park, Utah (Figure 1.7). From a human aesthetical perspective, we call them "nature's artworks."

As pointed out earlier, Maxwell's demon's work is goal-directed, and the result of its work is predictable; the order of the system will increase. That is, the gas in one part of the system will be hotter than in the other. In contrast, no "goal" motivates the actions of natural forces; hence the result of their actions is unpredictable. The randomness of the action of natural forces in a world dominated by entropy-increasing processes reduces the probability of the spontaneous rise of improbable structures in the inorganic world. Wasteful in terms of its colossal forces, inanimate nature has only a slim chance of producing improbable structures, and even when such structures arise, the degree of their complexity is only moderate. Despite the aesthetical values of "nature's artworks," it is virtually impossible that the process

Figure 1.7 Balanced Rock Reflection by Kevin Mikkelsen (from http://www.panoramio.com/photo/8294824).

of orogenesis (formation of mountains) could create Mount Rushmore. Nature's artworks are not a match for the ordered structures created by humans.

Artificial Improbable Structures

Despite the natural trend toward disorder, humans continually add new order on Earth by investing work and knowledge (information) into nature. Like Maxwell's demon, before starting to do or create something, any human artisan, designer, or constructor sets the goal and form the mental pictures of what he or she wants to create. Hence, in the beginning, there is the idea of doing something, followed by the work and knowledge to actualize the idea.

An essential difference between nature's artworks and artificial structures is that the first arise by chance and, thus, their form is unpredictable. For instance, the continued action of natural agents (temperature fluctuations, rain, snow, wind, etc.), gradually eroded a piece of rock until it produced the Balanced Rock—an unpredictable product of chance. Its shape and appearance is determined both internally (by the structure and texture of the original rock) and externally (by the nature of the

Figure 1.8 The Great Pyramid of Cheops.

eroding factors). By contrast, artificial structures, determined by the creator, are predictable; they are products of forces acting against the second law, in the direction of the decreasing entropy.

Human civilizations created highly ordered structures. There are not only "seven wonders of the world," but inhabited centers, roads, cultivated fields, industrial production centers, etc. For thousands of years, human civilizations around the world, to encourage better living conditions and to exploit nature for their own benefit and satisfy their aesthetic, intellectual, and religious needs, created a world of material culture full of structures that could never arise naturally: homes, palaces, pyramids, and other buildings.

Even the wildest human imagination could not dream of a pyramid sprouting spontaneously by a fortuitous combination of the stones they are built from. Human work and design have been indispensable in erecting these mystifying structures. The Pyramid of Cheops (Khufu) (Figure 1.8) was built more than 4,500 years ago. About 2.5 million stone blocks and 20 years of work of about 200,000 slaves were used to erect it. However, the building blocks and the labor of slaves were not all that was needed for erecting the pyramid; slave workers did not know how to build a pyramid that would lead the pharaoh's soul to the realm of the ancient Egyptian gods. A detailed construction plan by Hemon, the Khufu's architect and high official, was the first step in the pyramid's construction. What essentially took to build the Khufu (or any other artificial structure, for that matter) are building blocks (matter), work (free energy), and knowledge (information).

Similarly, work and information were used for transforming rock (matter) into Mount Rushmore National Memorial (Figure 1.6) and for building Stonehenge (Figure 1.9). What enables us to intuitively distinguish between natural artwork such as the Balanced Rock (Figure 1.7) and a work of art that is fashioned from rock? We viscerally recognize that Stonehenge (Figure 1.9) and the pyramids of Egypt or Mesoamerica are improbable structures that could not have arisen by themselves.

What makes it easy for us to understand that the first is a natural formation and the second is a product of human skills? It is not the degree of complexity, for the Balanced Rock structure is just as complex as a stone in the Stonehenge monument. Our intuitive ease at distinguishing an artificial structure from nature's artworks is a prescientific "thermodynamic lore" that makes us feel that Stonehenge stones cannot rise spontaneously at the top of other stones.

Figure 1.9 The Stonehenge monument in the county of Wiltshire, southern England.

Biological Improbable Structures

Living systems, from unicellulars to human beings, are also improbable structures of incomparably greater complexity. They represent the most complex structures that we know. Who would expect a cell to arise spontaneously from a pile of its building blocks (DNA, proteins, carbohydrates, lipids, water, etc.)? No one would claim the Great Wall of China spontaneously arose even if all the building blocks were available.

In clear distinction from nature's artworks, which are created by chance from the action of external agents, living systems arise predictably from built-in programs and the work of assembling them is done by forces intrinsic to them. While, like artificial structures, biological structures can be predicted from the work and information invested in creating them, they differ not only by the level of complexity but also from the source of the work and information used in building them. While artificial structures are products of the work and information that is extrinsic to these structures, biological structures are self-built and self-designed structures that do the work and generate the information necessary for erecting their structure.

Homeostasis

The Maintenance of Constant Internal Environment Is Prerequisite of Living Systems

Living systems exist as open systems in a state of nonequilibrium or dynamic equilibrium that continuously exchange matter and energy with the environment, while maintaining a relatively constant internal environment.

States of dynamic equilibrium are known to exist in natural systems as well. Examples include lakes whose water level and volume are kept relatively constant by a balanced input from rivers that flow into lakes, precipitation (rain, snow, hail, etc.), and outputs of water (i.e., by rivers that flow out of them, evaporation, leaking, etc.). The bed of the outflowing river may serve as a set point for maintaining

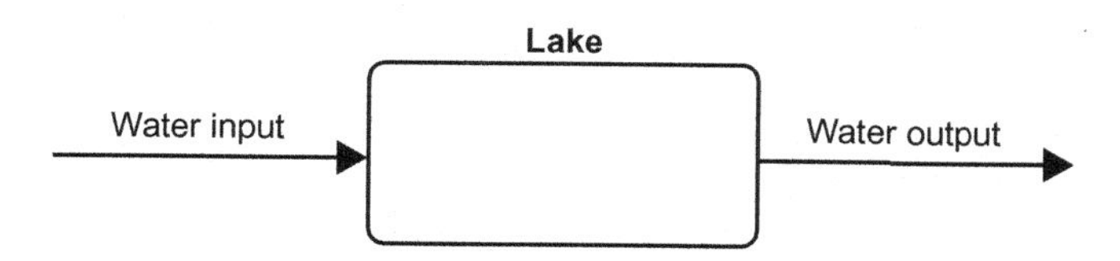

Figure 1.10 A diagrammatic representation of the open-loop control system of a lake.

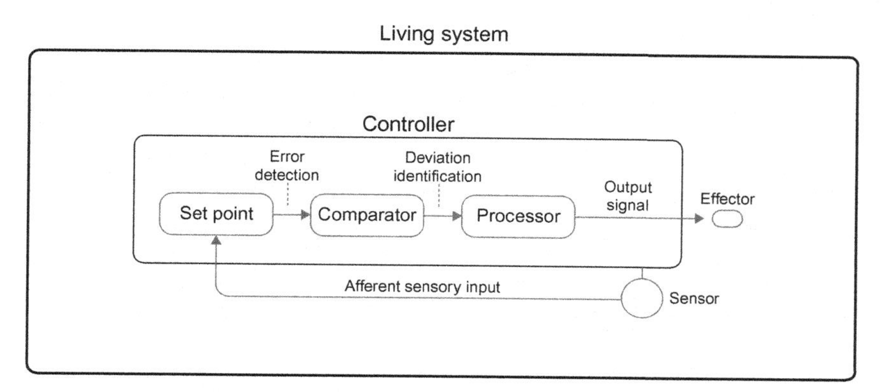

Figure 1.11 A diagrammatic representation of a closed-loop control system.

the relatively constant level of water in lakes. Yes, such lakes may be considered one-variable systems in a state of dynamic equilibrium. But there are essential differences between this and the dynamic equilibrium in living systems. First, the regulation of the dynamic equilibrium state of the lake as a water system is not a function of the lake itself, but of the level of the riverbed, which is extrinsic to the system, or at least is not determined by the system itself. This is an open-loop control system (Figure 1.10) in which the lake itself does not actively participate in maintaining a constant water level.

Living organisms are open systems exchanging matter, energy, and information with their environment, whose dynamic equilibrium is maintained by a closed-loop control system. The state of dynamic equilibrium of multiple variables in living systems is a function of a built-in control system, which determines the normal levels of variables, or their set points. While any change in the level of the riverbed may change the shape and size of the lake, changes in the environmental factors do not affect the size, shape, or nature of the system; the built-in regulator maintains the system variables within the normal ranges, so long as the environmental changes do not override the adaptive capacity of the organism. In an oversimplifed form, the closed-loop control system in living organisms looks as shown in Figure 1.11.

This leads us to Karl Ludwig von Bertalanffy's concept of equifinality. In a clear distinction from inorganic systems, where the initial conditions determine the final state of a system, in living systems, the final state may be reached even under different initial conditions. This is a state of "flowing equilibrium" (*Fliessgleichgewicht*), which implies maintenance of a constant ratio of the components of the system, despite variations in nutrition, as well as the ability of the system to restore its normal state after disturbances caused by external agents. Von Bertalanffy called such states *equifinal* (from the Latin *aequus*, "equal," and *finalis*,"pertaining to the end") and he invented the term *equifinality* to refer to the biological phenomenon (von Bertalanffy, 1950).

In order to sustain their vital functions, living systems need to create steady states (within certain limits) of their chemical and physical parameters, and they must maintain that state even under hostile environmental conditions. As shown in the preceding section, even without the influence of harmful environmental agents, the unavoidable degradation of structure determined by the thermodynamic forces of degradation will occur. In addition, the exceptional complexity of cells and multicellular organisms and the high improbability of their structure require an organism's internal environment be within ranges that allow normal functioning of the cell or organismic machinery in multicellulars.

The French biologist, philosopher, and author (dramatist) Claude Bernard (1813–1878) coined the term "*la fixité du milieu intérieur*" to describe the property of living systems to maintain a constant internal environment, which he considered to be "the condition for a free and independent life." Later, Walter Cannon (1871–1945), a Harvard physiologist, expanded the concept in his book *The Wisdom of the Body* (1932) and used the term *homeostasis* (from ancient Greek όμος (*hómos*), "similar," and στάσις (*stasis*), "stable") to describe the relatively constant state of the internal environment of living organisms (Cannon, 1963).

Homeostasis is loosely defined in different ways, but its classical meaning describes *both* the constancy of the internal environment (i.e., of the body fluids in animals or plants) and the biochemical and physiological processes that determine the constancy. In this broad meaning, it implies the constancy of the protoplasm in prokaryotic and eukaryotic unicellulars, including cells of multicellulars. Conventionally, in multicellular organisms, homeostasis is used to describe the chemical constancy of extracellular and body fluids (blood, lymph, etc.), which is necessary for the normal functioning of all the organism's cells. But each cell must also maintain a constantly normal internal environment. In line with my epigenetic theory of heredity and evolution, I use the term *homeostasis* in an even broader way, to include the maintenance of the integrity of the structure of both unicellular and multicellular organisms.

The control system (see the section "Control Systems" later in this chapter) is tasked with homeostasis. In multicellular animals, the control system closely monitors the homeostatic parameters and, based on various feedback loops, adjusts their level within the normal range, according to species-specific set points.

Most homeostatic parameters in animals are determined by the brain. The ubiquitous presence of the nervous system in the animal body allows the brain to receive current data on every part of the organism via afferent nerves. It compares the actual data with the set points to determine deviations from the norm, and, via efferent pathways, it sends instructions for restoring the normal levels of homeostatic parameters. In a simplified, generalized form, the control system of living systems (from unicellulars to multicellulars, both metazoan and metaphyta) consists of a controller that monitors the state of the system, compares it to the norm (set points), and sends instructions to effectors to maintain/restore the normal state.

The maintenance within the homeostatic ranges of thousands of variables in cells of unicellular and multicellular organisms is an incredibly complex undertaking. As typical examples of the physical parameters maintained at normal ranges under brain control are species-specific normal temperatures in warm-blooded animals and normal blood pressure. Thousands of chemical parameters, including the pH, are maintained at normal levels within the cells and in the body fluids. For example, glucose is a basic source of energy in animals. The amount of sugar we take with each meal would elevate the blood glucose to levels that are dangerous for the cell function and for human health. In normal people, this does not happen: our body senses immediately the higher-than-normal level of blood glucose and beta cells of the islets of Langerhans are instructed to secrete insulin, which transforms glucose into a glucose polymer (glucogen) that is stored in cells as a reserve source of energy. When the blood glucose level drops (hypoglycemy), autonomic nerves activate the secretion of the hormone glucagon, an antagonist of insulin, which, binding to its membrane receptor in liver cells via a complex pathway, releases glucose molecules from glucogen polymers, thus returning blood glucose to normal levels (Taborsky, 2010) (Figure 1.12). Glucose homeostasis is regulated by the autonomic nervous system that controls the secretion of insulin and glucagons by the pancreas, as well as the metabolic state of liver muscles and fat tissue (Thorens, 2011).

In all the cases studied as of yet, normal levels of homeostatic parameters in vertebrates are maintained by hormones according to signals that ultimately originate in the brain.

Maintenance of the homeostasis in a unicellular organism is less clear, and presently, we can speak only of speculative models of the control system and its controller (see the section "The Control System in Unicellulars" later in this chapter).

Control Systems

How do these complex and nonequilibrium structures, which are improbable from the viewpoint of physicists, arise in the real world? Improbable as they may be, living systems would perish of the increasing entropy that results from their own function, were it not for the antientropic mechanisms that they evolved in the course of evolution. They do not defy the second law because they cannot. But the antientropic mechanisms they evolved help them resist the thermodynamic forces of structural and energetic degradation temporarily until they grow up, mature, and produce offspring. These antientropic mechanisms only help living systems to buy the time necessary to perpetuate their structure in the offspring. Thus, while failing in the short term (the organism dies), the antientropic devices allow them to succeed in the long term by repeating through generations an endless “relay race” against the second law.

The biological antientropic devices are unique built-in control systems that maintain a state of dynamic equilibrium within living organisms and enable them to resist

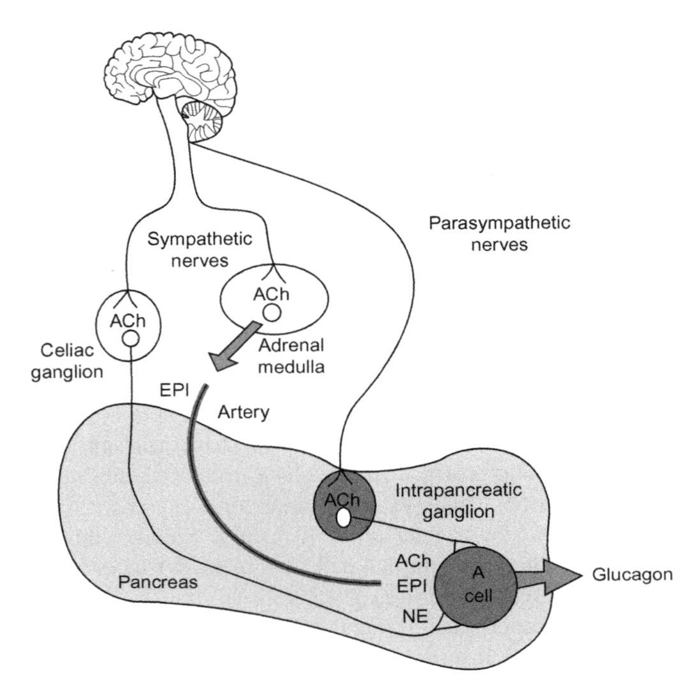

Figure 1.12 Three autonomic inputs to the islet alpha cell. Preganglionic sympathetic nerves travel in the spinal cord, releasing ACh from their terminals within the celiac ganglion. The released ACh activates postganglionic sympathetic neuronal cell bodies whose fibers innervate the islet alpha cells and release norepinephrine. Preganglionic parasympathetic nerves travel in the vagus, releasing ACh from their terminals within intrapancreatic parasympathetic ganglia. The released ACh activates postganglionic parasympathetic neuronal cell bodies, whose fibers innervate the islet alpha cells and release ACh. Preganglionic sympathetic nerves travel in the spinal cord, releasing ACh from their terminals within the adrenal medulla. The released ACh activates the chromaffin cells to release the sympathetic neurohormone epinephrine, which reaches the alpha cell via the arterial circulation. EPI, epinephrine; NE, norepinephrine (Taborsky, 2010).

and overcome the unavoidably increasing entropy of the system via feedback loops that control the output and stabilize their structure. Hence, the control system is an integral part of the living organism: "The existence and evolution of living beings is impossible to be imagined without a system of control and regulation" (Cabej, 1984). Even if by chance, or by human will and wisdom, a living structure arose in the absence of a control system, it would immediately perish, as do most fish on land. The emergence and evolution of control systems are inseparable from the emergence and evolution of life. Control systems increased the freedom of the whole (the organism) from the environment, at the expense of the freedom of the parts (organs, tissues, and cells) whose activity was placed under tight control and regulation.

Theoretically, control systems in living organisms may be integrated or separate for different parts of the living system. Integrated control systems are known to be operational in multicellular animals (Cabej, 2005, 2008, 2012, pp. 9–32). Such systems are better known in higher vertebrates, including humans. The picture of the control system in unicellulars is still blurred (see later in this chapter), while in plants, only separate controls are known (see later in this chapter), which can possibly be subordinate to a still-unknown integrated control system.

A prerequisite of an integrated control system is that it has to be able to monitor all parts of the organism by continuously receiving information on their state. This implies that the control system must have a pervasive presence throughout the living organism. A second imperative is that it has to compute in order to compare the actual state with the normal state, which is inscribed in the form of set points for various variables. This implies that the control system knows the normal structure of the system. Third, it needs to problem-solve; that is, to generate and send instructions throughout the animal's body for restoring the normal state.

While most of us may agree that coordination of the development of all the parts of organism is a necessity, the prevailing gene-centric view attributes this function to a genome rather than to a specialized center. Despite attempts to prove that a genetic program could account for the control of development, no model, however loose, has *ever* been presented on how this may happen. Even if one takes for granted that the genome is the control center in unicellulars, the question would naturally arise: which billions or trillions of cells of a multicellular animal would have the privilege of playing the role of the "genomic" integrated control system? How will this empowered genome receive information on the state of the system and send back instructions to restore the normal state in any parts of the organism?

There is absolutely no evidence, no hint, or even any hypothesis of a genome somehow controlling any systemic parameter in multicellulars. There is no evidence that the genome might control and regulate homeostasis in unicellulars. On the contrary, empirical evidence shows that genome (including its duplication and gene expression) is itself *regulated* rather than the *regulator* of homeostasis (see Section "The Control System in Unicellulars," later in this chapter).

We know where the crucial control center is in most metazoans (see later in this chapter, section "The Control System in Plants"). Beginning in the second half of the nineteenth century, experimental evidence progressively showed that in metazoans, a central system that controls and regulates vital functions (blood circulation, blood pressure, breath rate, and function of organs and organ systems) is operative and essential for their existence and evolution. Indeed, the study of the central mechanisms of control of animal functions is one of the main objects of animal physiology, and it is textbook knowledge that the maintenance of homeostasis and behavior in metazoans are functions of the nervous system. Through afferent pathways, sensory nerves send to the CNS sensory information about various parameters, which are compared to the corresponding set points; if deviations are detected, the brain sends chemical instructions via effectors to restore the normal level of the parameter.

In contrast with the clear picture of the control system in multicellular animals, the evidence on the control system in unicellular organisms and plants is still limited, poorly systematized, and awaiting scientific formulation and elaboration.

The Control System in Unicellulars

Both a unicellular and a multicellular organism are an architectural *magnum opus*, displaying incredibly high degrees of organization, and biology has not been able to answer two important questions that are crucial to understanding the nature of unicellular organisms. First, how do these microscopic beings determine the relative size, number, and spatial arrangement of organelles within the microscopic cell? We need to know where the "knowledge" necessary for producing and assembling the highly ordered structures comes from.

Second, how do they determine their adaptive behavior in search of sources of food and in avoidance behavior? They have to search for and find these sources and get access to them. Only rarely provided with food by luck, they have to go after food sources (it is the prophet that goes to the mountain, not the reverse). Simple as this may seem, their foraging behavior implies that the unicellular is capable of discriminating between the inorganic and organic debris and localize the latter. On its way to the source, it may come across a barrier and has to figure out how to circumvent it. Sometimes they have to use environmental cues such as light (phototaxis) or chemicals (chemotaxis) as leads to sources of food. Even their movement and necessary corrections in the direction of the source of food require precise calculations on the direction of the beating of cilia or flagella in ciliates or in determining the form, size, and direction of the beating of pseudopodia in amoeba. All these functions require adaptive changes in the structure of the cytoskeleton and microtubules of cilia and flagella and in the actin subunits of microfilaments. We need to know where all the calculations necessary for determining these adaptive changes in structure and behavior are made.

Both the spatial arrangement of organelles within the cell and the adaptive behaviors mentioned above are not randomly occurring events that require specific information to take place. Hence, they point in the direction of a specialized control center that receives information on the internal and external environment and by processing that information, it produces instructions (chemical signals) that, via effectors, determine the spatial arrangement of organelles and adaptive foraging behaviors. But essentially, these are the functions of a control system. From a theoretical viewpoint as well, it clearly seems that complex systems such as unicellular organisms could not subsist, let alone reproduce and evolve, without a control system. The question, however, would arise whether central or separate local systems of control are responsible for coordinating vital functions in unicellulars. We need to know where the control system and its controller are located within the cell of a unicellular organism. And the only rational approach in looking for the "controller" of these functions is to trace back the possible causal chain by sequentially following the described functions to their proximal causes and farther upstream.

The reductionist Zeitgeist still makes many of us focus on separate organelles rather than on the whole unicellular organism; we see functions and behaviors of unicellulars as products of specialized organelles, including chromosomes (for cell reproduction), ribosomes (for protein synthesis), cytoskeleton (for cell shape and transport of molecules throughout the cell), Golgi apparatus (for processing and secreting proteins from the cell), endoplasmic reticulum (for protein transport), cilia, flagella, and pseudopodia (for cell movement), etc. We are so accustomed to this view that we take it for granted that these organelles are self-controlled and self-regulated, even though the supporting evidence is nowhere.

Unlike the metazoans, for which we have a clear picture of the control system (especially for physiological and behavioral functions of animals), the picture of the control system of single-celled organisms is blurred. However, in recent decades, contours of the cell's control system are gradually beginning to emerge before our eyes.

As an example of a separate system, let us consider a control system that regulates the cell cycle (Alberts et al., 2002). The control system that regulates both DNA replication and cell mitosis consists of molecules of cyclin and Cdks (cyclin-dependent kinases), which form complexes of cyclin-Cdk. The main activators of Cdks are cyclins and cyclin-Cdk complexes that trigger sequential stages of the cell cycle. But in less complex cells, where the level of cyclins and inactivation of cyclin-Cdk complexes is determined by proteolytic enzymes of cyclin, the ultimate regulator of the cell cycle is external to the system that regulates production of these proteolytic enzymes, to which obviously the separate genomic control system is subordinate. The pending question then is: how do these enzymes know when to induce or suppress their synthesis according to the sequential stages of the cell cycle? Moreover, this control system of the cell cycle does not account for some of the critical events of the cycle, such as pole spindle formation and chromosome segregation.

Thus, although separate mechanisms of local control of the development and functioning of organelles within the cell would exist, a "supersystem" for controlling and coordinating the separate systems would be necessary for the unicellular organism to function properly. There is solid empirical evidence on a central control of functions in metazoans (including humans), and they are basic topics of animal physiology and animal behavior, respectively. There is also ample evidence of a central control of the animal organogenesis (Cabej, 2005, pp. 69 et seq, 2008, pp. 139 et seq, 2012, pp. 147 et seq). Since this mechanism in multicellulars will be briefly described later in this chapter, here I will only deal with the control systems in unicellular organisms.

Theoretical considerations aside, even facts such as the perfect coordination in space and time of the activity of cell organelles (e.g., ingestion and digestion of foods and excretion of waste in the environment), coordination of movements of appendages in locomotive behavior (phototaxis and chemotaxis of unicellulars), which involves the repatterning of the whole cell cytoskeleton and body, formation of pseudopods, coordination of thousands of cilia, and undulating motion of flagella, all of which point in the direction of the existence of a central control system.

The time for proclaiming the discovery of a central control system within the cell may not be on the horizon; hence, before we consider any speculative mechanisms of the central regulation of cell structure, function, and behavior, I find it appropriate to take a brief look at some facts and phenomena that represent counterinstances to the supposed view of the self-regulation of cell organelles, which might also suggest that a central control system is operative in unicellular organisms.

Cytoskeleton Induces DNA Replication

"The regulation of DNA replication cannot be explained at a genetic level alone," according to Casas-Delucchi and Cardoso (2011). More than a quarter of a century ago, it was observed that drugs doing the depolymerization of microtubules induce DNA synthesis, and the reverse is also true: drugs that stabilize the structure of microtubules block DNA synthesis (Thyberg, 1984). An increase in the number of microtubules and intermediate filaments in culture induces DNA and protein synthesis (Palmberg et al., 1985) and polymerization of microtubules is involved in cell proliferation (Ball et al., 1992).

A similar relationship between microtubules and the beginning of DNA replication is also observed in plants; dry tomato seeds are arrested at the G_1 phase of the cell cycle, but after imbibition, seeds show an increase in tubulin and DNA content that is immediately followed by seed germination (de Castro et al., 1995). Under mild heat stress, microspores in culture rearrange the microtubules of their cytoskeleton and start DNA synthesis, thus entering the cell cycle (Dubas et al., 2011).

These and other similar facts led investigators to the idea that microtubules of the cytoskeleton are involved in DNA replication and cell proliferation in eukaryotes.

Cytoskeleton Regulates Gene Expression

Rosette and Karin (1995) observed that experimental depolymerization of the microtubules of the cytoskeleton causes activation of the transcription factor NF-kappa B and induces expression of NF-kappa B-dependent genes. Cytoskeleton microtubules in the ciliated protozoan (*Tetrahymena thermophyla*) inhibit the expression of the gene for one type of β-tubulin, and it seems that microtubules of cilia inhibit both types of tubulin genes (Gu et al., 1995). The cytoskeleton is also involved in the localization and translation of specific mRNAs in HeLa cells (Hesketh and Pryme, 1991).

Evidence shows that cytoskeletal dynamics may regulate genome activity (Olson and Nordheim, 2010). Experimental reorganization of the cytoskeleton induces expression of the urokinase-type plasminogen activator (uPA) gene (Leeet al., 1993), and disruption of the actin microfilament structure (another type of cytoskeleton filament) in normal rat kidney cells with cytochalasin D induces expression of the PAI-1 (plasminogen activator type-1) gene (Providence et al., 1999). Similar microtubule effects on gene expression are also observed in apoptotic genes (Chen et al., 2003) and other genes in multicellulars (Bounoutas et al., 2011).

Microtubules of Spindle Poles Regulate Chromosome Segregation

In unicellular eukaryotes, after duplication of chromosomes, sister chromatids must be separated and the precise transport of the complete set of chromosomes to two "daughters" must occur. This is not an easy task: "The challenge for the cell is to compact the genome so that the DNA strands do not become entangled or broken

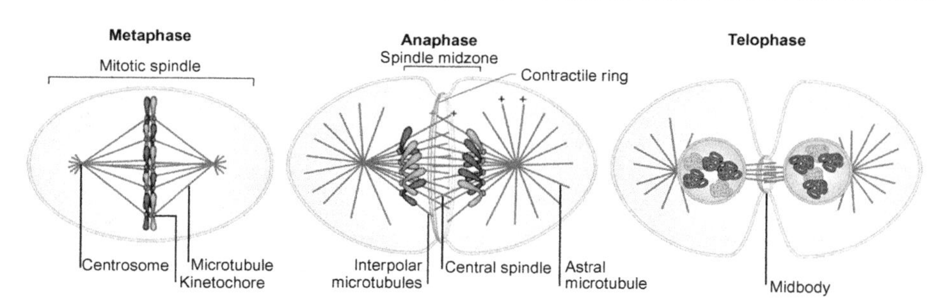

Figure 1.13 Assembly of the central spindle. Schematic diagrams of the distribution of microtubules and the chromosomes during cell division. In metaphase, the chromosomes align on the metaphase plate. At anaphase, the chromosomes move poleward, the central spindle assembles, and contractile ring assembly commences. In telophase, after cleavage furrow ingression, the contractile ring compresses the central spindle to form the midbody. Microtubule plus (+) ends are indicated (minus ends, which are positioned at the centrosomes, are not shown) (Glotzer, 2009).

during segregation and so that the appropriate genes are accessible to polymerases in the emergent mother and daughter cell" (Bloom and Joglekar, 2010). The machinery for the separation of sister chromatids and their transport to the daughter cells is the spindle apparatus consisting of the two microtubule-organizing centers (MTOCs) arising from the duplication of interphase centrosomes that, after detaching from the cell nucleus, relocate in one of the opposing poles of the cell. Microtubules, which are nucleated in centrioles, extend from spindle poles in the direction of the equatorial plate at which they attach to kinetochores. Spindles position themselves by astral microtubules nucleated from both of their poles and are attached to cortical sites (Glotzer, 2009; Grill and Hyman, 2005).

The force necessary for the segregation of sister chromatids is provided by motor proteins and by gradual, controlled depolymerization of the bipolar mitotic spindle microtubules. Kinetochores of sister chromatids are attached to microtubules at opposing poles, and hence, they move in opposing directions to form the complete set of chromosomes to each daughter cell (Figure 1.13). Almost all the microtubule-based force for the transport of chromosomes is provided by motor proteins, and microtubules determine the correct positioning of chromosomes in daughter cells: "The mitotic spindle supplies force, as well as positional cues, to the chromosome so that chromosome movements are consistent with the geometry of the dividing cell" (Bloom and Joglekar, 2010). The elongation of the mitotic spindle during anaphase leads to the formation of the central spindle, an array of antiparallel microtubules that is primarily responsible for cytokinesis (Glotzer, 2009).

Two decades ago, eukaryote cytoskeletal proteins (microtubules, actin filaments, and intermediate filaments) were discovered in prokaryotes. Soon after that, it was observed that the movement of daughter nucleoids (DNA-containing regions in prokaryotes) in two opposing poles requires the prokaryote actin homologue MreB (Gerdes et al., 2004) and β-tubulin homologue FtsZ (Shih and Rothfield, 2006).

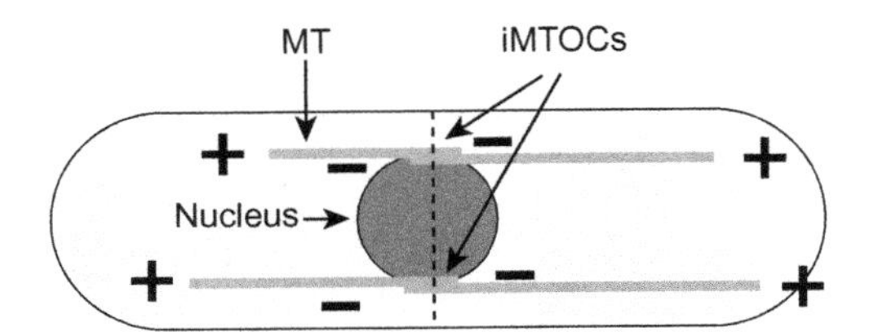

Figure 1.14 A model for nuclear positioning and interphase MT architecture in *S. pombe*. Microtubules (MTs) are organized from medial organizing centers (MTOCs) in multiple bundles with an antiparallel configuration and dynamic plus ends facing the cell tips and minus ends in medial-bundled regions. One MT bundle is attached to the nuclear envelope at the spindle pole body (SPB), and other MT bundles may be attached at additional sites. When an MT end contacts the cell tip, MT polymerization produces a transient pushing force that pushes the MT lattice and attached nucleus away from that cell tip. A balance of these pushing forces from these MTs may position the nucleus in the middle of the cell (Tran et al., 2001).

Cytoskeleton Regulates the Directed Release of Golgi Units from Nurse Cells via Ring Canals

In multicellulars, the transport of Golgi units from nurse cells to the oocyte is an active transport regulated by the cytoskeleton rather than a random diffusion process; Golgi units "move in a direct path toward the ring canals" that connect nurse cells with the oocyte and a conical actin basket at the nurse cell side of ring canals regulates their directed transport to the oocyte (Nicolas et al., 2009).

Cytoskeleton Determines the Position of the Nucleus in the Cell

The depolymerization of microtubules in yeast displaces the cell nucleus from its central position in the cell. Because the nucleus shifts its central position the septum is misplaced and cell division is unequal; that is, the resulting cells are different (Tran et al., 2000, 2001) (Figure 1.14).

Cytoskeleton Determines the Position of Organelles Within the Cell

Intracellular transport of molecules and organelles is responsible for their delivery to destination sites. Since the transport takes place primarily along microtubules, it is important that the free (plus) end of microtubules finds the correct destination site. Microtubules are dynamic polymers that continually growing or shortening in length to probe and explore many regions of the cell at random. This dynamic structure is conserved throughout the living world. In unicellulars, this allows cells to adapt their shape and form appendages (pseudopods, cilia, and flagella) and, in multicellulars, to determine the shape of tens to hundreds of different types of cells in the process of cell differentiation. It is interesting to observe that in unicellulars with permanent cell shapes depolymerizing drugs have little effect on microtubules (Heidemann et al., 1985). These cells can stabilize the polymer structure of their microtubules to a high degree when necessary. The process of differential stabilization of their structure is known as *selective stabilization* (Kirschner and Mitchison, 1986; Mitchison and Kirschner, 1985).

Positioning of organelles in the cell is species-specific, and it is maintained under varying internal and external environments. This implies that some kind of information is responsible for the strict determination of the positioning of organelles in the cell. In fission yeast, the position of mitochondria depends on microtubule dynamics not on motor proteins (Pon, 2011), and even the morphology of mitochondria and the Golgi apparatus depends on microtubules (Höög, 2003). The position of the Golgi apparatus in plant cells is determined by the organization of the actin cytoskeleton (Akkerman et al., 2011). It seems that microtubules somehow search and find the appropriate site for transporting and settling molecules, supramolecular components, and organelles. For example, actin filaments are responsible for the myosin-based transport of membrane organelles and the dynamics of these filaments is essential for their transport (Semenova et al., 2008) and for determining their destination. Microtubules are closely associated with the transport of peroxisomes and the depolymerization of microtubules causes peroxysomes to accumulate in the middle of the cell, preventing their transport (Rapp et al., 1996), whereas the actin cytoskeleton may be involved in determining peroxysome size, shape, number, and clustering (Schollenberger et al., 2010). In yeasts, the actin cytoskeleton regulates the partitioning of organelles and their movement to the bud (Catlett and Weisman, 2000). In the case of melanosomes, pigment granules where melanin is synthesized and stored (Wasmeier et al., 2008), their dispersion is related to the extension of microtubules that, via kinesins, move melanosomes away from the nucleus throughout the cell; and the reverse occurs when the microtubules shrink (Ikeda et al., 2011). Recent evidence shows that dispersion and aggregation of melanosomes is related to the increase and decrease (respectively) in the number of microtubules nucleated at the centrosome (Lomakin et al., 2011). This is proved by the fact that experimental inhibition of microtubule growth prevents the aggregation of melanosomes in the pericentriolar region (Lomakin et al., 2009). During pigment aggregation, it is observed that growing (plus) ends of microtubules capture melanosomes (Lomakin et al., 2009).

Microtubules are involved in exocytosis and in the transport of hormones, neurotransmitters, and neurotrophins through vesicles from the Golgi apparatus to the actin cytoskeleton at the site of cell membranes where these substances are released (Park and Loh, 2008) (Figure 1.15).

An examination of the mechanism of regulation of the vital functions of the cell and particular organelles may help shed some light on the presently dim picture of the control system in unicellulars. Adequate observational evidence shows all the stages of the cell cycle, both in unicellulars and multicellulars, are under the control and regulation of the cytoskeleton (see below and Chapter 2, section "Regulation of the Length of Microtubules – Key to Transport of Maternal Determinants in the Oocyte").

Cytoskeleton Controls the Formation of the Eyespot in Unicellulars

Over a century ago, Russian botanist Andrei Famintzin (1835–1918) observed that green algae from the shores of the Neva River in St. Petersburg contained a yellow

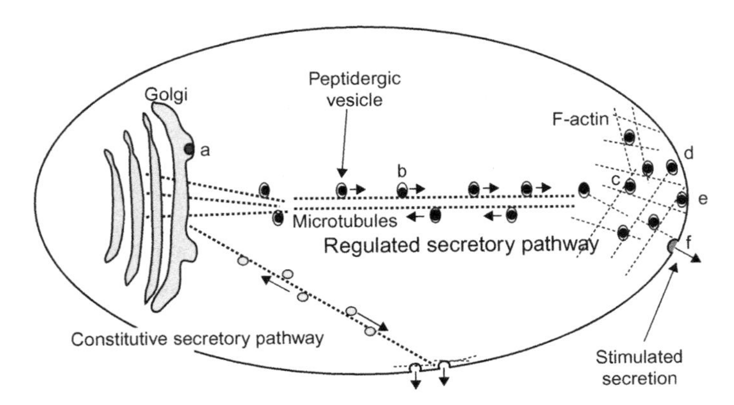

Figure 1.15 Steps for post-Golgi regulated secretory pathway (RSP) vesicle transport to the release site. Multiple steps are involved in transporting hormone-containing vesicles from the site of biogenesis at the trans-Golgi network (TGN) to the release site in the RSP. (a) Vesicle budding; (b) microtubule-based transport; (c) actin-based transport; (d) vesicle tethering; (e) docking; and (f) fusion with the plasma membrane. These steps share some commonality with the trafficking of constitutive secretory vesicles, but there are differences as well (Park and Loh, 2008).

spot in the middle of the body (Kateriya et al., 2004). This very simple photosensitive organelle, later observed in other unicellular organisms, is known as an *eyespot*. *Chlamydomonas reinhardtii* is a unicellular green alga with a single eyespot, which enables it to perceive the light and determine the direction to move in. This allows it to find places of optimal light intensity necessary for starch photosynthesis, its main source of energy. To reach this photosynthetically optimal place, the unicellular must adapt to the patterns of beating of the flagella. The phototactic response requires a particular asymmetric positioning of the eyespot (Mittelmeier et al., 2011) which, like the positioning of most of cell organelles, is determined by the centrioles or the MTOCs via the dynamic regulation of microtubule acetylation of the daughter four-membered (D4) microtubule rootlet (Boyd et al., 2011b) (Figure 1.16). The length of the D4 rootlet is the major determinant of the eyespot positioning along the anterior–posterior axis (Boyd et al., 2011a) and the equator (Boyd et al., 2011b).

Excitation of eye photosensitive pigments by light is followed by an electrical response and then by the adaptive change in the beating of the flagella. Finding the optimal intensity of light from a light source may require complex computation of the direction in which the unicellular has to move in and out of the possible adjustments it has to perform on the way to the intended destination.

Cytoskeleton Controls Locomotion in Unicellulars

Most unicellulars use cytoskeletal structures such as cilia and flagella for locomotion. Protozoans display a number of behaviors, such as adaptive responses to touch, light, and chemicals, and they can even learn from experience. These facts indicate

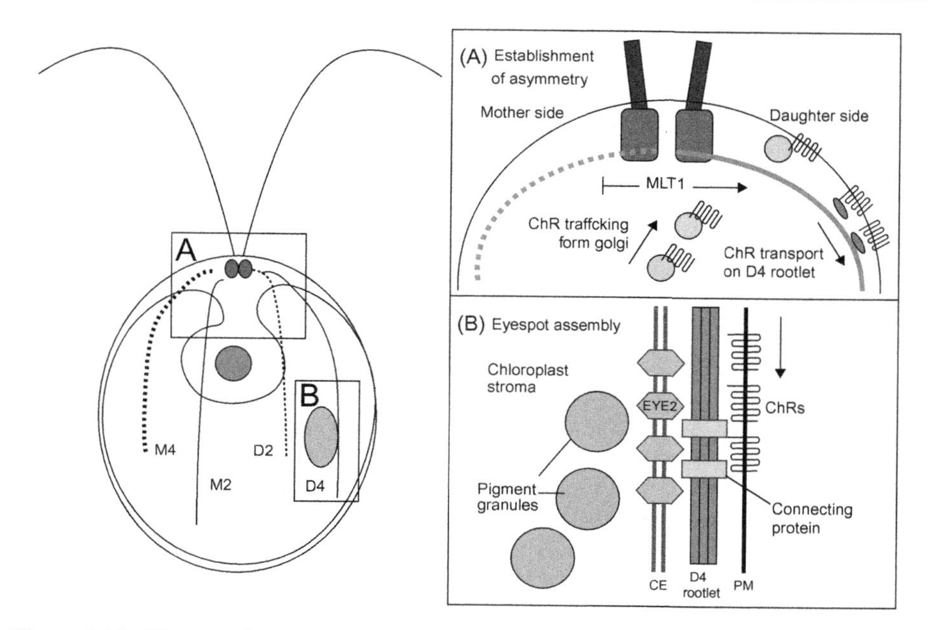

Figure 1.16 Eyespot placement and assembly in *Chlamydomonas*. Left: Diagram showing the cytoarchitecture of a *Chlamydomonas* cell. The basal bodies (small blue circles) nucleate the flagella and sets of two- and four-membered microtubule rootlets. The M2 and M4 rootlets are inherited from the mother cell, whereas the D2 and D4 rootlets are newly formed in the daughter cell. The asymmetrically placed eyespot (large orange ellipse) is associated with the D4 rootlet and positioned near the cell equator. (A) Simplified working model of asymmetric photoreceptor localization. Channelrhodopsin (ChR) photoreceptors traffic on endomembrane vesicles from the Golgi to the plasma membrane. MLT1 directs ChRs to the daughter side, where photoreceptors are transported by motor proteins (dark violet ellipses) along the D4 rootlet. (B) Simplified working model of eyespot assembly. ChRs form a patch in the plasma membrane. Rootlet or ChR-associated cues guide formation of a patch of EYE2 in either the inner or outer chloroplast envelope membrane, whose nucleates form eyespot pigment granule arrays. Specialized proteins in the eyespot establish and maintain connections between the chloroplast envelope and plasma membrane. CE, chloroplast envelope; PM, plasma membrane (Boyd et al., 2011b). (For interpretation of the references to color in this figure legend, the reader is referred to the web version of this book.)

that they receive, process, and store information from the environment. No evidence suggests that the genome is involved in these processes. The fact that the beating of cilias is perfectly coordinated (i.e., they beat successively in the same direction) indicates that electric impulses are generated and are sequentially transmitted to cilia throughout the cell. Ciliates resemble chemosensory neurons in their responses to stimuli (Hufnagel, 2008). Unicellular prokaryotes, such as bacteria, also display chemotaxis toward nutrients and are even capable of assessing levels of nutrients in the environment along with their stress level and that of other bacteria (Norris et al., 2011).

Cytoskeleton Controls Cytokinesis

Genetic studies show that chromosomes and centrosomes are not necessary for the positioning of the cleavage furrow and cytokinesis and these processes can take place even in their absence (Bucciarelli et al., 2003; Megraw et al., 2001; Zhang and Nicklas, 1996). The plane of cell division is dictated by the position of the central spindle during anaphase, and it is observed that perturbations of the microtubule cytoskeleton cause misplacement of the spindle and formation of two cleavage furrows (Werner et al., 2007). In unicellulars, cytokinesis is a well-coordinated and spatiotemporally precisely determined process that occurs during late mitosis but is not part of it. Production of two "daughter cells" is not the automatic result of DNA or chromosomes. It is the process of cytokinesis that equally partitions the duplicated structure of the mother cell to form two similar daughter cells. The process must occur at a precise time, beginning at the anaphase of the mother cell and ending at the beginning of the interphase of the two daughter cells. Division of the mother cell starts with the formation of the contractile ring and constriction of the cell membrane along the division plane, which forms the cleavage furrow, followed by the final separation of two daughter cells. This furrow must be created in a narrow and strictly determined region in order to ensure that both daughter cells are provided with the same organelles, nutrients, energy, and information necessary for an immediate independent life. The origin of information for cytokinesis in unicellulars is ultimately provided by microtubules of the bipolar spindle apparatus (Figure 1.17). Formation of the cell membrane at the site of separation of the dividing cells may also be controlled by a bundle of actin filaments of the cytoskeleton.

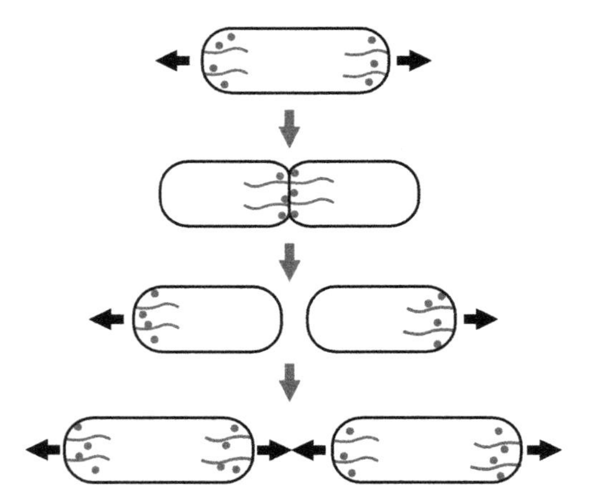

Figure 1.17 Diagram showing the redistribution of the actin cytoskeleton during the cell cycle. Prior to cytokinesis, actin accumulates at growing tips; during mitosis, it accumulates in the middle; daughter cells start to grow in a monopolar manner and transition to bipolar growth at the new end takeoff (Drake and Vavylonis, 2010).

The cytoskeleton is not unique to eukaryotic unicellulars. Bacteria also contain all three structural and functional homologues of eukaryotic cytoskeletal structures (microtubules, intermediate filaments, and actin filaments) (Muñoz-Espín et al., 2009), as well as the protein-building blocks of these polymers, tubulins, intermediate filament dimers, and globular actins. In prokaryotes like *E. coli* FtsZ, the prokaryote tubulin homologue is responsible for forming the Z ring, a constriction at the middle of the dividing cell. Depolymerization of the FtsZ ring is believed to provide the force necessary to complete cytokinesis (Errington et al., 2003; Nanninga, 2001) and the formation of septum by PBP3 (penicillin-binding protein) (Vollmer and Höltje, 2001). In the fission yeast, microtubules are the main factors determining organization of actin filaments and formation of the cytokinetic ring (Chang and Martin, 2009).

The cytoskeleton modifies its structure in response to the binding of integrins, hormones, growth factors, etc., to their respective cell-membrane receptors. This leads to changes in the cell shape and the relative position of cell organelles. The changes also extend up to the cell nucleus. The resulting phosphorylation of the fibrillar elements of the cytoskeleton and nuclear matrix induces the reorganization of the fibrillar network, which leads to the exposure and sequestration of specific regions of the chromosome-only enabling expression of specific genes (Ben-Ze'ev, 1991; Puck et al., 1990). This implies that the cytoskeleton may also be involved in the process of cell differentiation in mammals. In support of this, there is empirical evidence that the administration of different agents (insulin and IGF-I) in mammal cells induces different changes in the structure of cytoskeleton and in different results of their administration, implying activation of different signal transduction pathways (Berfield et al., 1997).

How does a unicellular coordinate, in space and time, the formation of its organelles and complex processes such as cell reproduction? How does a unicellular so finely tune its locomotion course in the direction of light or nutrients?

The above evidence shows clearly that all the processes of cell division, including duplication of chromosomes, their bidirectional separation and their placement to the to-be "daughter cells," directed cell locomotion, etc., are not determined either by the nucleus or by chromosomes. This short review of the regulation of all the stages of cell reproduction, as well as formation and precise asymmetric localization of the eyespot, shows that the key element in controlling all of the above is the cytoskeleton: all the evidence leads to the cytoskeleton and MTOCs.

There is ample evidence that the cytoskeleton regulates all the above vital functions of the cell, gene expression, DNA replication, chromosome segregation, cytokinesis, and assembly of the organelles in daughter cells. But what the cytoskeleton does cannot help us to understand how it does it; what we need is not a description but an explanation of these functions of the cytoskeleton. We need to know whether cytoskeletal structures are endowed with the capacity to receive data from the internal and external environments, and, by processing them, to make decisions and send instructions to organelles or be directly involved in executing these decisions. Again, the evidence at hand strongly suggests that microtubules may be endowed with such computational capabilities, which are very much unknown.

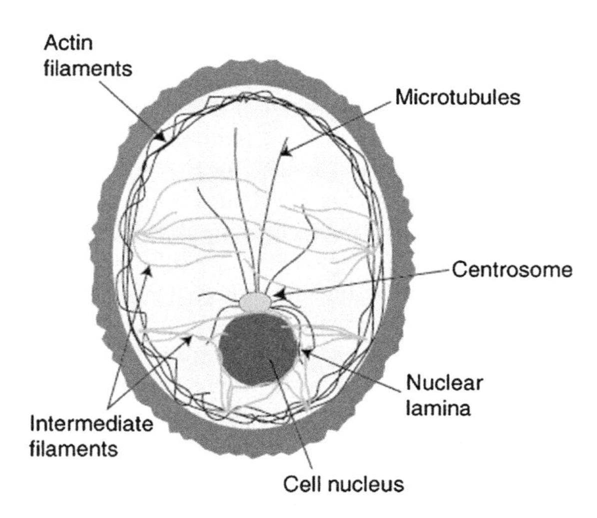

Figure 1.18 Organization of actin, microtubules, and intermediate filaments within a cell (from http://www.sparknotes.com/biology/cellstructure/intracellularcomponents/section1.rhtml).

The Structure of the Cytoskeleton

The cell-consuming cytoskeleton is a dynamic network of three types of protein filaments, microtubules, actin filaments (microfilaments), and intermediate filaments (Figure 1.18). Cytoskeleton occupies the whole volume of the cell. In many eukaryotic unicellulars, it determines the formation of pseudopodia, which are locomotor appendages. The term *cytoskeleton* was used first in 1903 by the Russian biologist Nikolai K. Koltsov (1872–1940).

The cytoskeleton determines a cell's shape and changes of that shape. It is responsible for cell motility. In eukaryotic unicellulars, it often forms locomotor appendages (pseudopods) by extending and contracting actin microfilaments, which make the movement of the cell possible. For locomotion, other unicellulars use cilia and flagella, which are also part of the cytoskeleton.

Microtubules

These are the longest cytoskeleton filaments, with an average diameter of 25 nm. These self-assembling hollow tubules are formed by 13 protofilaments of heterodimers of tubulin (α- and β-tubulin), globular protein molecules (Figure 1.19). Microtubules are also part of the structure of cilia and flagella. They are dynamic molecules that elongate or shorten, according to the needs of molecule transportation, organelle localization, etc. In animal cells, the minus end of the microtubule is enucleated to the centrioles of the centrosome (see later on their structure and function see Chapter 3, section "Epigenetic modes of cell differentiation"), while the plus end is free in cytoplasm. The polymerization/depolymerization of microtubules is faster at the plus end.

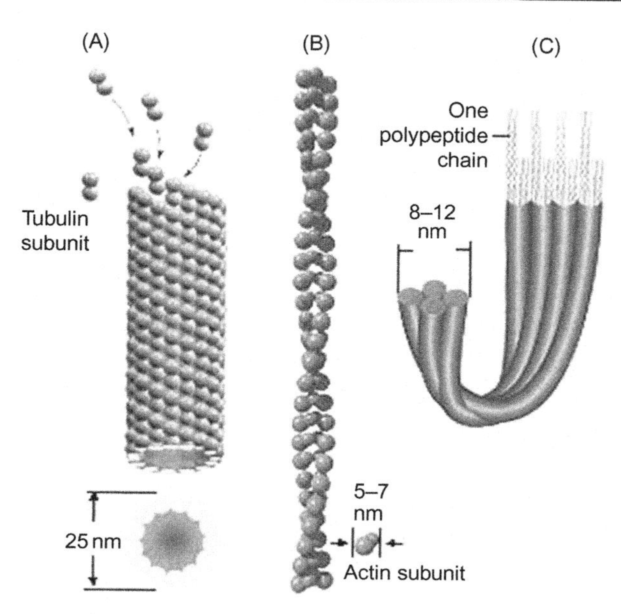

Figure 1.19 In these cells, actin filaments appear light purple, microtubules yellow, and nuclei greenish blue. (A) Part of a microtubule. (B) Part of a microfilament. (C) Part of an intermediate filament. (For interpretation of the references to color in this figure legend, the reader is referred to the web version of this book.)

Microtubules originate in the centriolar region of the centrosomes, but they may be nucleated in acentriolar MTOCs that may assemble and disassemble spindle poles in acentriolar cells (Moutinho-Pereira et al., 2009). Microtubules form mitotic and meiotic spindles during the cell cycle. Microtubules of a 9 + 2 structure also form cilia and flagella.

By participating in determining the shape of the cell, microtubules are involved in cell growth, apoptosis, and differentiation (Yujiri et al., 1999). They provide positional information, determine the position of organelles within the cell, and control the ordered transport of molecules throughout the cell with the aid of molecular motors, dynein, and kinesin that transport them in opposite directions. They carry out directed transport of molecules and organelles to their destination sites, generate pushing forces for central positioning of the cell nucleus, and mark sites of cell division (Picone et al., 2010; Piel and Tran, 2009; Tischer et al., 2010).

It is believed that the realization of the apoptosis program (programmed cell death) depends primarily on reorganization of the cytoskeleton rather than on activation of proteases and signaling molecules involved in apoptosis (Janmey, 1998).

In vitro experiments have shown that in the process of their self-organization, microtubules transport various particles, and that transport ends as soon as the process of self-organization is completed. Not only do microtubules transport the particles, but they also organize these particles and the content of particles, as seems to be the case with the transport of neurotransmitter vesicles along axons in the synapse (Glade et al., 2004). They determine the arrangement and movement of organelles (including the movement and separation of chromosomes during mitosis/meiosis) in the cell; they transport nuclei to the surface of the embryo during the blastoderm stage, thus determining the position of the nucleus within the cell.

Actin (Micro-) Filaments

These polymers of actin molecules are the thinnest (7 nm) (Figure 1.19) of all three types of cytoskeleton filaments. Actin polymerization generates a push force, whereas binding with myosin causes the actin cytoskeleton to contract. Actin filaments respond to the binding of a number of proteins with changes in the organization of the actin cytoskeleton. The actin cytoskeleton responds to a number of external signals (hormones, growth factors, etc.), and the mediators of these external influences are approximately 20 Rho guanosine triphosphatase (GTPase) enzymes (Sit and Manser, 2011).

Intermediate Filaments

As their name indicates, these filaments are intermediate in average diameter (12 nm) between microtubules and actin filaments (Figure 1.20). They are a family of fibrous proteins encoded by about 70 genes. They exist in monomer and dimmer forms and contribute to maintaining cell shape but do not participate in cell motility (Lodish et al., 2000).

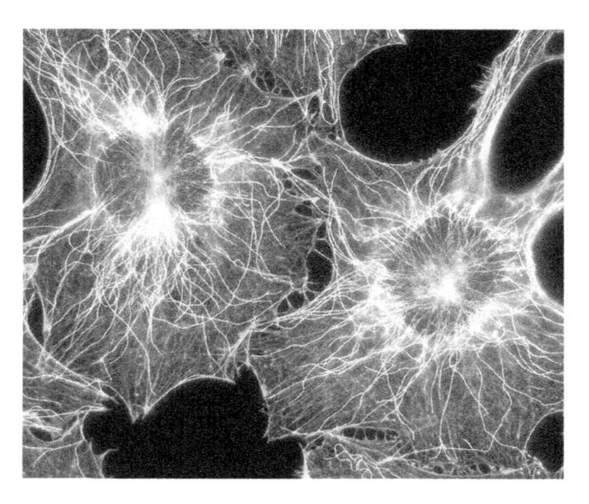

Figure 1.20 Structural organization of microtubule, microfilament, and intermediate filament (from http://www.tutorvista.com/content/biology/biology-iii/cell-organization/nonmembranous-cell-organelles.php).

Centrosomes as Microtubule-Organizing Centers

The discovery of centrosomes at the end of the nineteenth century is related to the work of Edouard van Beneden (1846–1910) and Theodor Boveri (1862–1915). In 1883, van Beneden described a new organelle in the middle of the cell, which Boveri called "centrosome" (1888) and "centriole" (1895). By the end of the nineteenth century, they independently developed the hypothesis that centrioles were centers of cell division (Heuttner, 1933).

Centrosomes are nonmembranous organelles, which, depending on the position of the nucleus, are mostly positioned along the middle of the cell, attached to the nucleus. Centrosomes consist of a spherical body of pericentriolar material (PCM), proteins embedded on a fibrous lattice. At the middle of the centriole is a pair of centrioles, hollow cylinders measuring about 100–150 nm in diameter and 210–400 nm in length (Delattre and Gönczy, 2004). Centrioles are arranged perpendicularly to each other and connected through coiled proteins (Figure 1.21). In the process of cell division, centrosomes form the poles of the spindle.

Centrioles are cylindrical structures formed by nine triplet sets of microtubules. During cell division, the centrosome divides and centrioles migrate to opposite poles, forming two centrosomes, one for each cell formed. At the start of the transition from the G1- to the S-phase of the cell cycle, centrioles begin to duplicate; next, each

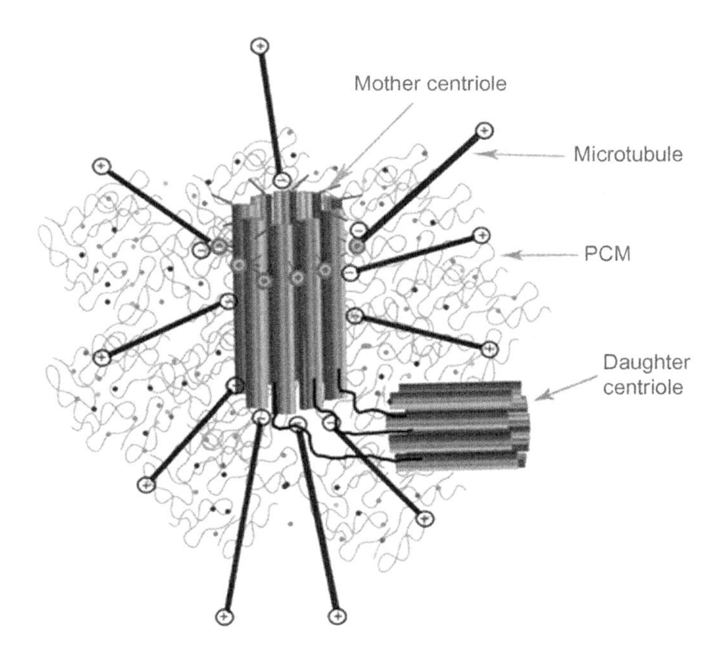

Figure 1.21 A schematic diagram of a typical mammalian centrosome composed of two centrioles (mother and daughter centriole) surrounded by a meshwork of PCM (Schatten and Sun, 2009).

of the two mother centrioles starts growing a daughter centriole perpendicularly to itself (Figure 1.21).

Both centrioles are connected through interconnecting fibers. The mother centriole is distinguished from the daughter centriole by distal and subdistal appendages. The centrosomal material consists of a fibrous scaffolding lattice with a large amount of coiled-coil centrosome proteins, and the centrosome's three-dimensional architecture is primarily maintained through specific protein–protein interactions. Microtubules are anchored with their minus ends to the centrosome core structure and microtubule growth is regulated by distal plus-end addition of tubulin subunits (Schatten and Sun, 2009).

Centrioles regulate the size of centrosomes (Conduit et al., 2010). Based on their biological role, centrosomes are also known as MTOCs. In animal cells, they are crucial to the spindle orientation and genome stability (Bornens, 2012).

Despite the dominant role that centrioles play in organizing the cytoskeleton, cells that lack centrioles are also known, a fact that suggests the existence of concurrent mechanisms (Moutinho-Pereira et al., 2009) of spindle formation (i.e., noncentriolar MTOCs. In higher plants, it is the nuclear envelope that acts as an MTOC for microtubule nucleation and formation of spindle poles.

During cell division, the centrioles separate and migrate to opposite poles of a cell and function as centers for spindle microtubule organization. Centrioles are also associated with the organization and development of undulapodia (cilia and flagella). They move to the periphery of the cell and act as basal bodies. Preexisting centrioles can replicate to form two pairs, one for each daughter cell.

Centrioles can move away from the middle of the cell and migrate to the periphery of the cell to convert into basal bodies (kinetosomes) (Figures 1.22 and 1.23). In the formation of cilia and flagella, the actin cytoskeleton plays the primary role (Vaughan and Dawe, 2011), as is indicated by the fact that pharmacological inhibition of actin impairs migration and the anchoring of mature basal bodies to the apical cell surface and formation of normal cilia (Boisvieux-Ulrich et al., 1990). In the cell membrane, the mother centriole converts to the basal body at the basis of a cilium and the basal body triplet microtubules transform into doublet microtubules of the axoneme (Kobayashi and Dynlacht, 2009) of cilia and flagella (Figure 1.22).

In 1991, one century after Boveri's theory of paternal inheritance of the centrosome in the round worm and sea urchin, Sathananthan et al. found that in human zygote, the paternal centrosome is functional alone (Sathananthan et al., 1991, 1996). In the soil nematode worm *Caenorhabditis elegans*, the paternal centriole is responsible for the asymmetric first zygote division and the varying fates of the resulting cells. Sperm centrioles play the key role in the asymmetric division of the zygote and in the following divisions, thus determining different cell fates (Vaughan and Dawe, 2011).

Can the Cytoskeleton Compute?

The cytoskeleton, and especially microtubules and actin filaments, are essential for nuclear processes taking place during the mitotic and meiotic divisions. They are also

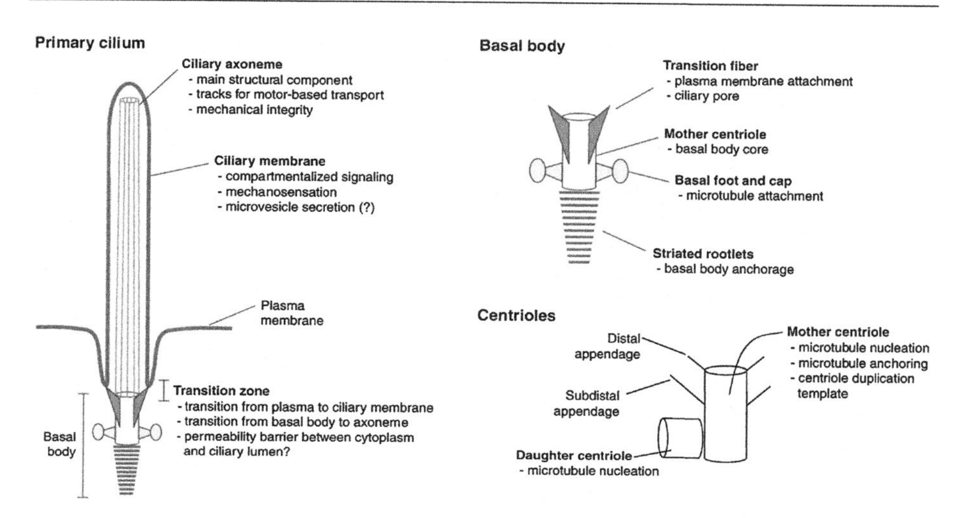

Figure 1.22 Major structural features and functions of primary cilia, basal bodies, and centrioles. The primary cilium is comprised of a basal body, an axoneme, and the ciliary membrane. Basal bodies are mother centrioles that have been modified by the addition of defining accessory structures, including transition fibers, basal feet and caps, and striated rootlets. The triplet microtubules of the basal body centriole give rise to the doublet microtubules of the ciliary axoneme at the region of the transition zone. Centrioles are typically found to be orthogonal pairs comprising a mother and a daughter, the former being associated with specialized functions (Seeley and Nachury, 2010).

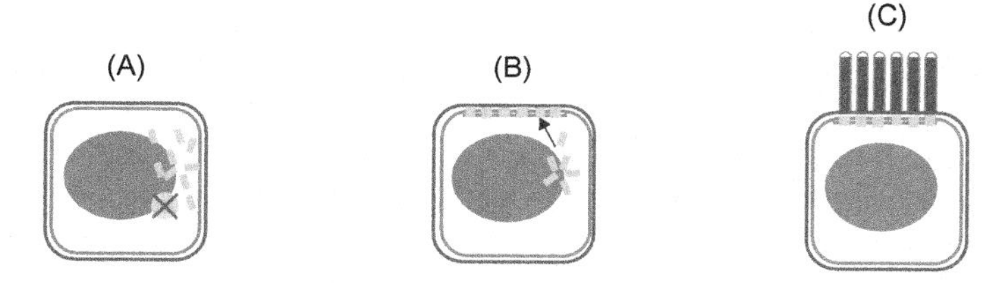

Figure 1.23 Centriole/basal body migration in terminally differentiated cells. The process of ciliogenesis produces thousands of motile cilia on many specialized, terminally differentiated cells. (A) Large numbers of basal bodies are formed within a single cell. Basal bodies migrate and dock with the cell membrane. (B) Movement requires actomyosin and is regulated by GTPase RhoA. (C) Cilia are assembled from the docked basal bodies (Vaughan and Dawe, 2011).

crucially involved in determining the position of organelles and for intracellular transport of vesicles and other supramolecular elements; they determine the shape of the cell and the dynamism of the cell shape. All these functions accomplished by microtubules and actin filaments are required rather than randomly occurring processes.

Under strictly experimental conditions, the behavior of an amoeba (formation of pseudopodia, direction of its movements, beginning of reproductive activity, and many physiological processes) is predictable. The amoeba's behavior is intended to adapt it to the experimental challenges of the external environment. But intended changes that bring about adaption imply that the system predicts* the adaptive result, figures the ways and takes the steps to achieve it. But, if "prediction is computation" (Pfaffmann and Zauner, 2001), where is the computation for determining the amoeba's behavior made?

The father of modern neuroscience, the Spanish scientist Santiago Ramón y Cajal, expressed the idea a century ago that the cytoskeleton is involved in conducting the nerve impulse and as early as 1951, the founding father of neurophysiology, Charles Scott Sherrington, seemingly anticipated the present trend of looking at the cytoskeleton as a potential "nervous system" of the cell in unicellulars:

> *Of nerve there is no trace. But the cell framework, the cyto-skeleton, might serve.*
> *Sherrington (1951)*

Indeed, neurons use the cytoskeleton to increase their potential computational capacity at 10^{16} bit states/s (Rasmussen et al., 1990).

There is no experimentally verified mechanism of computation in microtubules of the cytoskeleton. However, it is suggested that the cytoskeleton processes information in unicellulars and in nerve cells. It is known, for example, that learning is associated with increased production of tubulins (building blocks of microtubules) in the brains of baby chicks and baby rats and higher activity of microtubules (Hameroff, 1998; Tuszynski et al., 1998). It is thought that in neurons, microtubules may be utilized in the processing of electrical information and in cognitive processing (Priel et al., 2006). There is adequate evidence to firmly assert that centrioles, the MTOCs of the unicellulars, are organelles where such computation takes place. In this context, it is important to remember that the proper placement of organelles within the cell is also determined by the MTOC. Indeed, the improper positioning of centrioles causes *Chlamydomonas*' inability to move in the direction of the light source.

From the perspective of computational scientists, these unicellulars are biological micromachines, while the cytoskeleton appears to be their central system. Responding to intrinsic and extrinsic signals, the cytoskeleton reorganizes to generate responses that adapt the cell to the changed external or internal environment (Glade, 2008).

*While the conventional concept of prediction implies consciousness, today this term is homonymously used in biology and behavioral sciences to comprise unconscious adaptive behaviors of animals, including instinctive behaviors. This seems to be justified because all of them have in common input of the external stimulus, processing, and the output of the processing that manifests itself in the predictable behavior. Instinctive behavior, like all predictable behaviors, implies computation regardless of whether it results from the activation of evolutionary stable circuits or from circuits that form in response to specific external stimuli.

In a model by Glade (2008), the elementary units of the biological micromachine (cytoskeleton) are individual fibers (microtubules, actin filaments, and intermediate filaments) whose monomers (molecules of tubulin, actin, etc.) may act as molecular bits of information. Individual microtubules are connected by microtubule-associated proteins (MAPs) and other proteins, thus making communication between the fibers of the cytoskeleton network possible. In the process of their growth and depolymerization, microtubules leave chemical trails that enable information to be shared by neighboring microtubules. As a result, the dynamic skeleton is enabled to perform functions of a primitive brain and act "as an autonomous system sensible to external stimuli, conferring very complex behaviours to cells." (Glade, 2008). It seems to instruct the cell to react adaptively to external and internal challenges.

Microtubule functions and structure are closely related to MAPs. Axons and dendrites of nerve cells are very rich in microtubules. The growth of microtubules results from the polymerization of tubulin dimmers (α- and β-tubulin molecules) that is influenced by the biochemical environment, and especially by MAPs bound to microtubules.

In a model of the regulation of microtubule networks in the cell, MAPs convey cytoplasmic signals to microtubules, which also receive signals from the whole network of fibrous elements of cytoskeleton. MAPs in turn receive the output of the processing of cytoplasmic information from microtubules.

The growth of microtubules depends on the microtubule network's MAP-binding affinity. When the processing of signals in the network is good, MAPs bind to the microtubules. Increased binding activity in the network increases the stability of the network and signals that no change is needed. On the contrary, decreased stability of the network serves as a signal for change in the network structure. The adaptive self-stabilization (Figure 1.24) is based on a feedback mechanism on the structure of the

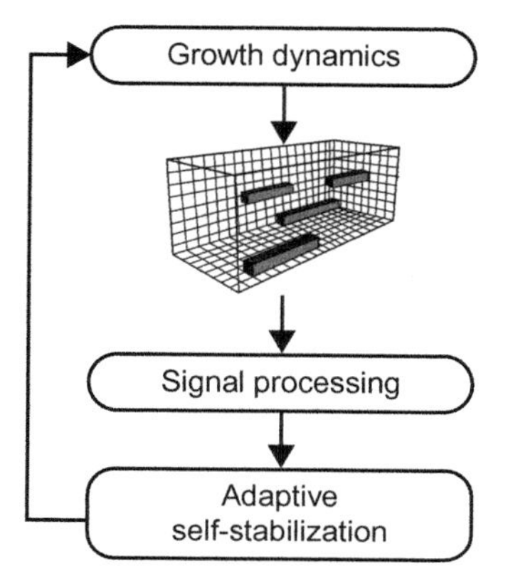

Figure 1.24 Cyclic flow of control. The growth dynamics and signal-processing modules share a common microtubule network representation. The adaptive self-stabilization module couples the growth dynamics to signal-processing performance (Pfaffmann and Conrad, 2000).

microtubule network and distribution of MAPs in the structure. The system displays some elementary learning and implies a sort of memory. The adaptive information processing needs to be experimentally verified, but the authors conclude that "[t]he fine structure of the cell is a natural medium for adaptive information processing" (Pfaffmann and Conrad, 2000).

In addition to its role in determining the shape and movements of the cell and the positioning of organelles within that cell, trafficking of vesicles and stimuli throughout the cell and evidence of processes of cell division show that, in mammal cells, the cytoskeleton is involved in processes of gene expression (Berfield et al., 1997; Puck et al., 1990). It is hypothesized that in neurons, at least, it is the cytoskeleton that signals the requirements for protein production to genes (Georgiev, *Quantum Mind Theories;*http://www.quantum-mind.co.uk/danko-georgiev-c169.html). In mammal cells, the cytoskeleton is in contact with both the extracellular matrix (ECM) and the nuclear matrix, which acts as a transducer of extracellular signals to the cell nucleus. Intermediary filaments (IFs) are intimately connected with structural elements of the nuclear matrix and are an integral part of the nuclear skeleton (Tolstonog et al., 2002).

In a model of regulation of microtubule networks in the cell, MAPs convey cytoplasmic signals to microtubules, which receive signals from the whole network of fibrous elements of cytoskeleton. MAPs, in turn, receive the output of the processing of cytoplasmic information from microtubules.

Some ECM molecules are connected to cell microfilaments via actin-associated proteins, thus forming an integrated system with the intracellular elements of the cytoskeleton (Ingber and Folkman, 1989).

In response to photostimulation with near-infrared light, some unicellulars extend pseudopodia to the side of the source of the light. It is argued that the extension cannot be performed by the cell's production of diffusible chemical signals which would indiscriminately affect pseudopodia throughout the cell. Hence, microtubules may relay the signal to the specific pseudopodia. Indeed, it is observed that after administration of antimicrotubular substances, the cell loses this ability to adapt by extending the right pseudopodia, although it can still move (http://www.basic.northwestern.edu/g-buehler/nerves.htm). Here is the interpretation of the phenomenon by the investigator:

> *After receiving the light pulses the centrosome destabilizes the radial array of microtubules which run towards the cell cortex which, in turn, will subsequently extend special pseudopodia to the light sources. Therefore, it seems that the observed destabilization is the signal that is propagated along the microtubles like along "nerves".*
>
> *http://www.basic.northwestern.edu/g-buehler/nerves.htm*

Based on experimental evidence, Gunter Albrecht-Buehler concluded that the centrosome is the organelle where the integration of light signals takes place and where microtubule signals to form pseudopodia originate (Albrecht-Buehler, 1998).

A controversial explanation, and far from experimentally verified, of a control system has been developed by physicist Roger Penrose and anesthesiologist Stuart

Hameroff. Protofilaments are composed of tubulin molecules in the form of tubulin dimmers of α- and β-tubulin monomers. Tubulin dimers may be the basic computational units of MTs (Penrose, 2003) that act as cellular automata for signal transmission (Hameroff and Penrose, 1996). MTs propagate and process signals, and these propagating signals appear to be relevant to the way that microtubules transport various molecules alongside them, and to the various interconnections between neighboring microtubules in the form of MAPs (Penrose, 2003).

In another model, Georgiev proposes that dipoles of water molecules within neurons break into dipole wave quanta that transmit information in water, influencing microtubule tubulin "tails," which transmit information along microtubules. "Collisions of waves formed by the tubulin tails are suggested to act as a computational gate for the control of cytoskeletal processes" (Georgiev, 2011), including the determination of MAP attachment sites in microtubules, which in turn transmit the output of the computations performed by the tubulin tails.

The Control System in Plants

The existence of an integrated control system in plants is not only a theoretical *sine qua non*, but it is also an *ex post factum* conclusion to be drawn from the precise control and regulation of plant reproduction, development, and growth. The high order and certainty of the occurrence of these processes speak unmistakably to their controlled and regulated nature.

Presently, however, knowledge about the plant control system is incomplete. We know of the intracellular mechanisms of gene expression, and there is abundant evidence on the extracellular signals, such as hormones, involved in gene expression and systemic networks. We can trace back the causal chain of gene expression upward through epigenetic changes in DNA (methylation) or histones (acetylation/deacetylation, etc.), consequential chromatine remodeling, and intracellular transduction pathways up to the binding of cell receptors by these hormones. But we do not know anything about causes behind the activation of hormonal pathways in plants. We do not know yet what determines the coordinated activation of these hormones and their pathways at the right time, at the right place.

The situation, in this regard, is by far more favorable in animals, where the causal chain is known beyond just hormones that bind to membrane receptors. In animals, the synthesis of hormones of the third tier (hormones produced by target endocrine glands such as thyroid, adrenal, and genital glands) is triggered by hormones of the second tier, secreted by the pituitary gland, such as thyroid-stimulating hormone (TSH), adrenocorticotropic hormone (ACTH), and gonadotropins (follicle-stimulating hormone (FSH) and luteinizing hormone (LH)). In turn, secretion of the pituitary hormones is induced by hormonal signals that "release" hormones secreted from a part of the brain called the hypothalamus, which represents the first tier of hormones in the hierarchy of control in animals. Further up, the top of the control hierarchy is occupied by higher centers of the brain.

The Plant Bauplan and Control System

The modern concept of the modular organization of plants by repetition of the same elementary units owes its origin to the observation of the founder of the discipline of morphology, Johann Wolfgang von Goethe (1749–1832), that flowers and fruits are repetitions of the foliage that "differ only in degree and not in kind." Goethe best expressed this observation in his famous dictum: "All shapes are similar, yet all unlike" (*Alle Gestalten sind ähnlich, und keine gleichet der andern*), which appears at the beginning of his poem on the *Metamorphose of Plants* (*Metamorphose der Pflanzen*) (1798) (Nickelsen, 2002).

We do not know whether, or how much, knowledge of the control system in animals can help reveal the plant's integrated control system. And this is for a good reason: essential differences exist in the Bauplans of both groups, and the degree of complexity and diversity of the animal structure is obviously greater. There are only three organ systems (root, shoot, and reproductive systems) in plants, whereas 11 organ systems (nervous, digestive, cardiovascular, urinary/excretory, muscular, skeletal, respiratory, reproductive, endocrine, immunitary, and integumentary systems) have evolved in higher animals.

At one time, plants were thought to consist of five morphological units (roots, stem, leaf, floral organs, and ovules), but the prevailing view now is that plants consist of only three units (roots, stem, and leaf) and according to the "heterodox" concept, the shoot and leaves are not distinct units; the leaf is seen as a partial shoot in the sense that "Each part is in essence an image of the whole" (Lacroix et al., 2005) and the plant itself is "a repetitively branched system alternatively composed of leaves and shoots, each being a shoot in different degrees of wholeness" (Classen-Bockhoff, 2001).

The number of architectural models in living plants is very small (23), and far less than would theoretically be predicted by the possible combinations of the elementary units (Barthélémy and Caraglio, 2007). The strict modularity of the plant architecture may be responsible for the majority of the relatively limited morphological inventiveness in the kingdom Planta, compared to animals (Figure 1.25).

Unlike animals, where adulthood marks the end of the development of tissues and organs, plants continue to produce new organs throughout their life. So while the purpose of the control system in animals is to maintain the structure and function of the organism upon reaching adulthood, in plants, it is also to produce new organs during the entire life.

Regardless of surprises that future research may bring, the modular organization and relative simplicity of plant structure, function, and behavior compared to animals, would tempt a biologist to predict that plant evolution may not have required a control system of the complexity and sophistication of that found in animals.

In the present fragmentary picture of plant control, hormones represent the principal extracellular signals and the higher step in the known hierarchy of control in plants. This is substantially different from what we observe in animals, from lower

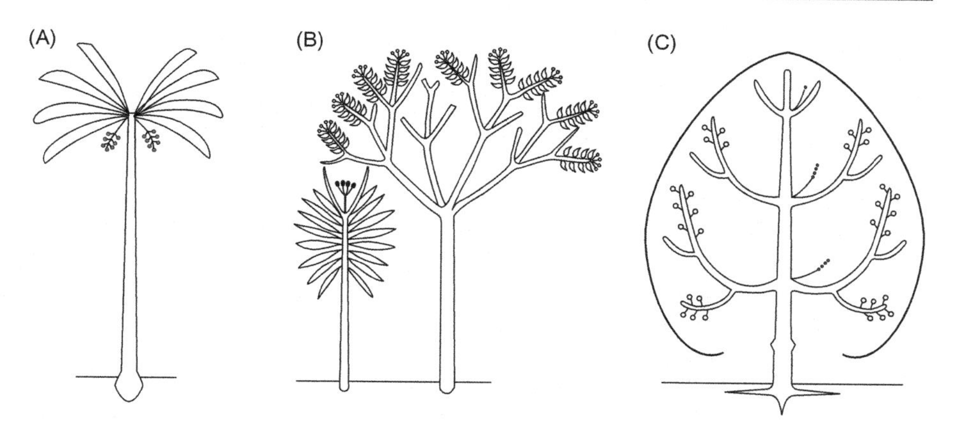

Figure 1.25 Three plant architectural models. Corner's model (A) concerns unbranched plants with lateral inflorescences. Leeuwenberg's model (B) consists of a sympodial succession of equivalent sympodial units, each of which is orthotropic and determinate in its growth. Rauh's model (C) is represented by numerous woody plants where growth and branching are rhythmic, all axes are monopodial and sexuality is lateral (Barthélémy and Caraglio, 2007).

metazoans to higher vertebrates. There are no known specialized hormone-releasing organs and cells in plants that are comparable to the endocrine glands in animals; even in lower metazoans, such as cnidarians, the secretory cells of the neural net are specialized in the production of hormonal substances (peptides). The level of hormonal control of development in plants corresponds to the level of hormonal control accomplished by hormones released by the target endocrine glands (thymus, thyroid, parathyroid, ovaries and testes, adrenal, and pancreatic gland) and nonendocrine organs (intestines, kidneys, etc.).

In metazoans, the controls at the supracellular organismic level form a strict hierarchy that in a simplified form looks as follows:

The *nonendocrine brain* axis (*endocrine brain* in invertebrates such as insects)
The *hypothalamus*
The *pituitary*
The *target endocrine glands*

There is no visible reason to believe that plants must have a hierarchic control system similar to the one seen in animals. However, what we are clearly missing in the plant control system is the controller, a center that is indispensable for continually monitoring the *system*, maintaining homeostasis and controlling the reproduction of plants and lifelong reproduction of their organs.

Recent experimental evidence has accumulated, and theoretical models have been developed on possible separate or even integrated control systems and controllers.

Hormones in the Plant Control System

The bulk of information on the plant control mechanisms at the supercellular level is related to the actions of plant hormones, known as *phytohormones*. Many phytohormones that regulate physiological and developmental processes in plants are known. In principle, their actions and mechanisms are similar to hormones in animals.

Especially relevant to understanding of the nature and mechanisms of hormonal action in plants have been studies on the key plant hormone auxin or indoleacetic acid (IAA), which among other things, is responsible for plant growth. Auxin moves from one cell to another through plasma membranes not just with the help of auxin transporters; in response to the perception of environmental stimuli, such as light and gravity, plant cells also release auxin to neighboring cells via special vesicles in a way reminiscent of the transport of neurotransmitters from the axon of one neuron to the dendrites of another in animals. Indeed, auxin functions not only as a hormone, but also as a morphogen and neurotransmitter. It is proposed that the quantal release of auxin may have consequences for plant development. Hence, its release requires tight control (Baluška et al., 2008).

When a shoot is tipped off, auxin moves to the lower side of the shoot, causing it to bend upward (negative gravitropism). The polarity of transport of auxin in the shoot is rootward (Lewis et al., 2011). When a shoot is cut, auxin stimulates the undifferentiated cells in the interior of the shoot to differentiate and form the apical meristem of the root. This takes place in various plant species known to reproduce vegetatively and is experimentally induced by dipping the cut surfaces into auxin solutions. The stimulating effect of auxin on cell growth is based on the fact that the hormone activates enzymes of the cell wall and causes acidification and loosening of the rigid cellular wall.

Unlike any other hormone in plants, auxin transport between cells is polar. The polarity of the movement of auxin from one cell to another is determined by the influx and efflux carriers of auxin in the plasma membrane. Auxin is also involved in its own transport through plant cells and tissues (Petrášek and Friml, 2009). The action of auxin depends not only on its synthesis and transport, but also on the cell's competence to react to auxin, which "seems also to be controlled in time and space" (Vernoux et al., 2010). Several auxin feedback loops acting at the cell level may contribute to the elegant patterning of the plant tissues and organs. It is proposed that auxin transport via the existing stem vasculature and auxin gradients set up through a feedback loop auxin/PIN1 (PIN1 - an auxin efflux carrier) enable the shoot apical meristem (SAM) to proceed with the initiation of leaves (Bayer et al., 2009; Leyser, 2010) (Figure 1.26).

Auxin is essential for lateral root initiation. This process starts with the accumulation of auxin in the central cells of the lateral root primordia, and later its concentration shifts to the tip of the primordia (Péret et al., 2009). The new organ anlagen (primordia) arises in the peripheral zone of the shoot meristem. Formation of organ anlagen occurs at the sites of auxin maxima that are determined by the PIN1 gene, which induces auxin efflux at the site of organ initiation at the shoot apical meristem (Bohn-Courseau, 2010).

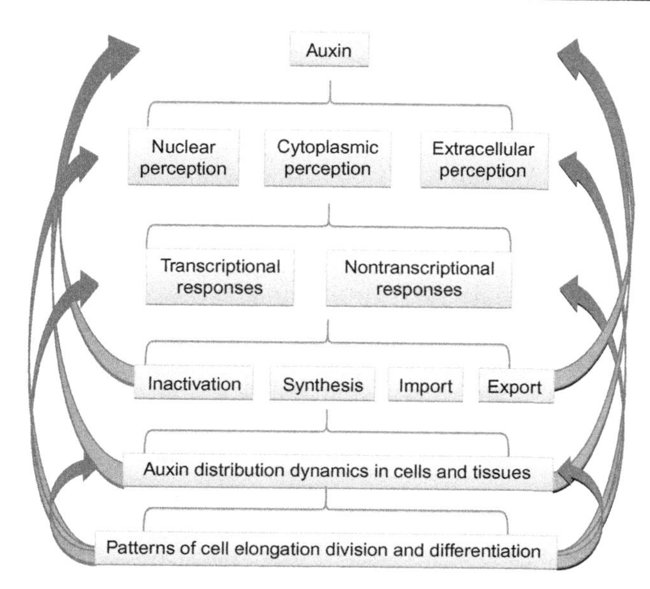

Figure 1.26 Feedback between the auxin action machinery occurs at multiple levels (Leyser, 2010).

Bud formation is a good example to illustrate the hormonal mechanism of organ formation in plants. In a simplified form, this mechanism illustrates the coordinated interactions of two hormones, auxine and cytokinine, in determining the development of the leaf bud (Figure 1.27).

One of the main functions of cytokinins is stimulating cell division and leaf expansion. Cytokinins bind to the histidine protein kinase receptor CRE1 and control the number of early cell divisions via an intracellular signaling pathway stimulating nuclear transcription activators (ARR1, ARR2, and ARR10) or repressors (ARR4, ARR5, ARR6, and ARR7) (Hwang and Sheen, 2001; Sakai et al., 2001; Schmuelling, 2002). The effects of cytokinins in some respects oppose those of auxin (they inhibit stem elongation and stimulate a plant's bushiness), and the balance of these hormones may control the size of the meristem (Beemster and Baskin, 2000).

Initiation of bud formation requires the export of auxin from the bud. The antagonistic action of auxin and cytokinin in bud formation is explained by the production of cytokinin, which increases the flow of auxin to the stem, which inhibits bud growth. The local synthesis of cytokinin is responsible for bud formation, although cytokinin is produced primarily in roots. Recently discovered hormones called *strigolactones* are also involved in bud formation; by dampening auxin transport, they inhibit bud growth (Crawford et al., 2010).

Gibberellins are another group of hormones involved in the germination and elongation of the stem, as well as floral timing and development (Schwechheimer, 2008). They are synthesized under the influence of external cues such as the lengthening of the day and the drop in temperature.

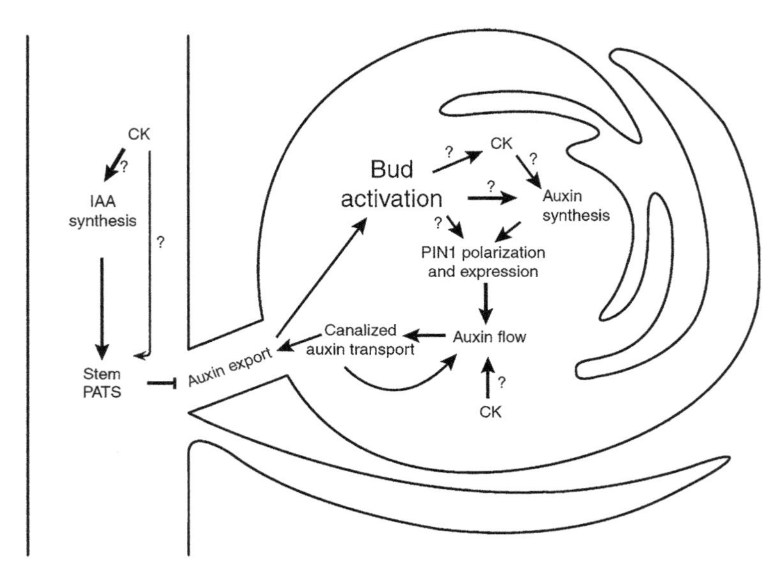

Figure 1.27 Bud activation and auxin transport. One model for the mechanism by which auxin regulates bud outgrowth relies on establishment of auxin export out of the bud as a prerequisite to allow bud outgrowth. An initial auxin flow towards an auxin sink promotes auxin transport canalization along this path. The polarization and upregulation of auxin transport feeds back to promote auxin flow further. The canalization of auxin transport into cell files connecting the auxin source (bud) to the auxin sink (main stem) allows sustained auxin export out of the bud. This means that auxin export out of the bud can be prevented if the main stem is a weak auxin sink. This can be achieved by increasing the amount of auxin transported in the polar auxin transport stream (PATS) in the main stem. Therefore, if cytokinin (CK) in the primary apex promotes auxin synthesis and/or PATS, this would inhibit bud outgrowth. On the other hand, if CK promotes auxin synthesis and/or auxin flow in the bud, it would promote bud outgrowth (Müller and Leyser, 2011). Symbols? indicate possible but unconfirmed involvement of various substances and processes.

Other important plant hormones are abscisic acid (a stress hormone), ethylene, jasmonates, brassinosteroids, etc.

Searching for the Plant Control System

Biologists are still puzzled that a relatively small number of hormones, many of them functionally redundant, enable plants to accomplish, during embryogenesis and in later life, many phenotypic outcomes and give rise to the observed diversity of forms in the plant kingdom.

This may be related to a number of genetic and epigenetic factors. In plants, hormonal signaling pathways are connected between them, thus giving rise to pathway networks where the combined activity of different pathways can lead to amplification, attenuation, or fine tuning of phenotypic outcomes (Kuppusamy et al., 2009). Different hormonal pathways may be involved in producing the same phenotypic result by using different genes or transcriptional programs, but in the process, they crosstalk and influence each other (Nemhauser et al., 2006).

Ample evidence has been accumulated on the control and regulation of gene expression and the role of epigenetic mechanisms, such as DNA methylation, histone acetylation/deacetylation, and chromatine remodeling. We have a detailed picture of signal transduction pathways in plants. However, this is only a partial picture; it elucidates the control at the cellular level. Knowledge of gene expression is of no help in understanding how tall a plant will grow, how long it can live, how often it will flower, how big its fruit will grow, how it will react to tissue injuries, etc. These are emergent plant phenomena that can neither be predicted by its cells or genome.

We need to know what determines the development of the plant multicellular structure, as well as what monitors its state, detects deviations from the norm, and sends instructions for restoring it for long periods of time, sometimes thousands of years.

The examples of plant control mechanisms described briefly and in simplified form above unquestionably show the progress in the field of molecular mechanisms of control at the plant cell and supracellular levels. The evidence at hand has shed light on some important aspects of the control mechanisms in plants, but they are far from anything that could be considered an actual control system in plants. All the evidence on hormonal regulation presented in this brief review has an essential flaw: hormonal pathways described so far take the regulation of first element of the pathway to be self-explanatory, as if the timing and switching on/off of the hormone synthesis do not require explanation. But the mechanism of its regulation cannot be excluded from the general scheme of the control system. Further work to piece together all the available facets into a comprehensive control mechanism that is comparable to the control system as we know it in animals will be necessary to fill the present gaps.

Searching for the Controller of the Control System in Plants

We have seen that the highly ordered structure of the cell in unicellular organisms is determined by a control system consisting of genetic and epigenetic mechanisms. Both genetic control by the genome and epigenetic control by the cytoskeleton, described earlier, are cell level controls and do not extend beyond the individual cells. Hence, they are not relevant to the multicellular structure of animals and plants. The development and maintenance of that structure requires memorization of that structure and special biological devices to direct its construction. Expressed more simply, this implies regulation in space and time of the arrangement of individual cells of different types in strictly determined patterns that allow the system to develop into an organism of its species and perform species-specific functions at the tissue, organ, and organismic levels. The question that we face in the case of plants is not whether a control system exists, but where it is. In this regard, there is no question that the situation for plants is different from the one in metazoans. The picture of the control system in plants, as can be seen from the few examples of hormonal control of growth of various organs, with a number of controls at the level of tissues and organs, is still fragmentary.

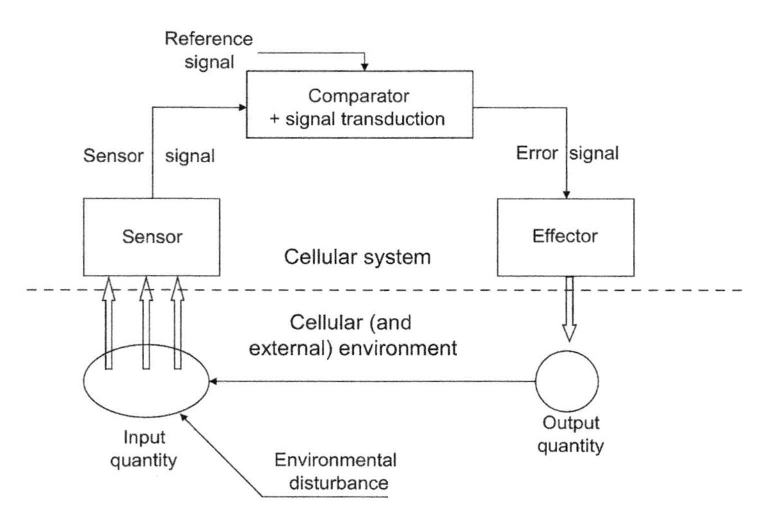

Figure 1.28 Basic control design of a system manipulating many aspects of plant behavior and incorporating negative feedback (Trewavas, 2006).

The heart of the control system in animals is the controller, the part of the control system that monitors the state of the system; it detects deviations from the norm, makes decisions, and sends instructions to restore the normal state. The controller enables self-correction and maintenance of the normal state. As it will be shown in the next chapter, in the animal's integrated control system, this supreme role belongs to the brain (the neural net in lower metazoans).

An integrated control system would imply the physical omnipresence of the elements of the control system in plant tissues and organs or, theoretically at least, an overarching mechanism of communication between the controller and the rest of the plant body, as well as an ability to transmit information on the state of the system to the controller. While one can theoretically argue the necessity of an integrated control system for the coordination of the development and maintenance of the plant structure and function, there are no substantial clues as of yet to track the whereabouts of the controller of the control system in plants. Interesting in this regard is a theoretical model of the control system in plants developed by Trewavas (2006) (Figure 1.28).

As shown in this diagram, the control system in plants is a closed-loop system. While the logic of the functioning of this control system is solid, the devil is in the details. Central to this system are the comparator and the reference signal. At the plant organism level, "the reference signal is currently unknown" (Trewavas, 2006). Neither is the comparator. The reference signal or the set point has to be part of the control system, rather than an input from the environment. The set point is an evolutionarily determined parameter of the species that unfolds in the process of development, rather than an environmentally provided instruction. In discussing the control system of animals, I came up with the idea that the determination of physiologically precise norms is a function of living systems, which is related to the emergent computational capabilities of the neuron, neural net, and CNS (Cabej, 2005, pp. 63–66). In explicit terms, we know that the set points for maintaining normal levels of glucose, glucocorticoids, mineralocorticoids, electrolytes in body fluids, water content, blood pressure, etc., are determined in brain structures. We are still far from being able to answer the question of where the set points are that determine the physiological levels of hormones and other parameters of plant fluids in plants.

Being a prerequisite of the maintenance and reproduction of any ordered structure, the control system implies mechanisms of monitoring and communicating with the rest of plant organism. The mechanisms of communication have been the Achilles' heel of our understanding of the nature of the control system in plants. However, undeniable progress has been made in this regard, especially during the last decade or so. Biologists on both sides of the Atlantic have provided substantial evidence on the existence of elements of a neuroid system and attempted to put it on a solid theoretical foundation.

Let us remember that neuroid phenomena evolved in unicellular ancestors of both multicellular plants and animals and that the nervous system plays the role of the controller in animals. In view of the common unicellular origin of animals and plants, these two facts may have some bearing on the issue of the control system in plants. Transmission of electrical signals between plant cells has been observed for a long time, and a number of environmental agents can induce action potentials in cell plants. It is interesting to point out that *Arabidopsis* possesses 30 percent of the nervous system-related genes of flatworms (platyhelminthes).

It is argued that "[n]eural aspects of biological systems are obvious already in bacteria and unicellular biological units such as sexual gametes and diverse unicellular eukaryotic organisms" (Baluška and Mancuso, 2009). Indeed, elements of electrical impulse conduction are observed in protozoans. Based on their complex morphology and physiology, the American ethologist James L. Gould called *Paramecium* "a promising nerve cell analogue" that has "all the complexity of real nerve cells and more" (Gould, 1982).

In plants, the situation with respect to impulse conduction is closer to the one we observe in sponges, which have no neurons or a nervous system but show coordination of activity and endogenous contraction rhythm for water exchange. Sponges have a nonneural control system (Cabej, 2012, pp. 421). They show an electrical conductance of the type observed in epithelial cells of lower metazoans, which is conserved in particular epithelia of higher animals. Sponges also possess many chemical messengers involved in their systemic contraction behavior, and even a number of synaptic genes.

Plants display electrical excitability (i.e., the ability to respond to various stimuli by electrical signaling); they can perceive light of various wavelengths, sense gravity, and even "smell" various volatile substances and determine their behavior accordingly. This gives rise to the well-known phenomena of phototropism, gravitropism, and "olfactotropism."

In 1991, an electric potential on the surface of the root apex of the *Lepidium sativum L.*, was reported for the first time. The authors concluded that "unevoked fluctuations of the potential ... are due to signals of an unknown nature" (Hejnowicz et al., 1991). Now the evidence of the existence of synchronized electrical activity in the transition zone of the root apex is believed to reflect the integration of internal and external stimuli for adapting plant physiology and metabolism to the changes occurring in the environment (Masi et al., 2009).

Plants generate electric signals in response to many environmental stimuli and stresses, such as extreme temperatures, diseases, and attacks by animals, and have evolved relevant defense mechanisms. These electrical signals are propagated locally or globally in the plant tissues and organs.

In response to damage caused by a caterpillar, a tomato leaf induces an action potential and propagates it to other leaves, which leads to synthesis in all of them of a substance that is toxic to the attacker. It is also believed that jasmonic acid (JA) produced in the wounded leaf may travel throughout the plant body (Sato et al., 2011) and trigger production of the toxic substance by all the leaves. However, the rapidity of the response by distant leaves contradicted evidence of the involvement of chemicals in the defense response. This, as well as the fact that electrical stimulation alone can induce the same response, demonstrates that the response results from the propagation of the action potential induced by the caterpillar attack (Leys et al., 1999; Stanković and Davies, 1997; Wildon et al., 1992). The action potential induced by the caterpillar attack travels to other leaves.

There is evidence that electric signals are involved in the regulation of growth, respiration and water uptake, and photosynthesis in plants (Koziolek et al., 2003).

Besides the role of action potentials and variation potentials, a new type of long-distance propagation that is transmitted systemically but varies in intensity from leaf to leaf that is defined as systemic potential (SP) has been recently demonstrated in plants and can be induced chemically or by cutting (Zimmermann et al., 2009).

There are two modes of the propagation of action potentials in plants. In lower plants and other plants such as *Dionaea* flytraps, they propagate in all directions, whereas in higher plants, they propagate along the plant axis via vascular bundles (xylem) that are used for water and nutrient transport (Brenner et al., 2006). The Venus flytrap, *Dionaea muscipula*, offers a curious example that illustrates the role of propagation of action potentials in determining plant behavior. At the apex of leaves, this plant has two half-open lobes that it uses as traps to catch insects, which are digested and used as food. Entrance of an insect into the inner surface of lobes is perceived by mechanosensory hairs as a stimulus and induces an action potential, which, via plasmodesmata (small channels connecting plant cells to each other), is conducted to all lobe cells, thus stimulating the motor cells of the midrib to close the lobes and trap the insect in less than a second (Figure 1.29).

Figure 1.29 The entrance of an insect in the inner surface of lobes induces the trap to close.

Even electrical stimulation alone, without mechanical stimulation, suffices to induce the trap to close (Volkov et al., 2008). It is interesting to point out that not only a plant like *Arabidopsis*, but even yeasts as well, have 30 percent of the nervous system-related genes of flatworms (platyhelminthes).

Despite the evidence on nonneural transmission of electrical signals in plants, we still are haunted by the question, "Where does the controller of the plant control system reside?"

References

Akkerman, M., Overdijk, E.J., Schel, J.H., Emons, A.M., Ketelaar, T., 2011. Golgi body motility in the plant cell cortex correlates with actin cytoskeleton organization. Plant Cell Physiol. 52, 1844–1855.

Alberts, B., Johnson, A., Lewis, J., Raff, M., Roberts, K., Walter, P., 2002. Molecular Biology of the Cell, fourth ed. Garland Science, New York, NY.

Albrecht-Buehler, G., 1998. Altered drug resistance of microtubules in cells exposed to infrared light pulses: are microtubules the "nerves" of cells? Cell Motil. Cytoskeleton 40, 183–192.

Ball, R.L., Albrecht, T., Thompson, W.C., James, O., Carney, D.H., 1992. Thrombin, epidermal growth factor, and phorbol myristate acetate stimulate tubulin polymerization in quiescent cells: a potential link to mitogenesis. Cell Motil. Cytoskeleton 23, 265–278.

Baluška, F., Mancuso, S., 2009. Deep evolutionary origins of neurobiology. Turning the essence of 'neural' upside-down. Commun. Integr. Biol. 2, 60–65.

Baluška, F., Schlicht, M., Volkmann, D., Mancuso, S., 2008. Vesicular secretion of auxin—evidence and implications. Plant Signal. Behav. 3, 254–256.

Barthélémy, D., Caraglio, Y., 2007. Plant architecture: a dynamic, multilevel and comprehensive approach to plant form, structure and ontogeny. Ann. Bot. 99, 375–407.

Bayer, E., Smith, R., Mandel, T., Nakayama, N., Sauer, M., Prusinkiewicz, P., et al., 2009. Integration of transport-based models for phyllotaxis and midvein formation. Genes Dev. 23, 373–384.

Beemster, G.T., Baskin, T.I., 2000. Stunted plant 1 mediates effects of cytokinin, but not of auxin, on cell division and expansion in the root of *Arabidopsis*. Plant Physiol. 124, 1718–1727.

Ben-Ze'ev, A., 1991. Animal cell shape changes and gene expression. Bioessays 13, 207–212.

Berfield, A.K., Spicer, D., Abrass, C.K., 1997. Insulin-like growth factor I (IGF-I) induces unique effects in the cytoskeleton of cultured rat glomerular mesangial cells. J. Histochem. Cytochem. 45, 583–594.

Bloom, K., Joglekar, A., 2010. Towards building a chromosome segregation machine. Nature 463, 446–456.

Bohn-Courseau, I., 2010. Auxin: a major regulator of organogenesis L'auxine: un regulateur majeur de l'organogenèse. C. R. Biol. 333, 290–296.

Boisvieux-Ulrich, E., Lainé, M.C., Sandoz, D., 1990. Cytochalasin D inhibits basal body migration and ciliary elongation in quail oviduct epithelium. Cell Tissue Res. 259, 443–454.

Bornens, M., 2012. The centrosome in cells and organisms. Science 335, 422–426.

Bounoutas, A., Kratz, J., Emtage, L., Ma, C., Nguyen, K.C., Chalfie, M., 2011. Microtubule depolymerization in *Caenorhabditis elegans* touch receptor neurons reduces gene expression through a p38 MAPK pathway. Proc. Natl. Acad. Sci. USA 108, 3982–3987.

Boyd, J.S., Gray, M.M., Thompson, M.D., Horst, C.J., Dieckmann, C.L., 2011a. The daughter four-membered microtubule rootlet determines anterior–posterior positioning of the eyespot in *Chlamydomonas reinhardtii*. Cytoskeleton 68, 459–469.

Boyd, J.S., Mittelmeier, T.M., Dieckmann, C.L., 2011b. New insights into eyespot placement and assembly in Chlamydomonas. Bioarchitecture 1, 196–199.

Brenner, E.D., Stahlberg, R., Mancuso, S., Vivanco, J., Baluška, F., Van Volkenburgh, E., 2006. Plant neurobiology: an integrated view of plant signaling. Trends Plant Sci. 11, 413–419.

Bucciarelli, E., Giansanti, M.G., Bonaccorsi, S., Gatti, M., 2003. Spindle assembly and cytokinesis in the absence of chromosomes during Drosophila male meiosis. J. Cell Biol. 160, 993–999.

Cabej, N.R., 1984. *Mbi Natyrën e Jetës* (On the Nature of Life). 8 Nëntori, Tiranë, pp. 32.

Cabej, N.R., 2005. Neural Control of Development. Albanet, Dumont, NJ, pp. 35–47.

Cabej, N.R., 2008. Epigenetic Principles of Evolution. Albanet, Dumont, NJ, pp. 29–50.

Cabej, N.R., 2012. Epigenetic Principles of Evolution. Elsevier, London/Waltham, MA, pp. 33.

Cannon, W., 1963. The Wisdom of the Body. W.W. Norton and Co., New York, NY.

Casas-Delucchi, C.S., Cardoso, M.C., 2011. Epigenetic control of DNA replication dynamics in mammals. Nucleus 2, 370–382.

Catlett, N.L., Weisman, L.S., 2000. Divide and multiply: organelle partitioning in yeast. Curr. Opin. Cell Biol. 12, 509–516.

Chang, F., Martin, S.G., 2009. Shaping fission yeast with microtubules. Cold Spring Harb. Perspect Biol. 1, a001347.

Chen, J-G., Yang, C-P.H., Cammer, M., Horwitz, S.B., 2003. Gene expression and mitotic exit induced by microtubule-stabilizing drugs. Cancer Res. 63, 7891–7899.

Classen-Bockhoff, R., 2001. Plant morphology: the historic concepts of Wilhelm Troll, Walter Zimmermann and Agnes Arber. Ann. Bot. 88, 1153–1172.

Conduit, P.T., Brunk, K., Dobbelaere, J., Dix, C.I., Lucas, E.P., Raff, J.W., 2010. Centrioles regulate centrosome size by controlling the rate of Cnn incorporation into the PCM. Curr. Biol. 20, 2178–2186.

Crawford, S., Shinohara, N., Sieberer, T., Williamson, L., George, G., Hepworth, J., et al., 2010. Strigolactones enhance competition between shoot branches by dampening auxin transport. Development 137, 2905–2913.

de Castro, R.D., Zheng, X., Bergervoet, J.H.W., De Vos, C.H.R., Bino, R.J., 1995. β-Tubulin accumulation and DNA replication in imbibing tomato seeds. Plant Physiol. 109, 499–504.

Delattre, M., Gönczy, P., 2004. The arithmetic of centrosome biogenesis. J. Cell Sci. 117, 1619–1630.

Drake, T., Vavylonis, D., 2010. Cytoskeletal dynamics in fission yeast: a review of models for polarization and division. HFSP J. 4, 122–130.

Dubas, E., Custers, J., Kieft, H., Wędzony, M., van Lammeren, A.A.M., 2011. Microtubule configurations and nuclear DNA synthesis during initiation of suspensor-bearing embryos from *Brassica napus* cv. Topas microspores. Plant Cell Rep. 30, 2105–2116.

Errington, J., Daniel, R.A., Scheffers, D-J., 2003. Cytokinesis in bacteria. Microbiol. Mol. Biol. Rev. 67, 52–65.

Georgiev, D.,2011. Quantum Mind Theories. Available from: <http://www.quantum-mind.co.uk/danko-georgiev-c169.html> (accessed 04.2012).

Gerdes, K., Møller-Jensen, J., Ebersbach, G., Kruse, T., Nordström, K., 2004. Bacterial mitotic machineries. Cell 116, 359–366.

Glade, N., 2008. Computing with the cytoskeleton: a problem of scale. Int. J. Unconv. Comput. 4, 33–44.

Glade, N., Demongeot, J., Tabony, J., 2004. Microtubule self-organisation by reaction-diffusion processes causes collective transport and organisation of cellular particles. BMC Cell Biol. 2004 (5), 3.

Glotzer, M., 2009. The 3Ms of central spindle assembly: microtubules, motors and MAPs. Nat. Rev. Mol. Cell Biol. 10, 9–20.

Gould, J.L., 1982. Ethology - The Mechanisms and Evolution of Behavior. W.W. Norton and Co., pp. 319.

Grill, S.W., Hyman, A.A., 2005. Spindle positioning by cortical pulling forces. Dev. Cell 8, 461–465.

Gu, L., Gaertig, J., Stargell, L.A., Gorovsky, M.A., 1995. Gene-specific signal transduction between microtubules and tubulin genes in *Tetrahymena thermophila*. Mol. Cell. Biol. 15, 5173–5179.

Hameroff, S., 1998. Did consciousness cause the cambrian evolutionary explosion?. In: Hameroff, S., Kaszniak, A., Scott, A. (Eds.), Toward a Science of Consciousness II: The 1996 Tucson Discussions and Debates. MIT Press, Cambridge, MA, pp. 421–437.

Hameroff, S.R. and Penrose, R., 1996. Orchestrated reduction of quantum coherence in brain microtubules: a model for consciousness. In: Hameroff, S.R., Kaszniak A.W., Scott, A.C. (Eds.), Toward a Science of Consciousness: The First Tucson Discussions and Debates, MIT Press, Cambridge, MA, pp. 507–539.

Heidemann, S.R., Joshi, H.C., Schechter, A., Fletcher, J.R., Bothwell, M., 1985. Synergistic effects of cyclic AMP and nerve growth factor on neurite outgrowth microtubule stability of PC12 cells. J. Cell Biol. 100, 916–927.

Hejnowicz, Z., Krause, E., Glebicki, K., Sievers, A., 1991. Propagated fluctuations of the electric potential in the apoplasm of *Lepidium sativum* L. roots. Planta 186, 127–134.

Hesketh, J.E., Pryme, I.F., 1991. Interaction between mRNA, ribosomes and the cytoskeleton. Biochem. J. 277, 1–10.

Heuttner, A., 1933. Continuity of the centrioles in *Drosophila melanogaster*. Z. Zellforsch. Mikrosk. Anat. 19, 119–134.

Höög, 2003. The 3D Architecture of Interphase Microtubule Cytoskeleton and Functions of Microtubule Plus End Tracking Proteins in Fission Yeast. Dissertation. Available from: <http://archiv.ub.uni-heidelberg.de/volltextserver/volltexte/2007/7515/pdf/thesis_final_with_summary.pdf/> (accessed 05.2012).

Hufnagel, L.A., 2008. Cortical ultrastructure and chemoreception in ciliated protists (ciliophora). Microsc. Res. Tech. 22, 225–264.

Hwang, I., Sheen, J., 2001. Two-component circuitry in *Arabidopsis* cytokinin signal transduction. Nature 413, 383–389.

Ikeda, K., Zhapparova, O., Brodsky, I., Semenova, I., Tirnauer, J.S., Zaliapin, I., et al., 2011. CK1 activates minus-end-directed transport of membrane organelles along microtubules. Mol. Biol. Cell 22, 1321–1329.

Ingber, D.E., Folkman, J., 1989. Mechanochemical switching between growth and differentiation during fibroblast growth factor-stimulated angiogenesis in vitro: role of extracellular matrix. J. Cell Biol. 109, 317–330.

Janmey, P.A., 1998. The cytoskeleton and cell signaling: component localization and mechanical coupling. Physiol. Rev. 78, 763–781.

Kateriya, S., Nagel, G., Bamberg, E., Hegemann, P., 2004. "Vision" in single-celled algae. News Physiol. Sci. 19, 133–137.

Kirschner, M., Gerhart, J., 1998. Evolvability. Proc. Natl. Acad. Sci. USA 95, 8420–8427.

Kirschner, M., Mitchison, T., 1986. Beyond self-assembly: from microtubules to morphogenesis. Cell 45, 329–342.

Kobayashi, T., Dynlacht, B.D., 2009. Regulating the transition from centriole to basal body. J. Cell Biol. 193, 435–444.

Koziolek, C., Grams, T.E.E., Schreiber, U., Matyssek, R., Fromm, J., 2003. Transient knockout of photosynthesis mediated by electrical signals. New Phytol. 161, 715–722.

Kuppusamy, K.T., Walcher, C.L., Nemhauser, J.L., 2009. Cross-regulatory mechanisms in hormone signaling. Plant Mol. Biol. 69, 375–381.

Lacroix, C., Jeune, B., Barabe, D., 2005. Encasement in plant morphology: an integrative approach from genes to organisms. Can. J. Botany 83, 1207–1221.

Lee, J.S., von der Ahe, D., Kiefer, B., Nagamine, Y., 1993. Cytoskeletal reorganization and TPA differently modify AP-1 to induce the urokinase-type plasminogen activator gene in LLC-PK_1 cells. Nucl. Acids Res. 21, 3365–3372.

Lewis, D.R., Negi, S., Sukumar, P., Muday, G.K., 2011. Ethylene inhibits lateral root development, increases IAA transport and expression of PIN3 and PIN7 auxin efflux carriers. Development 138, 3485–3495.

Leys, S.P., Mackie, G.O., Meech, R.W., 1999. Impulse conduction in a sponge. J. Exp. Biol. 202, 1139–1150.

Leyser, O., 2010. The power of Auxin in plants. Plant Physiol. 154, 501–505.

Lodish, H., Berk, A., Zipursky, S.L., Matsudaira, P., Baltimore, D., Darnell, J., 2000. *Molecular Cell Biology*, fourth ed. W.H. Freeman, New York, NY.

Lomakin, A.J., Semenova, I., Zaliapin, I., Kraikivski, P., Nadezhdina, E., Slepchenko, B.M., et al., 2009. CLIP-170-dependent capture of membrane organelles by microtubules initiates minus-end directed transport. Dev. Cell 17, 323–333.

Lomakin, A.J., Kraikivski, P., Semenova, I., Ikeda, K., Zaliapin, I., Tirnauer, J.S., et al., 2011. Stimulation of the CLIP-170-dependent capture of membrane organelles by microtubules through fine tuning of microtubule assembly dynamics.. Biol. Cell 22, 4029–4037.

Masi, E., Ciszak, M., Stefano, G., Renna, L., Azzarello, E., Pandolfi, C., et al., 2009. Spatiotemporal dynamics of the electrical network activity in the root apex. Proc. Natl. Acad. Sci. USA 106, 4048–4053.

Megraw, T.L., Kao, L.R., Kaufman, T.C., 2001. Zygotic development without functional mitotic centrosomes. Curr. Biol. 11, 116–120.

Mitchison, T.J., Kirschner, M.W., 1985. Properties of the kinetochore in vitro. II. Microtubule capture and STP-dependent translocation. J. Cell Biol. 101, 766–777.

Mittelmeier, T.M., Boyd, J.S., Lamb, M.R., Dieckmann, C.L., 2011. Asymmetric properties of the *Chlamydomonas reinhardtii* cytoskeleton direct rhodopsin photoreceptor localization. J. Cell Biol. 193 (4), 741–753.

Moutinho-Pereira, S., Debec, A., Maiato, H., 2009. Microtubule cytoskeleton remodeling by acentriolar microtubule-organizing centers at the entry and exit from mitosis in *Drosophila* somatic cells. Mol. Biol. Cell 20, 2796–2808.

Müller, D., Leyser, O., 2011. Auxin, cytokinin and the control of shoot branching. Ann. Bot. 107, 1203–1212.

Muñoz-Espín, D., Daniel, R., Kawai, Y., Carballido-Lopez, R., Castilla-Llorentea, V., Errington, J., et al., 2009. The actin-like MreB cytoskeleton organizes viral DNA replication in bacteria. Proc. Natl. Acad. Sci. USA 106, 13347–13352.

Nanninga, N., 2001. Cytokinesis in prokaryotes and eukaryotes: common principles and different solutions. Microbiol. Mol. Biol. Rev. 65, 319–333.

Nemhauser, J.L., Hong, F., Chory, J., 2006. Different plant hormones regulate similar processes through largely nonoverlapping transcriptional responses. Cell 126, 467–475.

Nickelsen, K., 2002. Alle Gestalten sind ähnlich, und keine gleichet der andern. Bilder von Pflanzenarten im 18. Jahrhundert. In: Müller-Wille, S. (Ed.), Sammeln – Ordnen – Wissen. Beiträge zu einem Festkolloquium aus Anlaß des 80. Geburtstages von Ilse Jahn, pp. 13–30. Available from: <http://www.philoscience.unibe.ch/documents/TexteHS09/Nickelsen2002.pdf/> (accessed 03.2012).

Nicolas, E., Chenouard, N., Olivo-Marin, J.-N., Guichet, A., 2009. A dual role for actin and microtubule cytoskeleton in the transport of golgi units from the nurse cells to the oocyte across ring canals. Mol. Biol. Cell 20, 556–568.
Norris, V., Zemirline, A., Amar, P., Audinot, J.N., Ballet, P., Ben-Jacob, E., et al., 2011. Computing with bacterial constituents, cells and populations: from bioputing to bactoputing. Theory Biosci. 130, 211–228.
Olson, E.N., Nordheim, A., 2010. Linking actin dynamics and gene transcription to drive cellular motile functions. Nat. Rev. Mol. Cell Bio. 11, 353–365.
Orgel, L.E., 1973. *The Origins of Life—Molecules and Natural Selection*. John Wiley & Sons, New York, NY, pp. 189.
Palmberg, L., Sjölund, M., Thyberg, J., 1985. Phenotype modulation in primary cultures of arterial smooth-muscle cells: reorganization of the cytoskeleton and activation of synthetic activities. Differentiation 29, 275–283.
Park, J.J., Loh, Y-P., 2008. Minireview: how peptide hormone vesicles are transported to the secretion site for exocytosis. Mol. Endocrinol. 22, 2583–2595.
Penrose, R., 2003. The Shadows of the Mind. Available from: <http://community.fortunecity.ws/underworld/continue/56/shadow.html>.
Péret, B., De Rybel, B., Casimiro, I., Benková, E., Swarup, R., Laplaze, L., et al., 2009. Arabidopsis lateral root development: an emerging story. Trends Plant Sci. 14, 399–408.
Petrášek, J., Friml, J., 2009. Auxin transport routes in plant development. Development 136, 2675–2688.
Pfaffmann, J.O., Conrad, M., 2000. Adaptive information processing in microtubule networks. Biosystems 55, 47–57.
Pfaffmann, J.O., Zauner, K-P., 2001. Scouting context-sensitive components. The Third NASA/DoD Workshop on Evolvable Hardware—EH. 12–14 July 2001, Long Beach, CA.
Picone, R., Ren, X., Ivanovitch, K.D., Clarke, J.D.W., McKendry, RA., Baum, B., 2010. A polarised population of dynamic microtubules mediates homeostatic length control in animal cells. PLoS Biol. 8 (11), e1000542.
Piel, M., Tran, P.T., 2009. Cell shape and cell division in fission yeast (Minireview). Curr. Biol. 19, R823–R827.
Pigliucci, M., 2008. Is evolvability evolvable? Nat. Rev. 9, 75–82.
Pon, L.A., 2011. Organelle transport: mitochondria hitch a ride on dynamic microtubules. Curr. Biol. 21, R654–R656.
Priel, A., Ramos, A.J., Tuszynski, J.A., Cantiello, H.F., 2006. A bio-polymer transistor: electrical amplification by microtubules. Biophys. J. 90, 4639–4643.
Providence, K.M., Kutz, S.M., Higgins, P.J., 1999. Perturbation of the actin cytoskeleton induces PAI-1 gene expression in cultured epithelial cells independent of substrate anchorage. Cell Motil. Cytoskeleton 42, 218–229.
Puck, T.T., Krystosek, A., Chan, D.C., 1990. Genome regulation in mammalian cells. Somat. Cell Mol. Genet. 16, 257–265.
Rapp, S., Saffrich, R., Anton, M., Jäkle, U., Ansorge, W., Gorgas, K, et al., 1996. Microtubule-based peroxisome movement. J. Cell Sci. 109, 837–849.
Rasmussen, S., Karampurwala, H., Vaidyanath, R., Jensen, K.S., Hameroff, S., 1990. Computational connectionism within neurons: A model of cytoskeletal automata subserving neural networks. Physica D 42, 428–449.
Rosette, C., Karin, M., 1995. Cytoskeletal control of gene expression: depolymerization of microtubules activates NF-xB. J. Cell Biol. 128, 1111–1119.

Sakai, H., Honma, T., Aoyama, T., Sato, S., Kato, T., Tabata, S., et al., 2001. ARR1, a transcription factor for genes immediately responsive to cytokinins. Science 294, 1519–1521.

Sathananthan, A.H., Kola, I., Osborne, J., Trounson, A., Ng, S.C, Bongso, A., et al., 1991. Centrioles in the beginning of human development. Proc. Natl. Acad. Sci. USA 88, 4806–4810.

Sathananthan, A.H., Ratnam, S.S., Ng, S.C., Thrin, J.J., Gianaroli, L., Trounson, A., 1996. The sperm centriole: its inheritance, replication and perpetuation in early human embryos. Hum. Reprod. 11, 345–356.

Sato, C., Aikawa, K., Sugiyama, S., Nabeta, K., Masuta, C., Matsuura, H., 2011. Distal transport of exogenously applied jasmonoyl–isoleucine with wounding stress. Plant Cell Physiol. 52, 509–517.

Schatten, H., Sun, Q-Y., 2009. The role of centrosomes in mammalian fertilization and its significance for ICSI. Mol. Hum. Reprod. 15, 531–538.

Schmuelling, T., 2002. New insights into the functions of cytokinins in plant development. J. Plant Growth Regul. 21, 40–49.

Schollenberger, L., Gronemeyer, T., Huber, C.M., Lay, D., Wiese, S., Meyer, H.E., et al., 2010. RhoA regulates peroxisome association to microtubules and the actin cytoskeleton. PLoS ONE 5 (11), e13886. doi: 10.1371/journal.pone.0013886.

Schrödinger, E., 1944. *What Is Life?* Cambridge University Press, Cambridge.

Schwechheimer, C., 2008. Understanding gibberellic acid signaling—are we there yet? Curr. Opin. Plant Biol. 11, 9–15.

Seeley, E.S., Nachury, M.V., 2010. The perennial organelle: assembly and disassembly of the primary cilium. J. Cell Sci. 123, 511–518.

Semenova, I., Burakov, A., Berardone, N., Zaliapin, I., Slepchenko, B., Svitkina, T., et al., 2008. Actin dynamics is essential for myosin-based transport of membrane organelles. Curr. Biol. 18, 1581–1586.

Sherrington, C.S., 1951. *Man on His Nature*, second ed. Cambridge University Press, Cambridge.

Shih, Y-L., Rothfield, L., 2006. The bacterial cytoskeleton. Microbiol. Mol. Biol. Rev. 70, 729–754.

Sit, S-T., Manser, E., 2011. Rho GTPases and their role in organizing the actin cytoskeleton. J. Cell Sci. 124, 679–683.

Stanković, B., Davies, E., 1997. Intercellular communication in plants: electrical stimulation of proteinase inhibitor gene expression in tomato. Planta 202, 402–406.

Taborsky Jr, G.J., 2010. The physiology of glucagon. J. Diabetes Sci. Technol. 4, 1338–1344.

Thorens, B., 2011. Brain glucose sensing and neural regulation of insulin and glucagon secretion. Diabetes Obes. Metab. 13 (Suppl. 1), 82–88.

Thyberg, J., 1984. The microtubular cytoskeleton and the initiation of DNA synthesis. Exp. Cell Res. 155, 1–8.

Tischer, C., ten Wolde, P.R., Dogterom, M., 2010. Providing positional information with active transport on dynamic microtubules. Biophys. J. 99, 726–735.

Tolstonog, G.V., Sabasch, M., Traub, P., 2002. Cytoplasmic intermediate filaments are stably associated with nuclear matrices and potentially modulate their DNA-binding function. DNA Cell Biol. 21, 213–239.

Tran, P.T., Doye, V., Chang, F., Inoue, S., 2000. Microtubule-dependent nuclear positioning and nuclear-dependent septum positioning in the fission yeast *Saccharomyces pombe*. Biol. Bull. 199, 205–206.

Tran, P.T., Marsh, L., Doyec, V., Inoué, S., Chang, F.A., 2001. Mechanism for nuclear positioning in fission yeast based on microtubule pushing. J. Cell Biol. 153, 397–412.

Trewavas, A., 2006. A brief history of systems biology. Plant Cell 18, 2420–2430.

Tuszynski, J.A., Brown, J.A., Hawrylak, P., 1998. Dielectric polarization, electrical conduction, information processing and quantum computation in microtubules. Are they plausible? Phil. Trans. R. Soc. Lond. A 356, 1897–1926.

Vaughan, S., Dawe, H.R., 2011. Common themes in centriole and centrosome movements. Trends Cell Biol. 21, 57–66.

Vernoux, T., Besnard, F., Traas, J., 2010. Auxin at the shoot apical meristem. Cold Spring Harb. Perspect. Biol. 1, 1–14.

Volkov, A.G., Adesina, T., Jovanov, E., 2008. Charge induced closing of *Dionaea muscipula* Ellis trap. Bioelectrochemistry 74, 16–21.

Vollmer, W., Höltje, J.V., 2001. Morphogenesis of *Escherichia coli*. Curr. Opin. Microbiol. 4, 625–633.

von Bertalanffy, L., 1950. The theory of open systems in physics and biology. Science 111, 23–29.

Wasmeier, C., Hume, A.N., Bolasco, G., Seabra, M.C., 2008. Melanosomes at a glance. J. Cell Sci. 121, 3995–3999.

Werner, M., Munro, E., Glotzer, M., 2007. Astral signals spatially bias cortical myosin recruitment to break symmetry and promote cytokinesis. Curr. Biol. 17, 1286–1297.

Wildon, D.C., Thain, J.F., Minchin, P.E.H., Gubb, I.R., Reilly, A.J., Skipper, Y.D., et al., 1992. Electrical signalling and systemic proteinase inhibitor induction in the wounded plant. Nature 360, 62–65.

Yujiri, T., Fanger, G.R., Garrington, T.P., Schlesinger, T.K., Gibson, S., Johnson, G.L., 1999. MEK kinase 1 (MEKK1) transduces c-Jun NH_2-terminal kinase activation in response to changes in the microtubule cytoskeleton. J. Biol. Chem. 274, 12605–12610.

Zhang, D., Nicklas, R.B., 1996. 'Anaphase' and cytokinesis in the absence of chromosomes. Nature 382, 466–468.

Zimmermann, M.R., Maischak, H., Mithöfer, A., Boland, W., Felle, H.H., 2009. System potentials, a novel electrical long-distance apoplastic signal in plants, induced by wounding. Plant Physiol. 149, 1593–1600.

2 Epigenetics of Reproduction in Animals

Newmann's Machines: Living Organisms Defy Human Imagination

The Aristotelian concept of the soul as the essential distinction between living and nonliving things dominated the European view of the nature of living organisms for many centuries after antiquity. In the seventeenth century, French philosopher René Descartes (1596–1650), within the framework of his dualist philosophy of mind and body, developed the idea that the human body works like a machine and the organism's physiological functions can be explained as operating in a way mechanically similar to clocks. In contrast with the Aristotelian view, Descartes believed that the human body is governed by the same physical laws as animals, plants, and inorganic matter, and saw the mind or soul as an exclusive attribute of human beings. His reductionist concept that animals and human bodies are governed by the laws of nature laid the theoretico-philosophical grounds for the scientific study of the living world.

A century later, in 1748, his compatriot, the physician and philosopher Julien Offray de La Mettrie (1709–1751) in *L'homme machine* (Man a Machine) further swung the pendulum toward the materialist view of biology. He erased any distinction between living and nonliving matter, proclaiming that both obey the same natural laws. The human mind and body can be explained with the same natural laws. Later, Charles Darwin's theory of natural selection provided strong evolutionary theoretical support to La Mettrie's idea that still dominates the biological thought that laws of physics and chemistry can explain all the functions of animals, including the human mind.

The analogy of living beings as machines stimulated many scientists and engineers to think of the possibility of building machines that could reproduce themselves just as animals and plants do. In 1966, the Hungarian–American mathematician John von Neumann came up with the idea that it is logically possible to build a self-operating and self-replicating machine (von Neumann, 1966) capable of manufacturing copies of itself when provided with parts. He called his computing machine the "universal constructor." In theory, his self-replicating machine could function even under conditions of perturbation-induced changes in various parameters. Then, by using an installed program in a memory tape, it would assemble a copy of its own and install and turn on a copy of the operating program in the daughter machine. The machine, thus, would enter a perpetual cycle of self-replication.

Building the Most Complex Structure on Earth. DOI: http://dx.doi.org/10.1016/B978-0-12-401667-5.00002-X

He showed that computing machines are fragile to perturbation and are designed to crash when errors occur. Living systems cannot afford to stop, for it would be a point of no return. Instead, they have evolved the following:

a. The ability to detect and localize the site of the error, process the afferent information on the nature of the error, and generate instructions to correct the error or repair the damaged cell or supracellular structure, all in the controller.
b. The robustness to function normally, even under conditions of perturbation.

Living systems evolved many components whose purpose is the system reliability. They are highly integrated, and they can function normally even after errors, if those errors are perceived as negligible.

In his theory, von Neumann posited that living organisms are very complex aggregates of their elementary parts, and from a thermodynamic perspective, they are highly improbable structures. "That they should occur in the world at all is a miracle of the first magnitude; the only thing which removes, or mitigates, this miracle is that they reproduce themselves" (von Neumann, 1966). He observed that a theoretical difficulty arises when it comes to the reproduction of artificial automata (self-replicating machines): "one gets a very strong impression that complication, or productive potentiality in an organization, is degenerative, that an organization which synthesizes something is necessarily more complicated, of a higher order, than the organization it synthesizes."

This is an essential difference between living organisms and artificial automata. But, according to von Neumann, the complexity is degenerative only below a certain low level. If the complexity is crudely measured by the number of components of a system, it is possible, he reasoned, to build an automaton of such a high level of complexity that can "construct other automata of equal or higher complexity." These machines would compete in shared environments, would be subject to natural selection, and may evolve similarly to living systems. Suggestions are made that in order to minimize effects of perturbations during the assemblage of the universal constructor, subsystems may be needed to construct automaton's "workplace." The universal constructor in the workplace will pick up parts and assemble them into a copy of itself in sequential cycles of self-reproduction.

The self-reproduction of the universal constructor can be seen as an artificial analog for unicellulars, which produce their twin organism in a complete form capable of independent life and self-reproduction. This holistic mode of self-reproduction does not fit into the model of the development of multicellular animals and plants, where parents control only part of the development of offspring rather than the whole development. Interestingly, William R. Buckley recently came up with an alternative model of self-reproduction reminiscent of the development of multicellular animals based on the production of gametes (eggs and sperm cells). According to his model, the universal constructor builds not a copy of itself, but rather an incomplete structure capable of producing its parts, and assembles them into a complete functioning machine capable of self-reproduction. In clear distinction from von Neumann's machine, where the daughter machine becomes functional only after it is fully assembled by the mother machine, Buckley's model has the daughter machine becoming operational and taking over its own development before her assembly is completed (Buckley, 2008).

Buckley's model displays an essential feature of the reproduction of multicellular animals: the zygote (egg cell in parthenogenetic organisms) develops according to instructions provided parentally to gametes until the phylotypic stage, when the controller of the embryo, the central nervous system (CNS) takes over not just further development, but the growth and organogenesis of the animal until adulthood as well.

The eumetazoan "living machine" builds neither the complete machine nor the machine in miniature, which by growing in size, becomes a fully operational and replicating "machine" of its kind; nor does the parent insert an operating program in it. Rather, it provides the epigenetic information for its Bauplan at the phylotypic stage. At this stage, long before the construction of the "living machine" is complete, the embryo possesses an awesome information-generating machine, the CNS, which starts functioning and generating information for building its own species-specific multicellular structure and its own operating program.

All the self-replicating machines proposed until today are vastly inferior to even the simplest unicellular organism so far as sophistication, self-sufficiency, and reliability are concerned. But this is not always adequately appreciated. The complexity of metazoans is often reduced to the mechanisms of protein synthesis. So, for example, in an intriguing book called *Decoding Reality*, Vedral (2010) compares living systems with von Neumann's machine consisting, in his view, of four basic components:

1. The protein-synthesizing machine.
2. The biological nanoengine (akin to the Xerox copier).
3. Enzymes that act as controllers switching the nanoengine on and off.
4. The DNA information set.

Vedral continues, "So we see DNA as key to this process, as it contains the blueprint of how each cell operates and replicates. Based on it, the constructor machine within our cells synthesizes amino acids, which in turn make up various proteins and new cells for our bodies."

Unfortunately, his comparison is not appropriate. First, he compares the living system to a protein-synthesizing machine, as if proteins were building blocks of the multicellular organism. Two centuries after Schwann and Schleiden, we know that the basic unit of all multicellular organisms is the cell, not the protein. Hence, the key to understanding the nature of multicellular systems is to understand how the complex multicellular structure consisting of billions or even trillions of cells of the most different types of structure and function arises from a single cell (egg cell or zygote). Multicellular organisms, and unicellulars as well, are not bags of proteins. They are supracellular structures where the spatial arrangement of cells of different types follows strictly determined and highly complex patterns. The information of trillions or quadrillions of bits used for erecting multicellular structures is larger by several orders than all the information contained in all the nucleotide sequences of the DNA.

Unlike unicellulars, metazoans do not produce copies of themselves. They produce specialized cells, eggs, and sperm cells equipped with epigenetic information, which enables them (or the zygote or egg cell in parthenogenetic organisms) to develop the Bauplan up to an early embryonic stage known as the *phylotypic stage*. At this stage, when the epigenetic information provided to gametes becomes exhausted, the nervous system would have developed. It is the first organ system to develop, and it serves the embryo as a primordial epigenetic programmer. From this point in time, the embryo becomes developmentally self-sufficient. It generates the epigenetic information necessary to control the further postphylotypic development of the embryo into an adult organism.

Biological Information

The Spatial Arrangement of Cells Needs a New Kind of Information

The reproduction of any biological system basically requires the following:

- Building blocks.
- Information on the spatiotemporal arrangement of the building blocks.
- Free energy for implementing that arrangement.

The building blocks of unicellulars are proteins and other molecules, supramolecular structures, and organelles. In multicellulars, the cell is the universal building block.

The cell structure in unicellulars, plants, and animals exhibits significant differences. Figure 2.1 shows a *Paramecium* as a representative of eukaryote unicellulars. A generalized structure of the plant cell and animal cell is presented in Figure 2.2.

The building blocks are the objects of special studies of cytology in unicellulars and of cytology, histology, and anatomy in multicellulars. The use and transformation of the free energy in the metabolic processes of cells and multicellular organisms is the object of study in biochemistry. Structure, metabolism, and biochemistry are known in adequate detail. Hence, we are going to only review the essentials of biological information responsible for the spatiotemporal arrangement of molecular structures, supramolecular structures, and organelles in unicellulars, and of billions or trillions of different types of cells in metazoans from which the breathtaking diversity of forms in the animal kingdom arises.

The erection of animal structure, by any account, requires huge amounts of information of various types, which in an all-embracing term we call *biological information*. This information is not only used in the process of reproduction via individual development but it also functions to maintain the structural, functional, and morphological identity of species during the life of the animal. To avoid the confusion that may stem from the present varying concepts of inheritance, we start with a narrow definition of biological inheritance: all structures, functions, behaviors, and life histories derived from the parent. The definition remains purely biological in that it ignores all extrabiological artifacts, such as the construction of nests, human material, and spiritual culture, which have found a home in discussions of biological inheritance.

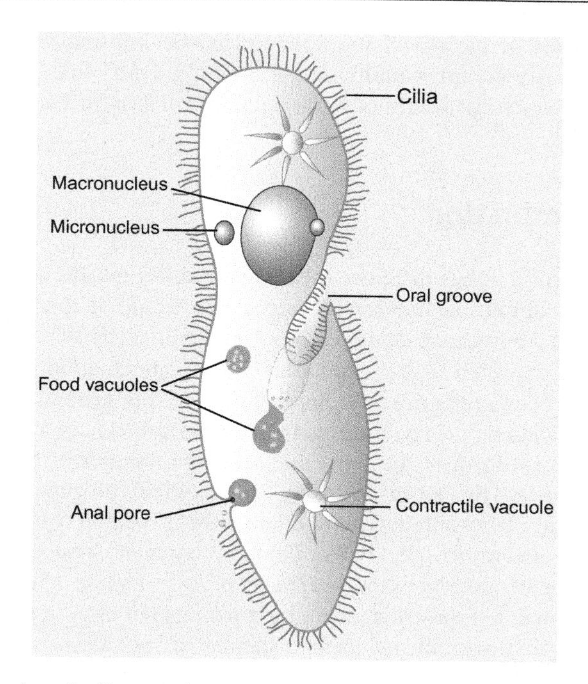

Figure 2.1 Drawing of a *Paramecium caudatum*.
Source: From http://sahsrojas.pbworks.com/w/page/3719748/svg%20paramecium.

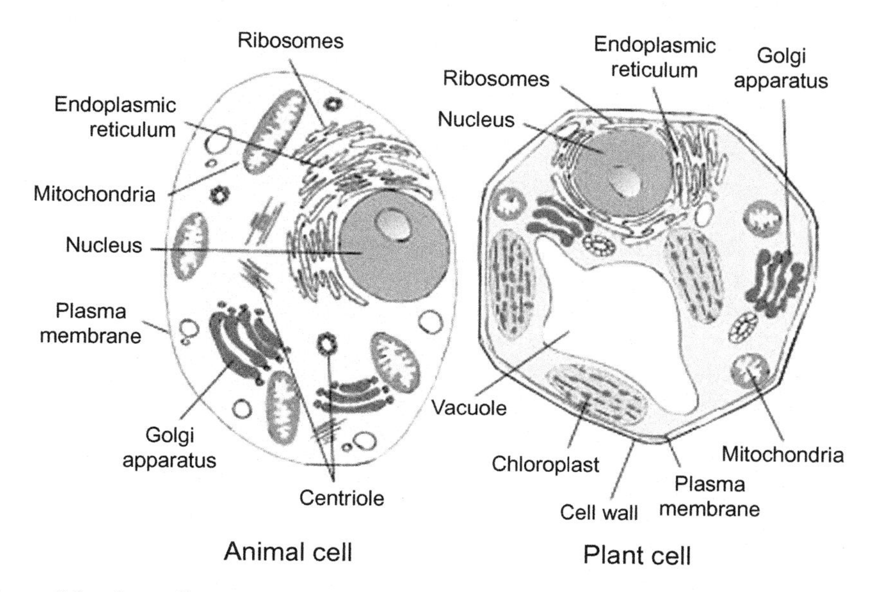

Figure 2.2 Generalized structures of an animal and a plant cell.
Source: From http://missbzscience.wikispaces.com/4%29+Cells.

Since at the heart of biological reproduction is the information that is used to erect the tremendously complex multicellular animals, I will first present a concise description of the nature and origin of biological information invested in the erection of animal structures.

Biological Information

Information is a term that has different meanings in different disciplines; there is no generally accepted definition universally applicable to all of them. The American mathematician and engineer Claude Elwood Shannon (1916–2001) pointed out: "It is hardly to be expected that a single concept of information would satisfactorily account for the numerous possible applications of this general field" (Shannon, 1993). In communications theory, there is no difference between a meaningful message and a nonsensical form of the same message; the number of bits transmitted is what matters primarily. This is not the case with biological information.

For many centuries, from antiquity until modern times, people believed that life may arise spontaneously and rather quickly from inorganic matter. Aristotle generalized the concept of the spontaneous generation in *History of Animals* (*ca.* 355 BCE), where he wrote that animals may not only arise from their parents but also "grow spontaneously and not from kindred stock; and of these instances of spontaneous generation some come from putrefying earth or vegetable matter, as the case is with a number of insects, while others are spontaneously generated in the inside of animals out of the secretions of their several organs." The theory of spontaneous generation prevailed among biologists in Europe until the seventeenth century and continued intermittently until the middle of the nineteenth century. With the advent of the scientific method in the seventeenth century, the hypothesis was put to the test by the Italian physician Francesco Redi (1626–1697), who in 1668 succeeded to experimentally refute the hypothesis of spontaneous generation. After biologist John Needham (1713–1781) revived the hypothesis in England, Louis Pasteur (1822–1895), by demonstrating that fermentation is caused by growth of microorganisms, rung the death knell of the hypothesis.

From the modern biological view, spontaneous generation encounters an insuperable difficulty: how can the *de novo* emerging organisms instantaneously acquire the huge amount of information for erecting their complex multicellular structure? Even when constructing a simple Lego structure, some instructions or information on what has to be built is necessary; it cannot arise by piling its building blocks randomly. Building a unicellular organism of millions of molecular and supramolecular building blocks requires an incomparably larger amount of information than a Lego unit does, and several orders greater amount of information would be required to erect a multicellular organism. Structures with such a small probability of arising spontaneously are known as *improbable structures*, and no improbable structure, be it a Lego, a car, a cell, or a multicellular organism, can arise spontaneously. Investment of some sort of information on how to build them is a prerequisite. The counterargument that, theoretically at least, improbable structures can arise spontaneously once (or even twice) in the case of living organisms is invalidated by the fact that these structures arise regularly in the process of the reproduction of living beings.

Genetic Information

Conventionally, *genetic information* is any sequence of bases in DNA (RNA in some viruses), but here I will consider the term to include only those sequences that, through processes of transcription and translation, can produce a polypeptide chain in the appropriate conditions that *per se*, or together with other polypeptide chain(s), can function as a functioning protein in the organism from which it derives. In this context, genetic information is representative of a polypeptide chain.

The conventional wisdom that assigns genes other functions above the molecular level is not warranted by the available biological evidence and is based on guesses. In strictly scientific terms, we still have no mechanism showing how a gene or a group of genes may determine the sequential stages of the development and evolution of a morphological, behavioral, or life history character. Studies on the consequences of gene mutations or lack/inactivation of genes in the development of such characters prove that these genes are necessary, but fall short of suggesting that they may be sufficient for the development of these characters; many nongenetic components are also necessary for the development of those structures, and in their absence, these structures do not develop (or develop with defects). More than half a century of unprecedented, intense research on the function of genes failed to produce evidence that any specific gene is responsible for the development of a supramolecular phenotypic character. No one has ever successfully demonstrated steps by which a gene or a number of genes, *per se*, lead to the formation of a morphological character.

The genetic information, thus, determines the sequences of amino acid residues in the polypeptide chains; hence, it represents "meaningful" or semantic information, as opposed to information in the communication theory, *sensu* Shannon.

Genes contain semantic information for protein biosynthesis alone, and recent developments in biology show that most proteins in vertebrates are produced by alternative splicing (AS), an epigenetic rather than genetic phenomenon (see the section "Alternative Splicing," later in this chapter). In this context, the idea that genes also represent morphological structures or contain information for their development is untenable. That could reasonably be claimed only if it was demonstrated that nongenetic factors are not involved in the process of the development of these structures, which obviously is not the case. Rejecting the idea that genes may contain semantic information for phenotypic characters, Shea (2007) believes that genes have what he calls "correlational information." But one could easily argue that epigenetic structures such as acetylated histones and chromatin marks in general, as well as DNA methylation/demethylation, also contain correlational information of the type he envisages.

From the opposite position, it has been argued that if genetic information is embodied in phenotypic structures, then the information of the relevant phenotypic structures, according to communications theory, will equal the information for the genetic structure that instructed the formation of the phenotypic structure. According to this theory, the information in the phenotype is equal and symmetrical to the information of the genotype and is such whether the information is in the receiver or the sender. This is causal information that shows that the state of the sender and the receiver reflects each other (Griffiths, 2000). However, the extrapolation from communications theory to biological information may not be warranted. The above propositions refute any idea of the primacy of DNA over phenotypic structures as a carrier of biological information and of the direction of the flow of information from DNA to phenotype.

It is curious to observe that as a reaction to the increasing evidence of the role of extragenetic factors in inheritance, many biologists adhere to a dichotomous concept of replicators and interactors, a view that gives genes and DNA a predominant role in inheritance, above and beyond other causal factors in development. According to this view, the genome has the monopoly of biological information, and biological entities are primarily little more than vehicles contributing to the immortality of replicators.

In response, the representatives of a new theoretico-philosophical trend known as the *developmental systems theory* (*DST*) came up with the "parity thesis," positing that all the elements (genetic and nongenetic) that interact during the developmental process are "parental resources." Accordingly, what really exist and determine the developmental and replication cycles of living systems are not replicators and interactors, but "developmental resources" (Griffiths and Gray, 1994). The sum of all developmental resources (i.e., of the elements involved in the developmental process) represent the developmental system. DST thus extended the biological contrivances of inheritance to include extragenetic elements. Developmental processes, according to this view, are cyclic series of events in which each cycle (developmental process) is a "causal consequence" of the preceding one (Griffiths and Gray, 1997). Not only the genome but also the other elements in the developmental system and developmental interactions are replicated (Griffiths and Gray, 1994). Indeed, DNA is not the only known biological structure that can copy itself; centrosomes and centrioles have their own self-replication machinery, and no DNA is involved in their replication (Balczon et al., 1995; Rattner and Phillips, 1973). The already classic experiments of Beisson and Sonneborn (1965) have shown that the *Paramecium* cortex reproduces itself exactly, without involving the genome.

However, Shea (2011) conditions the qualification of the developmental resources as carriers of inherited representations with whether they have the metafunction of "transmitting selected phenotypes down the generations." Nevertheless, he prudently admits that if his concept of the metafunction of the DNA is wrong, then "another account of inherited representation would be needed." In order for a developmental resource to be thought as part of an inheritance system, it should have semantic information in the form of "instructions" for building a structure. By denying this possibility for other "developmental resources," he ignores the unambiguous empirical evidence on transgenerational developmental plasticity. Recall that even according to his own criterion, epigenetic variations in the offspring have to be "carried down the generations," and transgenerational developmental plasticity falls in the category of what he calls "inherited representations." Transgenerational developmental plasticity (discussed further in Chapter 5) arises epigenetically and involves no changes in genes.

Focusing on the well-known fact that ostriches prenatally form calluses on the skin of parts of their body that come in contact with the ground (i.e., breast, pubis, and rump), Shea, like many other biologists, believes that the genotype has both information with indicative content, which tells the organism where the abrasion will take place postnatally, and information with imperative content for producing skin thickenings. Analogous to the old genecentric concept of "masters and servants," he believes in the existence of genetic representations or tokens and consumers specialized in "reading" these genomic representations or the genetic information and producing respective phenotypic outputs (Shea, 2012). But again, in examples of transgenerational developmental plasticity, there are no "genomic representations or tokens" to read, and both "indicative" and "imperative" information come from epigenetic sources (see Chapter 4, section Transgenerational developmental plasticity - Insights into the nature of evolutionary morphological change). For instance, locusts of species *Schistocerca gregaria* (Forskål), in response to crowding, develop several new morphological, behavioral, physiological, and life history characters that are transmitted to the next generation, although no changes in genes are involved in this phase transition.

From an informational point of view, in cases of developmental plasticity, genes themselves serve as channels of upstream epigenetic information, as proposed by Griffiths and Gray (1994).

Is There Any Genetic Program in the Genome?

Conventionally, a genetic program implies the information contained in the genome, presumed to determine patterns of gene expression leading to cell differentiation and sequential events, and ultimately leading to the development of all phenotypic traits. The program specifies *where* and *when* the new organ will develop, *which* cells will participate in the development of the organ, and *how* different types of cells will be arranged in the intricate patterns of tissues in the developing organ.

The route from genes to morphological characters is convoluted and, in many respects, still escapes us. This is the reason why we have no hypothesis or model yet of the sequential steps of the development of an organ. There is no evidence that any part of the genome (a gene or a number of genes) codes or determines steps for developing a brain, a heart, or even a strand of hair. It is true that genes or gene networks are involved or responsible for developing such characters, but their involvement is not linear: a number of different types of cells have to be differentiated and arranged in specific spatial patterns in order to form the organ. But both cell differentiation (Christophersen and Helin, 2010) and dedifferentiation (return of differentiated cells to pluripotency) (Takahashi and Yamanaka, 2006), as well as activation of gene regulatory networks (Cabej, 2012, pp. 23–24, 39–80), are epigenetically rather than genetically determined.

Since the normal development of the phenotype requires precise implementation of the program in time and space, the genome must also receive updated information on the implementation of a preceding step of the developmental program, release the relevant signal to implement the next step, and so on. Moreover, the genome has to restrict the effects of specific circulating inducers on the developing organ and block their effects on the rest of cells of the animal body. Relevant supporting evidence is still absent.

Even if, for the sake of argument, one ignored all of the above difficulties and took for granted that the genome can do all of the above, another insuperable difficulty looms: which of the genomes will function as directing center of the developing organ?

1. All the genomes of the myriad cells of the organism;
2. Genomes of cells participating in the development of the organ; or
3. The genome of a particular cell that sends instructions for the development to all the cells of the presumed organ.

If 1 is true, the embryo would hardly avoid a "developmental chaos."

If 2 is true, the genome of these cells has to be (initially at least) different from the genomes of the rest of the cells of the embryo, which is irreconcilable with the basic tenet of molecular genetics that the genome in all somatic cells of the body is identical.

If 3 is true, this would be a master genome, which is refuted by 2.

Let us again suppose that the above difficulty is also surmountable. Then another insurmountable difficulty pops up from the nature of individual development: during the intrauterine life, many mammal species, including humans, form quadrillions of specific neural connections. Obviously, not only tens of thousands of genes, but also the entirety of the few billion nucleotides in the genome of these species, are negligible as a source of information for determining the huge number of neural connections.

Given that there is no evidence that a gene, a number of genes, or the genome does determine the sequential signals or chemical instructions leading to the development of the phenotype, one cannot reasonably believe that the development of an organism, from a single cell (zygote or egg) to the complex structure of an adult metazoan organism, results from the implementation of any genetic program.

From a genecentric view, it is incomprehensible how from an initial developmental program different programs emerge that are operational in tens to hundreds of different types of cells in a metazoan organism. The cell differentiation that leads to these modified programs is determined and regulated epigenetically (Juliandi et al., 2010; Maruyama et al., 2011) rather than genetically: "Cellular differentiation is a well-orchestrated epigenetic program by which the developmental potential of the cells is progressively restricted" (Maruyama et al., 2011).

Epigenetic Information

In 1942, English scientist Conrad H. Waddington invented the term *epigenetics* to describe the interaction of genes with their environment and the phenotypic result of this interaction, as opposed to a one-to-one correspondence between the genotype and phenotype. Before and after that, only scant evidence existed on inherited nongenetic changes in phenotypic traits. So, for example, the German zoologist Richard Woltereck (1877–1944) observed that under cultivation, the Danish strain of the small crustacean *Daphnia cucullata* produced offspring with helmets, and this *Dauermodifikation* (long-lasting modification) lasted for up to 40 generations before reverting to the ancestral type. He attributed these inherited changes to environmental agents and said that they fell into the category of the "norm of reaction" rather than the inheritance of new characters.

Adequate and unambiguous evidence of inheritance of acquired characters in living organisms has begun reviving up only in recent decades. Numerous described cases of polyphenisms in invertebrates, transgenerational developmental plasticity, and especially discoveries of the inheritance of epigenetic changes in DNA and chromatin made epigenetics a key biological discipline and a focal point of modern biological investigation.

In the mid-1960s, Tracy Morton Sonneborn (1905–1981) of Indiana University, along with Janine Beisson, conducted some important experiments on *Paramecium*. Although the results of these experiments found their way into the *Proceedings of the National Academy of Sciences* in the United States (Beisson and Sonneborn, 1965), their experiments and conclusions did not get the attention they deserved; but this is hardly surprising for ideas that challenge gene theory at its apogee. After mating (conjugation), *Paramecium tetraurelia* sometimes fails to separate completely, and particular ciliary rows of one individual's cortex are incorporated into the recipient's cortex with rows of cilia in reverse polarity (180° rotation). For the recipient cell, the part of the cortex with reverse cilia polarity is an "acquired character." According to conventional genetic knowledge, it would be expected that the character would not appear in the offspring and later generations because it arose not from specific changes in the genome. Contrary to expectations, the progeny of recipients inherited the donor's reverse (posterior–anterior) polarity. The cilia of the grafted cortex waved in the opposite direction of the host's own cilia, i.e., they were not in the host's control. The same occurred when paramecia were experimentally grafted with pieces of cortex from donor paramecia; again, the host transmitted to the offspring not its own but the donor's polarity of cilia. Experiments showed clearly that determinants of cortical inheritance were neither genes nor the cytoplasm, but the donor's cortex (Figure 2.3).

Experiments were construed so that no contamination of the transplant (donor cortex) with DNA or liquid cytoplasm occurred, thus excluding the possibility of nuclear or cytoplasmic DNA involvement in the transmission of the character to the offspring. So no genetic material (genes, DNA, or RNA) was involved in the inheritance of the donor-derived reverse ciliary polarity; only the cortex was. Observations of *Tetrahymena pyriformis* in later studies showed that the number of ciliary meridians in different individuals often is different. It was experimentally demonstrated that these different "corticotypes" are hereditary traits and are determined not by genes or DNA (i.e., not by the nucleus), but by the cortex itself (Nanney, 1966).

The nongenetic cortical inheritance, which was also known as "cytotaxis" and "structural memory," was later observed in other ciliates. Sonneborn concluded:

> *For all cortical traits examined, development is hereditarily determined by existing and self-reproducing cortical arrangements: the genes (or DNA) doubtless control synthesis of the molecular building blocks, but not their site of assembly or the position, orientation, and number of assemblies.*
>
> *Sonneborn (1970)*

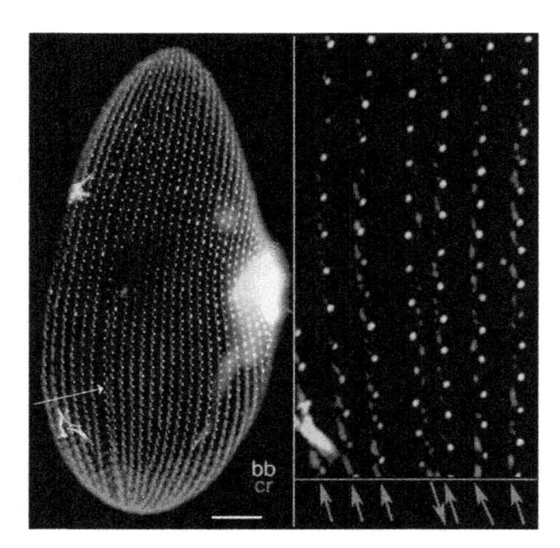

Figure 2.3 Visualization of an inverted ciliary row. In the left panel, the arrow points to a discontinuity in the cortical pattern. In the right panel, the enlargement around this discontinuity shows the direction of the ciliary rootlets, which run downward and to the right of the observer for inverted rows and run upward and to the left of the observer for normally oriented rows. Bar: 10 mm.
Source: Courtesy of F. Ruiz (Beisson, 2008).

And his former student, Nanney, stated:

> *A cell possesses (at least) two classes of structural information: information linearly coded in nucleic templates, either nuclear or cytoplasmic, and information encoded in supramolecular organization.*
>
> *Nanney (1966)*

Based on present knowledge, we may interpret cortical inheritance and the "structural memory" of *Paramecium* and other ciliary unicellulars as functions of centriole/basal bodies, i.e., of their capability to self-replicate genome independently.

Despite its great theoretical importance and implications for biology, cortical inheritance was overshadowed by the overwhelming success of the genecentric view. It now seems to be escaping that shadow.

The discovery of the DNA methylation in 1975, independently by the British biologist Robin Holliday and his student John Pugh (Holliday and Pugh, 1975) and by Arthur D. Riggs (Riggs, 1975) in California, triggered a snowball effect on epigenetic research.

Neural Origin of Epigenetic Information in Epigenetic Structures

Under the umbrella of epigenetic information are understood changes in DNA methylation/demethylation (Figure 2.4), histone marks leading to chromatin remodeling, and changes in the patterns of miRNA expression.

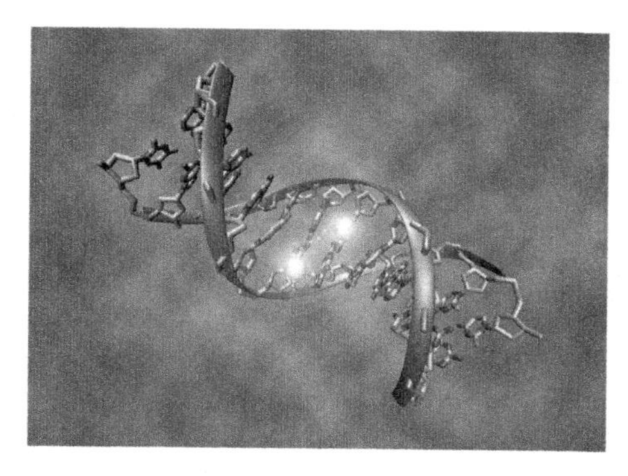

Figure 2.4 The crystal structure of a short DNA helix, which is methylated on both strands at the center cytosine.
Source: From Christoph Bock (Max Planck Institute for Informatics), 2006. Available from: http://en.wikipedia.org/wiki/File:DNA_methylation.jpg

The epigenetic information contained in methylated/demethylated DNA and histone marks have control over the expression of genetic information and epigenetic changes in DNA and histones are considered to represent an "epigenetic inheritance system" (Shea, 2007). But, if by the term *system*, we commonly mean the entirety of the parts that function as a whole to perform a particular function, then neither DNA methylation/demethylation nor histone acetylation/deacetylation and so on represent "systems." DNA methylation and chromatin remodeling are no more than epigenetically changed structures. No parts of an epigenetic inheritance system are visible in DNA methylation and histone marks, and the following discussion is aimed at revealing the parts of the epigenetic inheritance system, if at all.

If there is an "epigenetic inheritance system," then one must identify interacting parts of the system and possible flow of the epigenetic information into the system. This may also show whether these epigenetic information-carrying structures are genuine sources of the information they contain or they are just media or channels through which the epigenetic information is transmitted to, and embodied into, biological structures.

The simplest approach to this problem is to proceed from the knowns, which are DNA methylation and histone acetylation, to the unknowns, i.e., to the inducers of methylation and acetylation in the DNA and chromatin histones. The fact these epigenetic changes in DNA and chromatin histones are nonrandom and generally adaptive implies that the epigenetic information is invested in the right place in the genome or chromatin. Any inquisitive mind would ask: how does the organism manage to induce epigenetic changes exactly to the right nitrogen bases and histone molecules out of the millions of such bases and histone molecules during the processes of gametogenesis, zygote formation, and individual development. We need to know how these epigenetic structures come into being and what determines their occurrence at exactly the sites where they are needed. Where does the information for producing these adaptive changes in the structure of the genetic and nongenetic material come from?

A noncausal explanation would be that this is a property of living organisms, one that has evolved in the course of evolution. Such an answer merely circumvents a legitimate question.

Any causal approach to the problem should start with a close examination of the proximal causes, i.e., with an understanding of the mechanism of inducing epigenetic changes in DNA and histones. This approach could show whether any source of the epigenetic information exists upstream of the DNA and chromatin, and if this proves to be the case, in a step-by-step approach, to trace back the flow of the epigenetic information to its ultimate source. If we succeed in going that far upstream, then we may reveal the causal chain that leads to the production of epigenetic structures, which might be the very epigenetic system of heredity we are looking for.

DNA Methylation/Demethylation

This is the crucial element of gene imprinting, X-chromosome inactivation, stem cell differentiation, reprogramming, carcinogenesis, chromatin remodeling, and so on (Doi et al., 2009; Hewitt et al., 2011). DNA methylation is present in plants and *Animalia*. It is species-specific, and in multicellulars, it is transmitted to daughter cells. Experimentally induced DNA methylation has sometimes been transmitted to subsequent generations in multicellulars (Vandegehuchte et al., 2010).

The methylation of DNA is a function of DNA methyltransferases, which in turn are activated by extracellular signals; hormones such as glucocorticoids (Biswas et al., 1999; Laborie et al., 2003), estradiol (E2), (Lai et al., 2009); and by a transcription factor (TF), Egr-1 (Ebert et al., 1994). Recall that glucocorticoids produced by adrenal glands and the E2 produced by gonads are secreted in response to signals from the pituitary (adrenocorticotropic hormone (ACTH) and follicle-stimulating hormone (FSH), respectively), that in turn are secreted in response to brain signals, respectively corticotrophin-releasing hormone (CRH) and gonadotropin-releasing hormone (GnRH). The TF Egr-1 also is regulated cerebrally by the hypothalamic GnRH (Rachel Duan et al., 2002) and by glucocorticoids (Loizou et al., 2006).

Let us illustrate the role of the nervous system in inducing epigenetic changes in DNA with two examples. The neural activity resulting from stressful stimuli, via the protein Gadd45b (growth arrest and DNA-damage-inducible, beta), causes DNA demethylation in hippocampal and suprachiasmatic nucleus neurons (Ma et al., 2009). The causal chain from stressful stimuli to DNA demethylation and epigenetic modification of genes looks as follows:

> Stressful stimulus→processing of the stimulus in the stress neurocircuitry→induction of Gadd45b synthesis→DNA demethylation→expression of neurogenesis genes.
>
> Within 2 h, fear conditioning decreases DNA methylation (hypomethylation) in particular exons of the *bdnf* (brain-derived neurotrophic factor) gene in hippocampal neurons of adult rats (Lubin et al., 2008).

In a number of cases, the neurally induced DNA methylation is transmitted to subsequent generations and leads to the inheritance of acquired characters in metazoans. The illustrating example is the methylation of the *bdnf* gene. Licking and grooming (LG) is a characteristically inherited behavior in rats. In this respect, there are high-LG and low-LG rat mothers. Maternal high-LG behavior during the first postnatal week is processed in the brain cortex of rat puppies where it activates a signal cascade that leads to DNA demethylation and secretion of the neurotransmitter serotonin in their hipocampal neurons. The binding of serotonin by its receptor triggers a signal transduction pathway that leads to the activation of the gene *NGFIA* (nerve growth factor-inducible A) or *egfr*, inducing DNA demethylation and opening of the chromatin, thus enabling expression of the glucocorticoid receptor (*GR*) gene (Figure 2.5). As mothers, these puppies display the same high LG for their puppies. However, when these puppies are reared by low-LG mothers they show their offspring the same low-LG behavior of their adopted mother rather than the high-LG behavior of their biological mother. The reason for the transmission of the acquired character (low-LG behavior) is that the processing of the insufficient maternal LG in the brain cortex leads to DNA methylation and insufficient production of GR (Szyf et al., 2007).

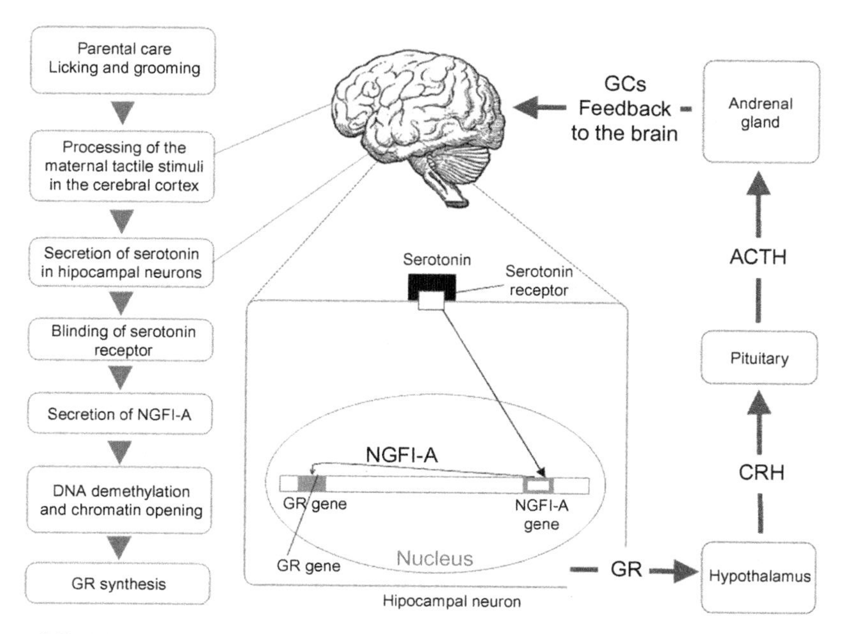

Figure 2.5 The mechanism of transmission of the high-LG behavior in female rats. GR, glucocorticoid receptor; CRH, corticotrophin-releasing hormone; ACTH, adrenocorticotropic hormone; GC, glucocorticoid.

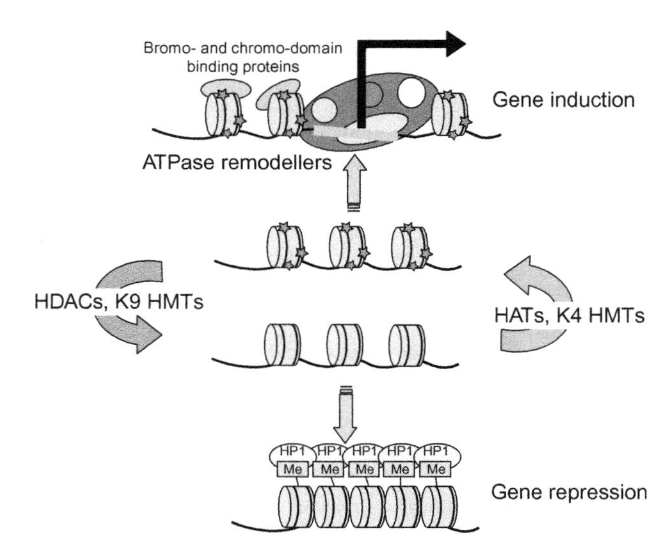

Figure 2.6 The histone switch. Targeted modifications under the control of histone methylases (HMTs), HATs, and HDACs alter the histone code at gene regulatory regions. This establishes a structure that contains bromo- and chromo-domains that permits recruitment of ATP-dependent chromatin remodeling factors to open promoters and allow further recruitment of the basal transcription machinery. Deacetylation, frequently followed by histone methylation, establishes a base for highly repressive structures, such as heterochromatin. Acetylated histone tails are shown as brown stars. Methylation (Me) is shown to recruit heterochromatin protein 1 (HP-1) (Adcock et al., 2006).

Histone Modification and Chromatin Remodeling

A number of hormones such as thyroid hormone, retinoic acid, steroids like E2 and testosterone, and other lipophilic hormones (Sonoda et al., 2008) bind to nuclear receptors, dissociating them from histone deacetylases (HDACs). The resulting molecular complex recruits one or more of a family of about 300 coregulators (coactivators and corepressors), including histone acetyltransferase (HAT). In this form, the TF complex acetylates specific histones, thus remodeling chromatin, binding to hormone specific response elements, and enabling transcription of specific genes (Hinojos et al., 2005; Lonard and O'Malley, 2007) (Figure 2.6). In the absence of these nuclear hormones and other inducers, the nuclear receptors remain associated with HDAC and prevent gene expression by compacting or "closing" chromatin (Gamble and Freedman, 2002).

Protein hormones, which due to the larger molecular size cannot penetrate the cell membrane, bind to the tramsmembrane portion of specific cell membrane receptors, activating signal transduction pathways that end with histone acetylation/phosphorylation/chromatin remodeling and the expression of specific genes. So, for example, the pituitary FSH binds its membrane receptor in ovarian granulosa cells, activating a signal cascade that leads to phosphorylation and acetylation of histone H3 and to

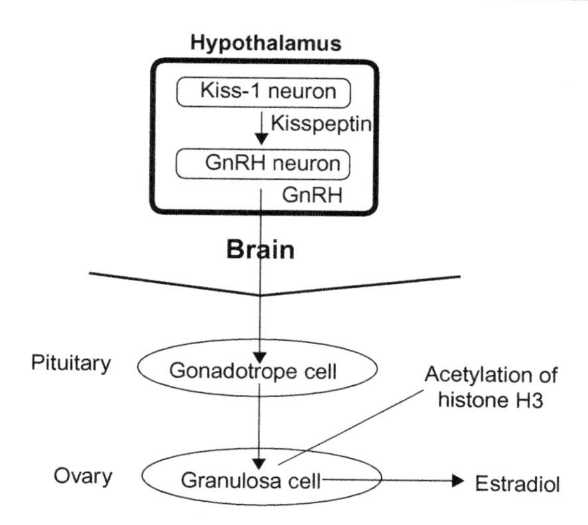

Figure 2.7 Neural control of the signal cascade that leads to acetylation of histone H3 and, consequently, induction of FSH-responsive genes and the synthesis of E2 by granulosa cells surrounding the oocyte. Note that the epigenetic information necessary for H3 acetylation flows from hypothalamic neurons via the FSH-secreting pituitary cells to granulosa cells, where it induces FSH-responsive genes.

the transcription of FSH responsive genes necessary for granulosa cell differentiation (Salvador et al., 2001). But the epigenetic information that flows down to the granulose cell histone originates farther upstream at the hypothalamic neurons that, by secreting GnRH, induce the pituitary to secrete FSH (Figure 2.7).

In a number of cases, histone acetylation and chromatin remodeling are induced directly by chemical signals released by nerve endings. This is the case with the regulation of expression of the synaptic nAChR (nicotinic acetylcholine receptor) genes in the neuromuscular junction (NMJ) (Figure 2.8). Axon terminals of the motor neurons release agrin (Herndon et al., 2011; Ravel-Chapuis et al., 2007) and neuregulin (Ravel-Chapuis et al., 2007), which activate the receptor muscle-specific kinase, which in turn triggers signal transduction pathways that lead to the activation of GA-binding protein (GABP). The latter recruits HAT p300, which induces histone acetylation causing chromatin to "open" and enabling the expression of AChR membrane receptor genes.

All the examples presented earlier in this chapter show that the epigenetic information embodied in methylated DNA and in modified histones is provided to them by external factors rather than being generated in these structures. It is transmitted to them via neurohormonal signal cascades along the hypothalamic-pituitary-target endocrine glands axes that ultimately start in the CNS and via intracellular signal transduction pathways (Cabej, 2005) or via chemical signals released by nerve endings that also originate in the CNS. It seems that the ultimate upstream source of information that flows through neurohormonal cascades is the CNS.

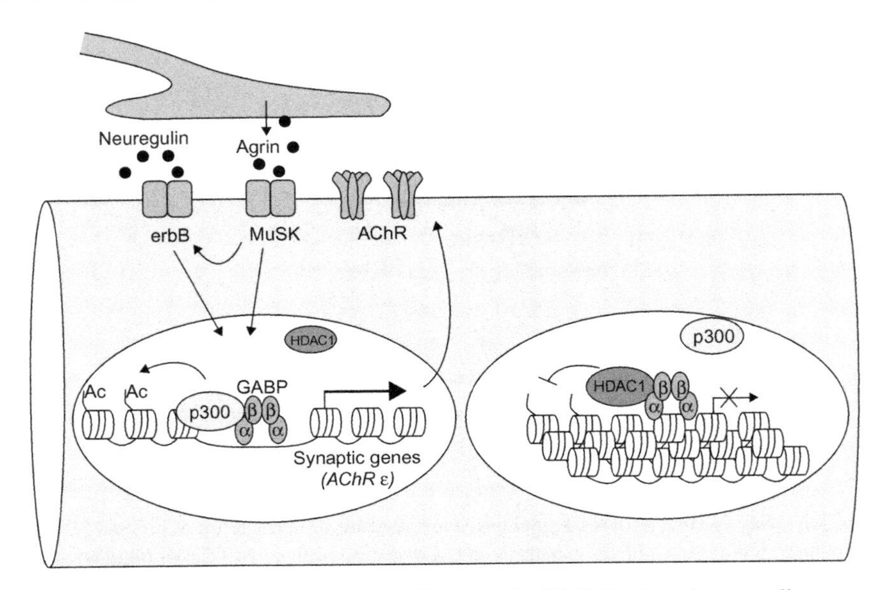

Figure 2.8 A model for synaptic gene regulation at the NMJ. Agrin and neuregulin accumulate in the basal lamina of the synaptic cleft and activate their muscle receptors to induce the local activation of intracellular signaling pathways, which in turn activate the TF GABP. In addition, they induce histone hyperacetylation and hyperphosphoacetylation, which participate in chromatin decondensation. The recruitment of p300 on synaptic genes by GABP in subsynaptic nuclei favors chromatin hyperacetylation and decondensation. Conversely, in extrasynaptic nuclei, GABP recruits the histone deacetylase HDCA1 on synaptic gene promoters, thereby promoting chromatin compaction (Ravel-Chapuis et al., 2007).

In a generalized form, the neurohormonal induction of changes in epigenetic structures (DNA methylation, histone acetylation) or the flow of epigenetic information looks as follows:

> External/internal stimuli→processing of the stimuli in neural circuits→secretion of specific hypothalamic-releasing hormones→secretion of specific pituitary stimulating hormones→secretion of specific hormones by target endocrine glands→changes in specific epigenetic structures (DNA methylation and histone acetylation).
> It is obvious from this simplified signal cascade that the epigenetic information necessary for inducing changes at specific sites of DNA and histones is generated in the CNS by processing of external and internal stimuli.

Alternative Splicing

Alternative splicing is an epigenetic mechanism of eukaryotes that utilizes the genome to generate protein forms that are not possible through genetic mechanisms. In plants and animals, more than 30% of genes are alternatively spliced (Xiong et al.,

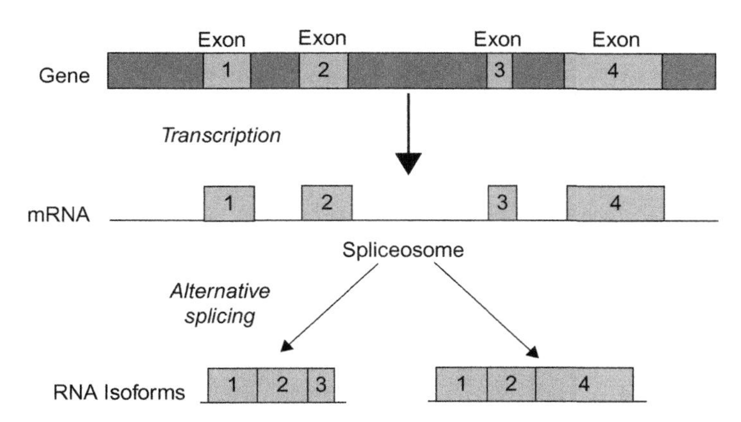

Figure 2.9 Demonstration of AS: In DNA, the genetic information that includes the code for making a protein is located in fragments (exons, shown by green boxes), which are interrupted by noncoding fragments (introns, shown by blue boxes). By the process of AS, the introns are removed and the exons spliced together in different combinations, generating different mRNAs that are decoded (translated) into distinct proteins. (For interpretation of the references to color in this figure legend, the reader is referred to the web version of this book.) *Source*: From Kashyap and Tripathi (2008).

2012). However, many AS variants in plants may be nonfunctional (McGuire et al., 2008; Tress et al., 2007). It is thought that "the role of AS as a mechanism for expansion of function proteome diversity in plants is very limited" (Severing et al., 2009).

AS is a function of an epigenetic mechanism for the combinatorial manipulation of linear DNA information to increase diversity of the proteome. AS combines multimeric blocks of pre-mRNA (*exons*) to construct new proteins that otherwise would be impossible to build genetically. A crude estimation of the number of proteins generated by AS in millions of animal and plant species suggests that it may exceed by far the number of proteins that evolved via genetic mutations during almost 4 billion years of the existence of life on Earth.

Both common sense and intuition tell us that AS and AS-derived proteins evolved via an intrinsic information-generating mechanism, which in all likelihood may be identical to the mechanism that performs AS in extant organisms. Hence, the study of the AS mechanism is important not only for our understanding of cell functions but it is also of paramount evolutionary theoretical importance. AS implies selective excision of introns and ligation of exons of pre-mRNA transcripts in specific ways (Figure 2.9).

Splicing Machinery

AS is the function of spliceosomes, which are nuclear organelles, that remove noncoding introns from pre-mRNA and recombine exons to produce protein isoforms. Spliceosomes are dynamic macromolecular organelles composed of five types of small nuclear ribonucleoproteins (snRNPs) as well as a number of other proteins

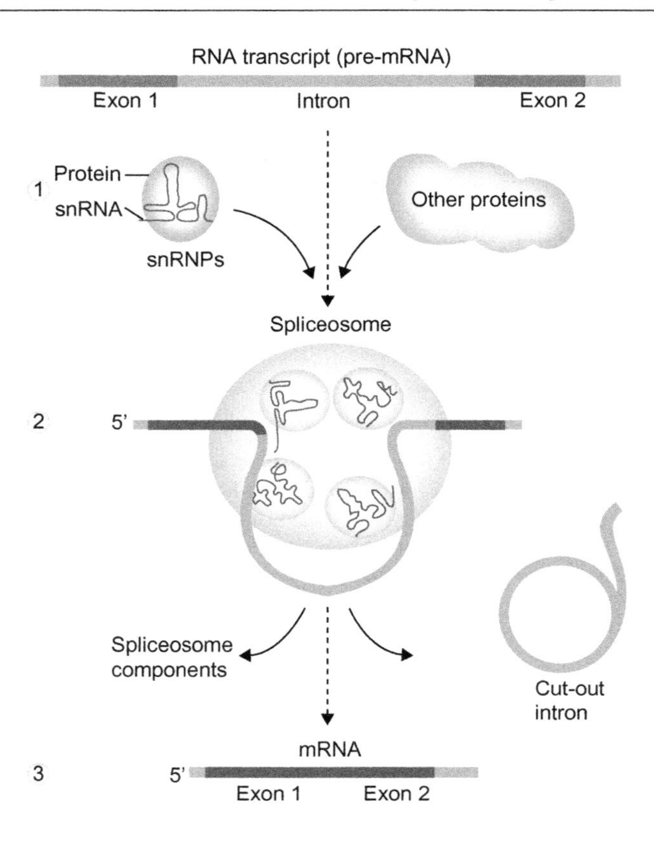

Figure 2.10 Diagram of removal of an intron from the pre-mRNA.
Source: From AP Chapter 17, "From Gene to Protein," http://www.quia.com/jg/1269935list.html.

(Oesterreich et al., 2011). They are roughly the size of ribosomes and are crucially involved in the AS process. The spliceosome is activated upon binding a pre-mRNA transcript. In this active form, it removes introns and joins the exons of the pre-mRNA to build a protein (Figure 2.10).

Epigenetic Determination of AS

Introns in the pre-mRNA have splice site sequences at both ends, but they are too degenerate to sufficiently always recognize correct splice sites. There is evidence that chromatin remodeling via histone acetylation/deacetylation helps in determining splice sites. A clear correlation is observed between the epigenetic marks (DNA methylation/demethylation and histone acetylation/deacetylation that lead to chromatin remodeling) and AS. Epigenetic signatures such as differential methylation and histone acetylation in the exon regions indicate that AS may be regulated epigenetically (Kornblihtt et al., 2009; Zhou et al., 2012).

Histone acetylation and deacetylation, along with histone methylation, and histone modifications generally, are involved in the assembly and dynamic rearrangements of the spliceosome (Gunderson et al., 2011). Modifications to the histone structure facilitate the binding of chromatin-binding proteins and serve to recruit splice factors. The deacetylation in the gene body not only prevents internal initiation of transcription but also restricts spliceosome assembly to the exons alone.

Histone acetylation alterates the transcription rate and, just like histone methylation, it is believed to influence AS by recruiting spliceosomal subunits. For example, one of the subunits of the spliceosome, U2snRNP, is recruited by the HAT Gcn5 (Hnilicová and Staněk, 2011), and polypyrimidine tract binding protein, which is associated with chromatin, by binding pre-mRNA, modulates the AS of several genes (Hnilicová and Staněk, 2011; Sharma et al., 2008). Histone H3K9me3 (H3 trimethyl Lys9) is a regulator of AS (Hnilicová and Staněk, 2011), whereas HDACs influence selection of splice sites (Hnilicová et al., 2011).

Recent evidence shows that, along with AS, especially during the development, alternative transcripts are generated by a novel mechanism, the "alternative transcription," which precedes AS and consists of the production of different pre-mRNAs by changing the start and/or end of transcription (Pal et al., 2011).

The experimental evidence of the role of histone modification and chromatin remodeling in the context of the evidence on the neural origin of information for changes epigenetic structures (histone modification/chromatin remodeling) presented earlier in this chapter, suggest that the nervous system may play a crucial role in AS of pre-mRNAs in metazoans.

At present, we are far from a real understanding of the control system in unicellulars and plants; hence, the ultimate source of information for the regulation of AS in these groups is still a mystery.

Epigenetic Information and Signal Cascades

We obviously oversimplify when we speak of interactions between genes and the external environment. The relationship is neither direct nor linear; more often than not, both lack physical contact to interact. In between are a number of separating biological macro- and microstructures. A plant or an animal organism is under the constant action of various environmental variables, but genes, within the cell nucleus, are well protected against their action. Even if genes were to come in contact with external agents, they would not respond adaptively by switching off or on, if they could respond at all. Notwithstanding, genes respond adaptively to external or internal stimuli, such as changes in the temperature of the environment, humidity, length of the day, day–night and seasonal cycles, changes in the social environment, presence of predators, or changes in the level of hormones and other chemicals in body fluids.

How do genes know to adaptively respond to specific environmental agents with whom they have no physical contact? Speaking metaphorically, genes are unilinguals; they understand the language of biology, but they are not proficient in the external environment language. The fact that genes respond adaptively to external influences, which they do not "see" and do not "understand," suggests that the external influences reach genes interpreted in another language. An interpreter of the environmental influences comes into play, and, via communications channels, provides them with instructions on how to respond in the familiar biological language.

The interpreter is the nervous system, which processes environmental data to generate the epigenetic information provided to genes in the form of chemicals or instructions for switching genes on and off and determining the spatiotemporal patterns of gene expression.

Now, let us try to outline the pathway from these environmental genetically unintelligible variables to genes. This is a neural pathway because it could not be anything else. The nervous system is the only system that, owing to its omnipresence down to the cell level, can receive data on both the external and internal environment and, via afferent pathways, send them to the CNS. The first thing the CNS does with the wealth of the incoming data is to filter them or separate "the wheat from the chaff"—to separate the meaningful or significant data from the insignificant data.

Ignoring the "insignificant" data, the CNS takes the rest for challenges to work out, which it labels as stimuli (Latin for *goad*). Once qualified as a stimulus, the environmental variable is processed in a specific neural circuit. The processing ends with a neural output in the form of an electrical/chemical signal assigned for activating a specific signal cascade.

In order to be taken as a stimulus in metazoans, an internal/external agent must exceed a certain threshold as measured by a set point determined in the brain. Accordingly, a stimulus may be functionally defined as any variable that, above or below an intrinsically determined set point, causes an adaptive response.

Confusion often arises in relation to the nature of external stimuli from the causal and the informational standpoints. Because external stimuli precede in time the intragenerational and transgenerational changes, they are considered to be causes of those changes. For the appearance of an adaptive change, stimuli are necessary but not sufficient conditions. An approaching fox may be taken as a stimulus by a bird, who immediately decides to take flight, while a horse may ignore it. The decisions are made in the brains of the bird and the horse based on predicted dangers or rewards. The cause is, thus, within the animal itself, rather than in the external stimulus, although the external stimulus is a necessary condition for the animal's decision.

That different organisms may respond differently to different stimuli suggests that the external agent *per se* is not information or instruction for the organism. It is the interpretation of that stimulus in the brain that generates the epigenetic information that via signal cascades instructs target cells to adaptively respond to the external stimulus. From the communications theory view, the processing of the stimulus in the brain reduces the uncertainty and increases the probability of the activation of a specific gene to 1, by selectively activating only one of a number of available signal cascades (Cabej, 2012, pp. 23–24, 39–80). Since the processing of the stimulus and the generation of the instruction take place at a subconscious level, the instruction has a purpose, but not a goal.

In an adaptive response to the lengthening of the photoperiod, higher temperature, and so on, many animals in temperate climates display reproductive behavior. These environmental agents are not instructions to genes; *per se*, they are unintelligible or neutral rather than instructions to genes for reproductive behavior and physiology. Neither sunlight nor temperature can induce the expression of specific genes. Only manipulation of these environmental data in the nervous system provides them with meaning, translates them into semantic information, and makes them intelligible to the genes. This neurally derived information via signal cascades reaches genes in the form of instructions or commands to switch them on or off. The environmental data translated into semantic information in the brain are useful and pragmatic to the organism; the neurally derived epigenetic information represents adaptive solutions to the challenges posed by environmental factors.

The function of the nervous system in its relationship with the environment is to structure the environmental data so that they serve as instructions to the target cells, tissues, and organs. The epigenetic information is produced in a species-specific language, in a sense that what is "semantic" information for a species may be nonsemantic or "meaningless" for another. Thus, it is in the nervous system that environmental data are converted into meaningful information. Thus, the epigenetic information generated in the course of the processing of the environmental data in the CNS is specific, as opposed to the genetic information, where the universal triplet code provides its characteristic universality. The amount of semantic information provided to the receiver via a signal cascade may be evaluated by the change in the probability of attaining the desired result before and after the reception of the information.

The behavior of the CNS in converting meaningless environmental data into information that is intelligible to its cells and genes can reasonably be characterized as a goal-directed behavior (i.e., intended to ultimately adapt the organism to changes in the environment). But such a statement raises an apparent problem: is the animal organism aware that it is manipulating the environmental data to produce information for adapting itself to the changing conditions in the environment? Obviously, the organism is not aware. But a counterquestion would be: is it necessary for an animal to be aware of all adaptive choices it makes? An amoeba does not need to be aware that its debris-engulfing routine is necessary for its survival, but debris engulfing is a clearly purposeful behavior; even when an amoeba corrects its course to reach a source of food, its behavior is goal-directed. If goal-directed behavior evolved as early as in primitive unicellulars, should one doubt (or be surprised) that goal-directed behaviors are conserved in higher invertebrate and vertebrate organisms?

The epigenetic information generated in the CNS by processing external environmental stimuli is intended to adapt the organism to changes in the environment. It is selectively provided to the cells/organs that have to adapt in the organism alone. The precisely restricted action of circulating inducers to relevant cells/organs alone is a formidable challenge that no genecentric-prone approach can explain.

Metazoans evolved a binary neural control of gene expression, which makes possible local action of global circulating inducers. The system uses at least two mechanisms of restriction:

1. Local innervation induces the expression of relevant receptors in target cells so that circulating inducers can express specific genes in target cells alone.
2. Local innervation induces the expression of specific genes by releasing local gene inducers.

Let us consider only the signal cascade of the steroid hormone estrogen, which is responsible for folliculogenesis, i.e., the growth and development of follicles. It starts with hypothalamic neurons secreting neurohormone GnRH. Its secretion stimulates the pituitary to synthesize gonadotropins, an FSH, and luteinizing hormone (LH). These hormones induce sex glands to produce E2, which is essential for folliculogenesis and the production of egg cells. How does the organism restrict the expression of E2 in the ovary and, more precisely, in the granulose cells surrounding the ovarian follicle?

We know that this is possible from the release by the ovarian sympathetic nerve endings of the neurotransmitters norepinephrine, vasoactive intestinal peptide (VIP), or both. This stimulates granulosa cells to express the FSH receptor (the mediator of the gonadotropic functions of the pituitary FSH; see Figure 2.7). The local innervation only "targets a circumscribed subpopulation of ovarian cells" (Mayerhof et al., 1997), the granulosa cells in the vicinity of developing follicles. Neonates that underwent sympathectomy showed stunted folliculogenesis, reduced production of E2, and delayed ovulation (Lara et al., 1990; Riboni et al., 1998).

Figure 2.11 shows a schematic model of regulation of the muscle growth in insects. As it shows, systemic signals for stimulating (Ilp neuropeptides) and inhibiting (PTTH→ecdysone) myogenesis originate in the insect's brain.

In *Manduca sexta*, the larval dorsal external oblique (DEO1) muscle consists of five muscle fibers, but as an adult, the insect remodels the muscle radically by eliminating four muscle fibers and allowing only one to remain. During ecdysis, muscle growth in insects stops because of ecdysone secretion, which inhibits muscle growth, but nevertheless, the surviving muscle fiber continues to grow. This is again neurally regulated: the terminal arbor of the motoneuron innervating five larval DEO1 fibers recedes from all but one of them, the surviving one, which then becomes the adult DEO1. The remaining branch of the motoneuron induces the muscle fiber to express the receptor ecdysone receptor B1 (EcRB1), which stimulates muscle growth (Hegstrom and Truman, 1996). The already uninnervated muscles express the isoform EcRA of the ecdysone receptor, which stimulates programmed cell death, degeneration, and elimination of four larval muscle fibers.

Whether ecdysone will perform its antimyogenic action depends on the type of receptor that will be expressed by the muscles. And the expression of EcRB1, which is necessary for muscle growth, or EcRA, which leads to the inhibition of muscle growth by ecdysone, depends on local innervation.

The binding of ligands (e.g., hormones and growth factors) to the cell surface (integral extracellular) receptors marks the beginning of an intracellular cascade

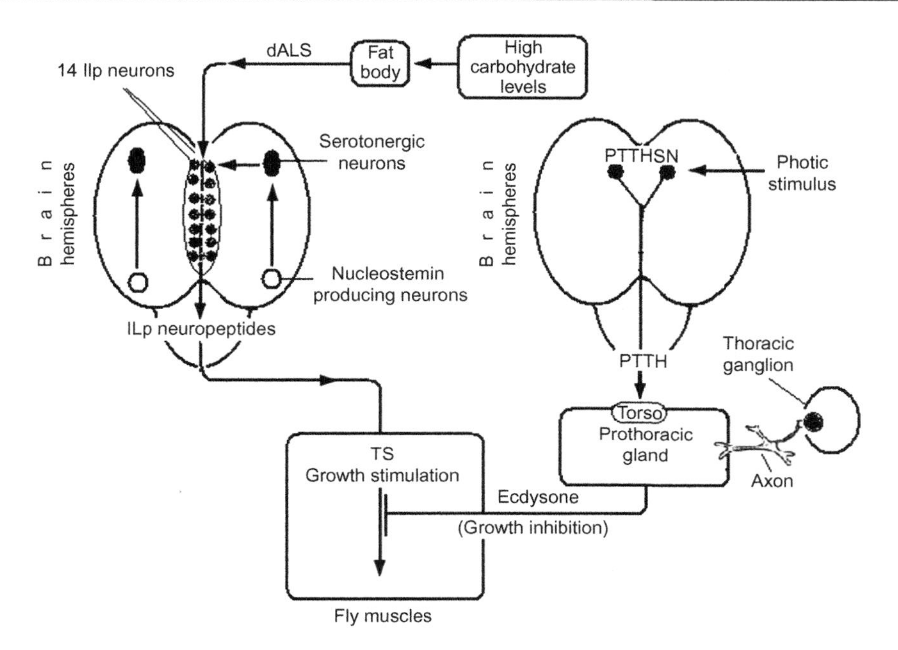

Figure 2.11 A simplified model of the global neural regulation of muscle growth in insects via antagonistic effects of IS and ecdysone. Left: the insulin-like peptides pathway. Right: the PTTH-ecdysone pathway. **dALS**, acidlabile subunit, a fat body-derived glycoprotein; **Ilp**, insulin-like peptides; **PTTH**, prothoracicotropic neurohormone; **PTTHSN**, prothoracicotropic hormone secreting neuron; **IS**, insulin signaling.

of signals known as the *signal transduction pathway*. Among the most important extracellular receptors are G protein-coupled receptors (GPCRs), receptor tyrosine kinases (RTKs), and integrins (Figure 2.12).

GPCRs are a group of cell surface receptors that bind various ligands (e.g., hormones and neurotransmitters) in eukaryotes. The binding of the ligand changes the conformation of the GPCR and activation of a G protein, leading to the activation of a second messenger pathway. RTKs are a large family of transmembrane enzymes that serve as receptors for a number of ligands such as hormones, growth factors, and cytokines. Upon binding a specific ligand, the receptor dimerizes and phosphorylates. In this phosphorylated form, the receptor activates a transduction pathway. Integrins are transmembrane receptors that specialize in the transduction of extracellular matrix (ECM) signals of fibronectin and collagen.

The activation of a signal cascade is triggered by a "perceived" difference between a variable's actual value and its set point. The last element in the cascade finalizes the cascade's purpose, which is the activation of a specific gene or a number of genes, the controlled variable in engineering terminology. According to this concept, the CNS interacting with the epigenetic control system of the cell determines the level of thousands of gene products in body fluids and in cells throughout the animal's body. The interaction of the integrated control system and the individual

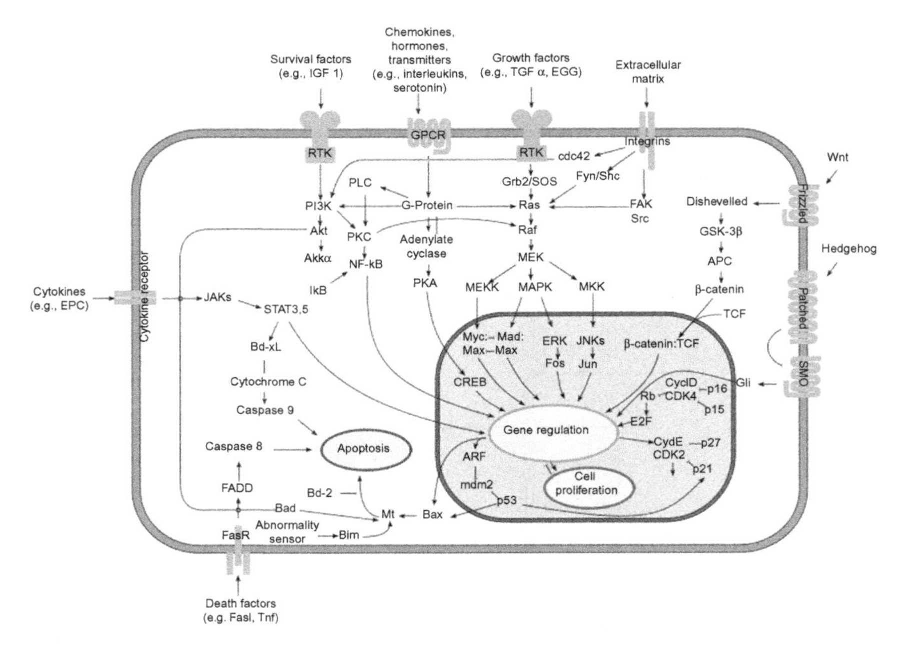

Figure 2.12 An overview of major signal transduction pathways.
Source: From http://en.wikipedia.org/wiki/File:Signal_transduction_v1.png.

cells' epigenetic control of gene expression enable the organism to maintain homeostatic values of variables despite the disturbances caused by the interference of external and internal agents.

The last element of the signal cascade acts as a ligand for the receptor molecule to which it binds. When the ligand has a low molecular weight (e.g., steroids and retinoic acids), it passes through the cell and nuclear membranes, where it binds to a specific nuclear receptor. The ligand changes the conformation of the nuclear receptor, transforming it into an active form of the TF that binds to specific regulatory sequences by inducing the expression of a specific gene. However, since they are too big to allow them to travel through the cell membrane, most protein ligands (e.g., hormones and growth factors) bind to their specific cell membrane receptors. The binding of the ligand changes the conformation of the intracellular part of the receptor molecule, thus activating it and triggering sequential events in the elements of the transduction pathway that ends with the activation of one gene or a number of genes (Figure 2.13). The receptors serve the cell as antennae for capturing extracellular signals.

All major transduction pathways in metazoans are activated by extracellular signals such as hormones, growth factors, ECM proteins, and neurotransmitters, which in turn are neurally regulated. The overwhelming majority of these signals are the final elements of signal cascades originating in the nervous system. All hormones of

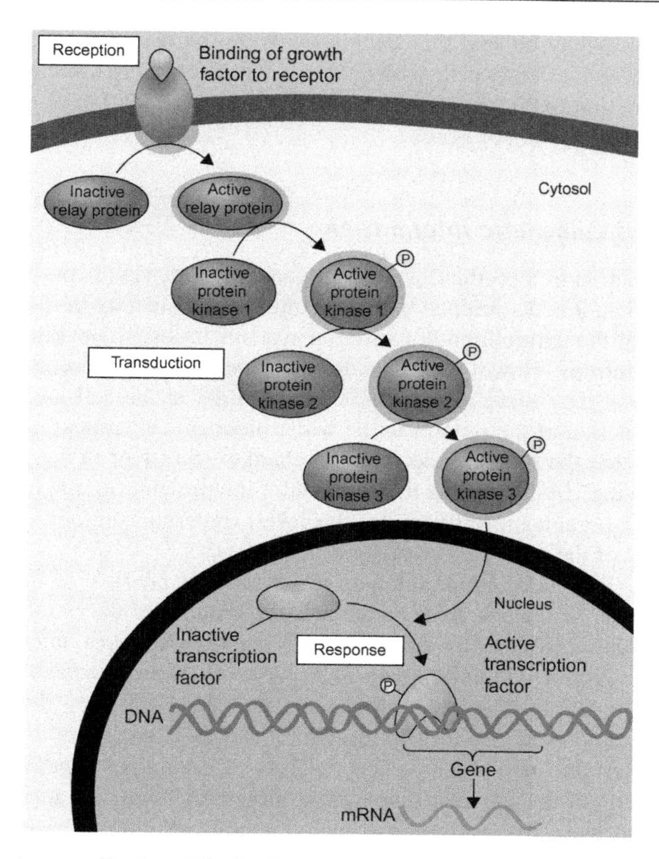

Figure 2.13 A generalized model of a signal transduction pathway.
Source: From http://biotecnologieindustriali.campusnet.unito.it/do/corsi.pl/Show?_id=2a54.

the target endocrine glands (thyroid, thymus, pancreas, adrenals, ovaries, and testicles) are induced by "stimulating" hormones released by the pituitary, which in turn are induced by respective neurohormones secreted by the brain. Many growth factors such as epidermal growth factor, transforming growth factor (TGF), and fibroblast growth factor are induced by various hormones of the target endocrine glands, but all of these hormones are induced by the respective pituitary stimulating hormones, which, in turn, are induced by corresponding neurohormones of the hypothalamus (see Cabej, 2005, 2012, pp. 23–24 for relevant sources). Wnt-4, a member of the family of secreted glycoproteins of the Wnt family, is necessary for the ductal side branching in the mammary gland (Brisken et al., 2000; Robinson et al., 2000), but Wnt-4 secretion is stimulated by the pituitary prolactin, which in turn is inhibited/induced neurally by the opioid system (Aurich et al., 2001; Soaje and Deis, 1997; Soaje et al., 2002). Proteoglycans, a major component of the ECM, are regulated by local innervation (Brandan et al., 1992).

In summary, it may be said that the signals from the nervous system are passed along a cascade to the target cell, where the last signal of the cascade starts an intracellular transduction pathway, which generally results in induction of the expression of a specific gene or group of genes.

Genetic Versus Epigenetic Information

Genetic information is a product of random changes in genes (errors in the replication of DNA), i.e., it is determined by the thermodynamic factors of disorder and is a manifestation of the natural trend of material systems to lose their original order and increase their entropy. However, the random change, which represents loss of order at the DNA level may serve as an acquisition of order at the cell level, i.e., it may improve the function of the protein in the wider biochemical context of the cell. It is in this context that the random, meaningless change—the error in the genetic text—acquires "meaning" by improving the function of the protein. So, the loss of order at a lower level of organization (the molecular level) represents an increase of order at the higher level of the cell.

By contrast, epigenetic information is not a random product of thermodynamic forces, conditions, or factors. Its generation is the product of the work, a product of computation (data or information processing) in specific structures, microtubules (and probably other structures) in unicellulars, and specialized neural tissue in metazoans.

Genetic information produces order at the molecular level by determining the sequence of amino acid residues in protein molecules, but there is no evidence that it can determine the order or the patterns of spatial arrangement of different types of cells in the intricate histological structures of organs in multicellulars. Development of these patterns requires huge amounts of data in the form of epigenetic information.

Just like genetic information is continually transmitted to effectors to compensate for lost proteins, epigenetic information is continually transmitted to target tissues and organs to induce them to produce new components and cells to compensate for relevant losses, thus maintaining homeostasis or restoring normalcy to the unavoidably degraded biological structures.

As opposed to genetic information, which exists or is embodied materially in the form of the sequence of nucleotides in DNA, epigenetic information is computed in a structure, such as a nerve cell or neural circuit, and probably in the cytoskeleton of all groups of living organisms, including unicellulars (see the sections "The Control System in Unicellulars" and "Can the Cytoskeleton Compute?" in Chapter 1). It is transmitted in the form of commands to particular structures, via neurohormonal algorithms (signal cascades) or directly by the information-generating structures, to produce a particular phenotypic result. Embodied in the structures of DNA, genetic information is transmitted by the template mechanism of transcription involving a one-to-one correspondence of nucleotides between molecules of DNA and mRNA and a three-to-one correspondence between mRNA nucleotides and amino acids in protein molecules during the translation process.

Epigenetic information is not encoded into a specific structure; hence, it cannot be transmitted in template mode. Instead, it uses a communicative mode of transmission. There is no physical correspondence between the neural circuit and the individual elements of the cascade or the biological algorithms, let alone the relevant phenotypic character. Epigenetic information happens, emerges in neural circuits, and is transmitted when needed. It emerges in the form of electrical/chemical signals released as the output of the computation of external/internal stimuli in specific neural circuits. It results from the computational activity of the emergent configurations neural circuits rather than any permanent structures. It is related to such emergent configurations, but it is different from them. Chemical signals (epigenetic information) released by the activity of neural circuits serve as commands that activate one of several available signal cascades; namely, the one that can produce the specific change in the target cells. (For an extensive discussion on the nature and origin of epigenetic information see Cabej (2012, pp. 39–80).)

In the organismic hierarchy of control, the genetic control at the molecular level is subordinate to the epigenetic control, but the latter is not independent of the former because the genetic system influences the epigenetic control via feedback loops.

Reproduction in Unicellulars

The mechanisms of reproduction in prokaryote unicellulars are asexual: binary fission and budding. Eukaryote unicellulars use asexual and sexual reproduction (conjugation and sporulation).

Reproduction in Single-Celled Prokaryotes

Prokaryotes (from the ancient Greek προ (pro), before, and καρυόν (karion), kernel), include bacteria. The earliest group of unicellular organisms evolved about 3.8 billion years ago (Cooper, 2000). They are divided into two main groups, *Archaea* (*Archaeobacteria*) and *Bacteria* (*Eubacteria*). Prokaryotes have no nucleus, and the genetic material (DNA) is less organized in chromatin and chromosomes than it is in eukaryote unicellulars. They have ribosomes and cytoplasmic satellite DNA and cytoskeleton, but lack membrane-bound organelles such as plastids and mitochondria. Having no separate sexes, bacteria reproduce asexually.

Asexual Reproduction

Asexual reproduction is the primary mode of reproduction in unicellular prokaryotes. The whole mitotic division in bacteria, and unicellular organisms in general, is under the epigenetic control of the cytoskeleton (see the section "Control Systems in Unicellulars" in Chapter 1).

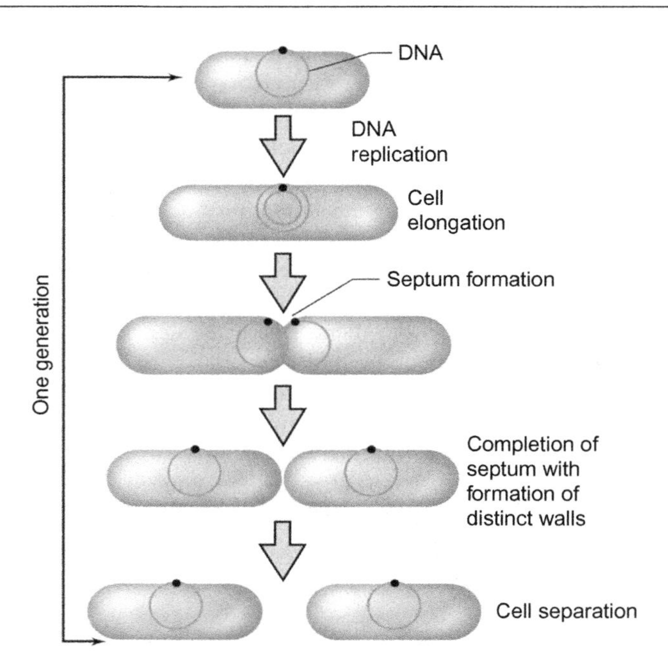

Figure 2.14 Asexual reproduction (binary fission) in bacteria.
Source: From mtsu32.mtsu.edu:11289/cellcycle_web/sld006.htm. Accessed in June 2012.

Binary Fission in Bacteria

Having no separate sexes, bacteria reproduce asexually. Binary fission is their main mode of reproduction (Figure 2.14). Bacterial DNA is condensed in a circular chromosome in a region called the *nucleoid*, which in the bacterium *Escherichia coli* represents one-third of the cell volume (Harrington and Trun, 1997). The cytoplasm also divides equally between the two resulting cells, so both cells are similar.

Budding in Bacteria

Many bacteria use budding for reproduction. The mother cell produces a small bud on one of its ends. The bud grows on the mother's body and, depending on the species, separates when it grows to a size comparable to its mother's size (Figure 2.15). Other forms of asexual reproduction in unicellulars include multiple fission and spore formation (sporulation).

Whether one talks of sexuality in bacteria depends on the definition of sexuality; if sexuality implies separate sexes, and the exchange of genetic material is a prerequisite of sexual reproduction, then bacteria could hardly be considered sexually reproducing organisms (conjugation is not a prerequisite of bacterial reproduction). It is true that during conjugation, bacteria exchange genetic material (DNA), but the exchange is not part of bacterial reproduction, and there are no genuinely separate

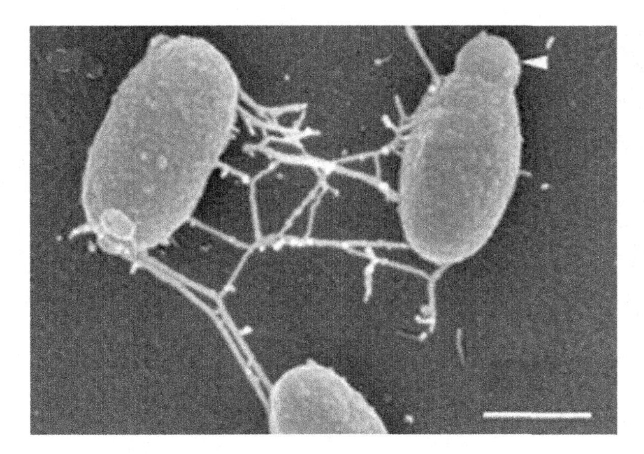

Figure 2.15 Budding in bacterium *Formosa agariphila* (from *Formosa agariphila* sp. nov., a budding bacterium of the family Flavobacteriaceae isolated from marine environments, and amended description of the genus *Formosa* IJSEM 2006, 56, 161–167).

sexes in bacteria. From this view, the two other processes of the lateral gene transfer, the bacterial transformation and transduction, are even less likely candidates to be distinct modes of sexual reproduction. Parasexuality (Otto, 2008) may be a better term to describe these phenomena of horizontal gene transfer in prokaryotes.

Reproduction in Single-Celled Eukaryotes

Eukaryotes (from the ancient Greek ευ (eu), "good, true," and κάρυον (karion), kernel) are characterized by the presence of a nucleus, a number of chromosomes in which DNA is organized in form of nucleoproteins, and by a number of membrane-bound organelles. Single-celled eukaryotes belong to two main groups: Protista and unicellular fungi. Eukaryote unicellulars may have evolved approximately 2 billion Mya, the photosynthetic algae evolved 1600–1500 Mya, and red algae, the oldest taxonomically identifiable eukaryote, evolved 1200 Mya (Hedges et al., 2004).

Single-celled eukaryotes reproduce asexually and sexually. Unicellular eukaryotes reproduce sexually or asexually. Asexual reproduction in single-celled eukaryotes involves mitosis, i.e., duplication of chromosomes and cytoplasm to produce "twin cells" in the process of cell division (Figure 2.16).

Asexual Reproduction in Ciliates

Ciliates have two nuclei, one small (micronucleus) and the other larger (macronucleus).

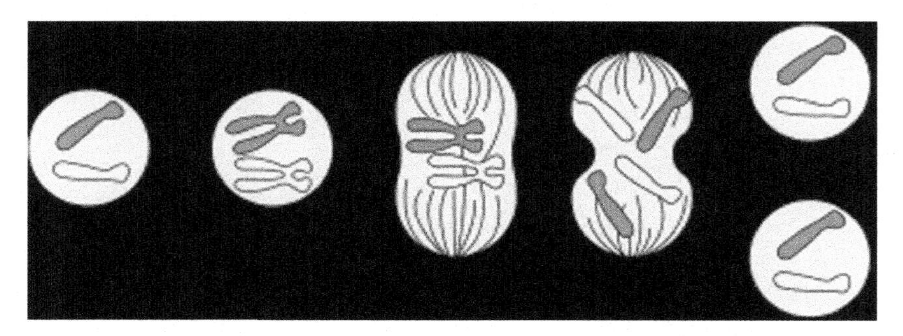

Figure 2.16 Asexual reproduction in eukaryotes. Mitosis divides the chromosomes in a cell nucleus.
Source: From http://en.wikipedia.org/wiki/Mitosis.

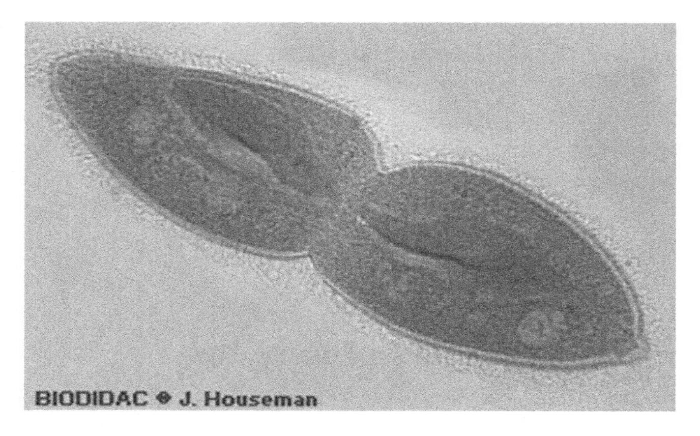

Figure 2.17 Binary fission in *Paramecium.*
Source: From http://biodidac.bio.uottawa.ca/thumbnails/filedet.htm?File_name=OLIH023P&File_type=GIF.

Like other unicellular eukaryotes, asexual reproduction in ciliates is a type of binary fission, where the nucleus is duplicated and each nucleus goes with one of the parts of the transversally divided body.

Binary Fission in Paramecium

In the case of a unicellular ciliate, such as *Paramecium*, binary fission consists of mitosis of the micronucleus, and each of the divided macronucleus halves goes to one of the resulting twin cells (Figure 2.17).

Distinct from other unicellular eukaryotes, paramecia have two nuclei: one polyploid macronucleus with numerous copies of genes, which is essentially involved in

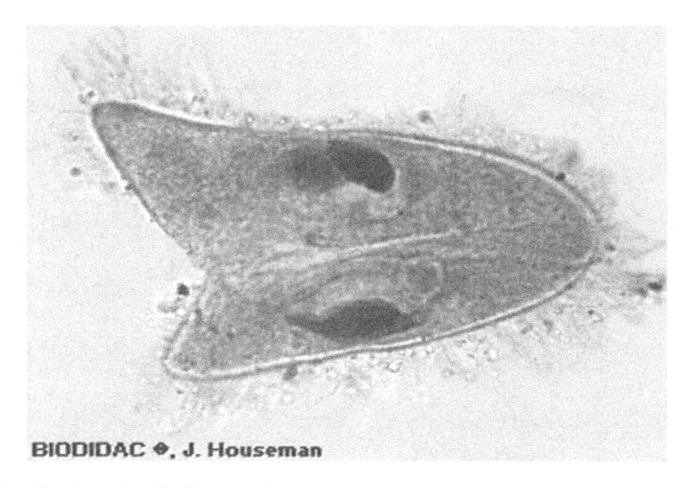

Figure 2.18 Conjugation in *Paramecium*.
Source: From http://biodidac.bio.uottawa.ca/thumbnails/filedet.htm?File_name=OLIH020P&File_type=GIF.

cell division, growth, and metabolism, and one micronucleus, which is activated only during conjugation.

Sexual Reproduction in Paramecium—Conjugation

Sexual reproduction in *Paramecium* occurs only when environmental conditions deteriorate. Sexual reproduction involves the exchange of genetic material via cytoplasmic bridges during conjugation (Figure 2.18). It is a relatively complex process: two paramecia join laterally to form a cytoplasmic bridge. The micronuclei of both conjugants divide by meiosis (producing haploid micronuclei). Each conjugant then donates a micronucleus to its partner. After exchanging genetic material, conjugants separate and both organisms produce a new macronucleus to replace the original macronuclei that disintegrate during the process. Thus, half of the new nuclear apparatus derives from the conjugant. The paramecia now are ready for asexual reproduction through binary fission.

Reproduction in Fungi

Fungi are a separate kingdom of eukaryote organisms with a nucleus, chromosomes, and organelles. Yeasts, molds, and mushrooms belong to this group. Like animals, they are heterotrophic organisms that use plant- or animal-derived materials as sources of food, and their cell walls consist of chitin. Similar to plants, they have cell walls and vacuoles and produce spores.

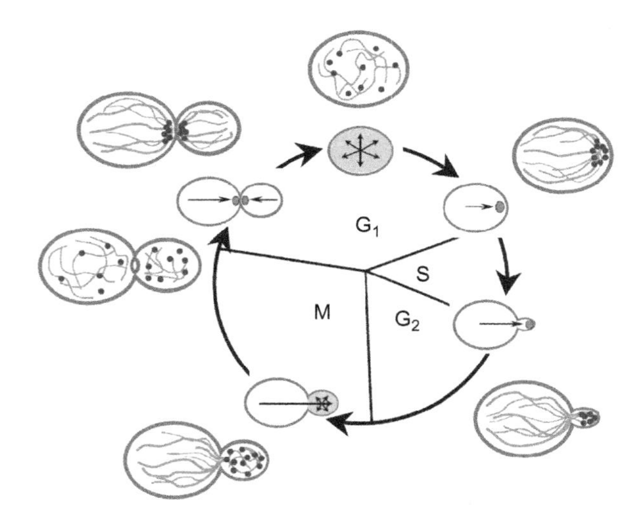

Figure 2.19 Cell polarity in budding yeast is established by a polarized actin cytoskeleton throughout the cell cycle. A cap (blue) of regulatory and cytoskeletal proteins establishes the polarity of actin cables (beige) and cortical patches (brown). Tight localization of the cap orients actin cables. Actin cables then guide secretory vesicles to the cap, where they accumulate (also shown in blue) and fuse, thus polarizing growth (arrows). During isotropic growth, the proteins of the cap are more diffusely distributed, cortical patches are isotropically distributed, and actin cables form a meshwork. A fourth cytoskeletal structure, a cytokinetic ring, mediates cell division (bright pink) (Pruyne and Bretscher, 2000). (For interpretation of the references to color in this figure legend, the reader is referred to the web version of this book.)

Asexual Reproduction in Fungi

Reproduction by Budding in Yeasts

Reproduction by budding in yeasts is known in greater detail than in any other fungi. The three major phases of the yeast life cycle are cell division by budding, mating between haploid cells, and sporulation of diploids (Chant, 1999). Formation of buds in yeasts is an epigenetically controlled process. It starts with the establishment of cell growth polarity, and the centerpiece of the regulation of this polarization for bud formation is the yeast's actin cytoskeleton (Chant, 1999), as opposed to animal cells, where the establishment of the cell's asymmetry depends on microtubules. Actin polymerization helps protein compounds to move and actin cables (actin long fibers) direct the vectorial transport of organelles to the bud (Catlett and Weisman, 2000; Moseley and Goode, 2007) (Figure 2.19). Actin cortical patches, small actin-containing structures, initially determine the budding site and later reorganize to form two rings in the neck of the bud, where they are believed to help in septation and cytokinesis. They may not only facilitate the anchoring of cables to the cortex but also be the sites of cable formation (Amberg, 1998).

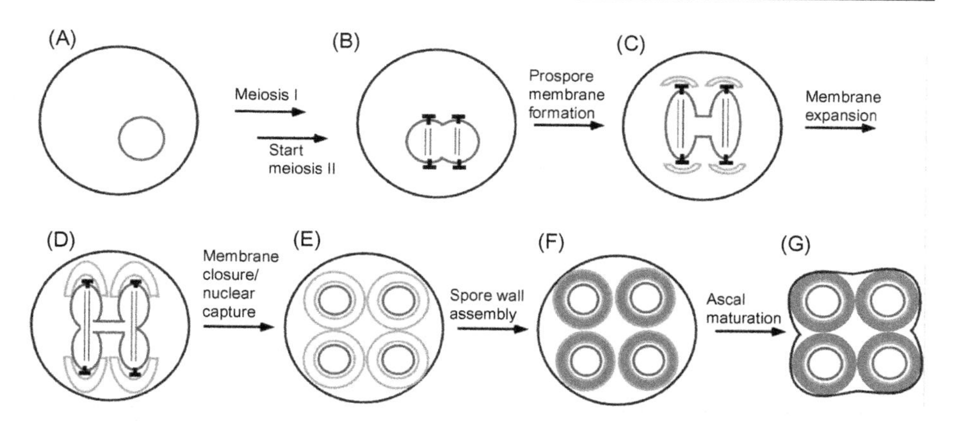

Figure 2.20 (A–G) Overview of the stages of spore and ascus formation. In the presence of a nonfermentable carbon source, diploid cells starved for nitrogen will undergo meiosis. During the second meiotic division, the spindle polar bodies (SPBs, indicated as ⊤), which are embedded in the nuclear envelope (shown in red), become sites for formation of prospore membranes (shown in green). As meiosis II proceeds, the prospore membranes expand and engulf the forming haploid nuclei. After nuclear division, each prospore membrane closes on itself to capture a haploid nucleus within two distinct membranes. Spore wall synthesis then begins in the lumen between the two prospore membrane-derived membranes. After spore wall synthesis is complete, the mother cell collapses to form the ascus (Neiman, 2005). (For interpretation of the references to color in this figure legend, the reader is referred to the web version of this book.)

Reproduction Through Sporulation

Sporulation (yeast gametogenesis) is the main reproduction mode for most fungi. Formation of spores may occur asexually or sexually by the union of male and female gametes. It is in many respects similar to gametogenesis in mammalian cells (Figure 2.20).

Sexual Reproduction in Yeasts

Sexual reproduction is less widespread in yeasts than asexual reproduction by budding. It is prevalent during unfavorable environmental conditions. Under stressful conditions, diploid yeast cells reproduce via sporulation (sexual reproduction), forming haploid spores of type a and α, which again can conjugate to produce diploid cells, and so on (Figure 2.21).

Sexual Reproduction in Green Algae (*Chlorophyta*)

Green algae is a group of about 7,000 unicellular photosynthetic eukaryote species. Most of them are free-living unicellular organisms, but others form colonies, such as

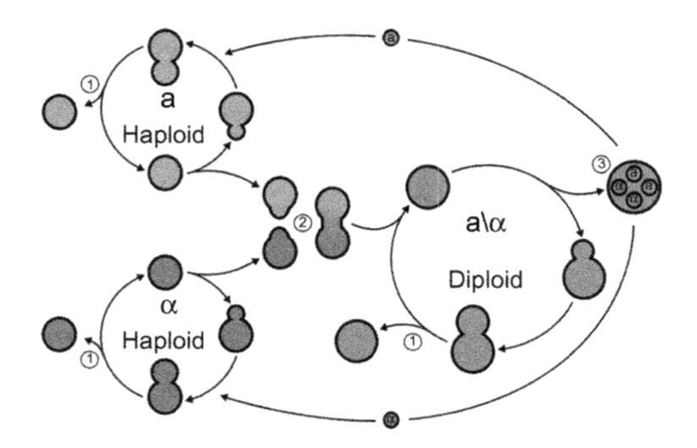

Figure 2.21 The life cycle of a yeast cell. This diagram shows budding in 1, conjugation in 2, and spore in 3.
Source: From *Developmental Biology Interactive*, http://www.devbio.biology.gatech.edu/?page_id=1226.

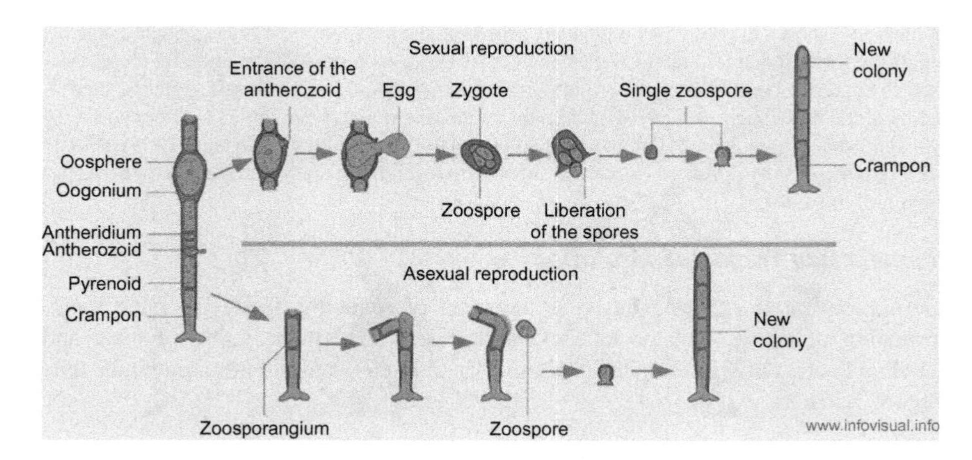

Figure 2.22 Sexual and asexual reproduction of green algae.
Source: From *The Visual Dictionary*, http://www.infovisual.info/01/022_en.html.

the well-known Volvox. The typical green alga has chloroplasts, central vacuole, and cellulose cell walls.

Asexual Reproduction in Green Algae

Asexual reproduction in green algae occurs in the form of fission or budding or by the production of zoospores in the zoosporangium. Asexual zoospores swim and form new colonies (Figure 2.22).

Sexual Reproduction

Sexual reproduction in green algae is common and may be isogamous (gametes both motile and same size), anisogamous (both motile and different sizes, with the female being bigger), or oogamous (female nonmotile and egglike; male motile). Many green algae have alternating haploid and diploid phases. The haploid phases form gametangia (sexual reproductive organs) and the diploid phases form zoospores through reduction division (meiosis). Some do not experience alternation of generations.

In algae, sexual reproduction involves the conjugation or joining of two haploid individuals in which one individual provides genetic material to the other.

The gametophyte produces haploid gametes by mitotic division, which unite to form a diploid zygote that develops into a sporophyte. The sporophyte then undergoes meiotic division to give rise to haploid spores, which grow into gametophytes. In this way, the gametophyte and sporophyte generations alter with each other.

Asexual Reproduction in Metazoans

Reproduction by budding or broken-off pieces of sponges occurs in some species of metazoan but is not widespread. Gemmules, or "internal buds," are groups of sponge cells wrapped in spongin capsules. They overwinter in a dormant state and then germinate in warmer weather to develop into new sponges.

Sponges display another variant of budding called *internal budding* or *gemmulation*. The buds (gemmulae) form within the body wall and are released when the sponge dies. Again, this form of reproduction seems to be complementary to sexual reproduction. Some sponge species form gemmulae that are resistant to cold and drought. Sponges can also reproduce by regenerating broken-off pieces.

Reproduction by Fragmentation

Fragmentation implies that the branching of sponges may break apart into storms (Leong and Pawlik, 2010). Being sessile organisms, sponges rely heavily on fragmentation for dispersion over long distances, while larvae produced by sexual reproduction contribute to establishing new populations in the neighborhood (Maldonado and Uriz, 1999).

Sexual Reproduction in Sponges

Sponges are hermaphrodite metazoans that produce both male and female gametes. Both gamete types are produced by choanocytes, but in most species, female gametes derive from transformation of archeocytes. Sperm cells carried by water currents

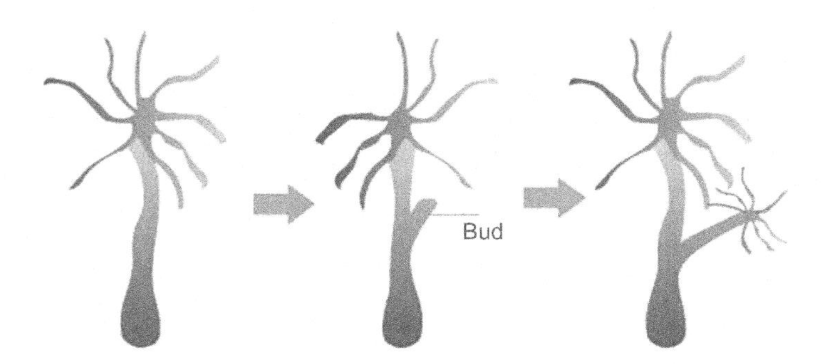

Figure 2.23 Budding in *Hydra*.
Source: From http://www.tutorvista.com/content/biology/biology-ii/reproduction/asexual-reproduction.php.

are captured by choanocytes, which transport them to the egg cell. The fertilized egg produces a larva from which a sponge gradually develops.

Reproduction in Eumetazoans

The transition from unicellularity to multicellularity required new modes of reproduction. Sexual reproduction is the most widespread form of reproduction, which is used by more than 99% of vertebrate species. A relatively small number of species reproduce by budding, internal budding, and fragmentation. They start from a single reproductive cell (egg or zygote) that is capable of developing into an organism of their own kind. Let us start with a glimpse on these "primitive" forms of biological reproduction.

Some lower invertebrates are reproduced both sexually and asexually. Some *Hydra* species switch from sexual reproduction in summer to asexual reproduction by external budding in winter (Figure 2.23). Budding is the most common form of asexual reproduction in eumetazoans: the mother produces in its body a group of specialized cells from which another individual of the same species develops. This form of reproduction is observed in the females of some cnidarians, tunicates, and flatworms.

Asexual Reproduction in Eumetazoans

Reproduction by Fragmentation

The fascinating reproduction mode of self-fragmentation is most commonly observed among flatworms, especially among some free-living freshwater planarians and a number of echinoderms. In an apparently "suicidal" behavior, they constrict and divide their body at some point behind the pharynx, after which each part

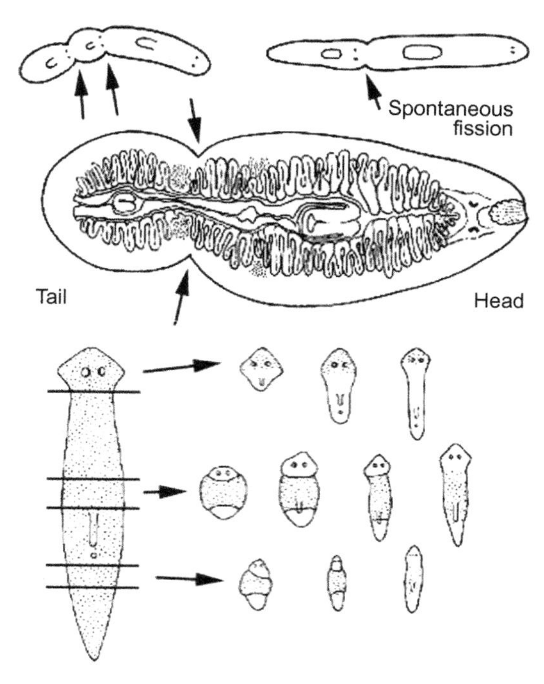

Figure 2.24 Regenerative capacity of the freshwater planaria *Dugesia tigrina* after transverse cutting.
Source: From http://www.rzuser.uni-heidelberg.de/~bu6/Introduction03.html.

regenerates the missing part. This in turn leads to the formation of two fully grown planarians (Figure 2.24). When split, these worm fragments develop the missing part.

Reproduction through fragmentation occurs in sponges, cnidaria, and some worms, as well as in some lower plants.

Reproduction by Budding

Reproduction by budding is characteristic for cnidarians and other lower invertebrates. The new individual starts as a protrusion in the parent's body (Figure 2.25).

The fact that reproduction by budding in lower invertebrates is not the exclusive mode of reproduction, but rather an adaptive form of reproduction under changed conditions in the environment, suggests that it might have been a complementary adaptation rather than the "primitive" form of reproduction.

Sexual Reproduction—the Prevalent Mode of Reproduction of Multicellulars

Sexual reproduction in eumetazoans is a complex physiological process that often requires strict coordination in space and time of the development, maturation,

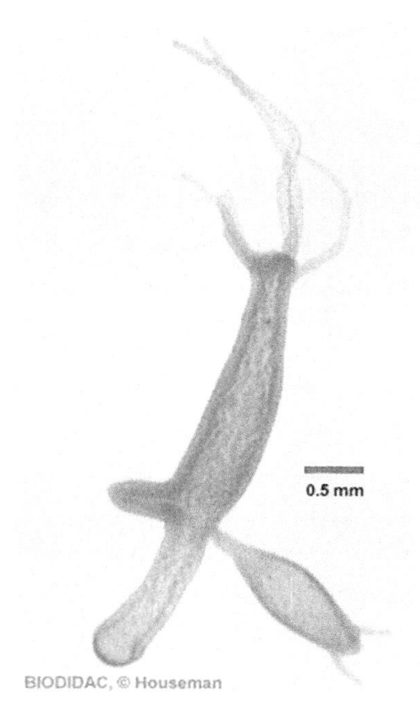

Figure 2.25 Budding in *Hydra*.
Source: From http://biology.about.com/library/weekly/aa090700a.htm.

release, and union of egg and sperm cells. As already mentioned, metazoans do not produce copies of themselves, and apparently, this is impossible from the developmental point of view. They produce reproductively specialized cells and egg and sperm cells. With the exception of the hermaphroditic organism, these cells are produced from both male and female individuals; hence, the name *dioecious* (from Greek δύο (*dyo*), two, and οίκος (*oikos*), house, from proto-Indo-European **weiḱ*, to settle) for the species with separate sexes and their mode of reproduction via the union of egg and sperm cells. The great majority of animal species are dioecious and reproduce sexually, i.e., via the union of the gametes of opposite sexes, the egg cell and sperm cell, produced by the female and the male parent, respectively.

A smaller number of metazoan species develop from eggs that *per se* are capable of developing into organisms of the kind that produces them. These animal species are known as *parthenogenetic* (from Greek, παρθένος (parthenos), virgin, and γένεσις (genesis), origin, creation] and this mode of reproduction is called *parthenogenesis*. Curiously, there are some salamander species that, in adaptive responses to environmental circumstances, may switch between the dioecious and parthenogenetic modes of reproduction.

The principal mechanism of sexual reproduction in metazoans was discovered by the end of the nineteenth century, when biologists in several countries were independently investigating cell nucleus. In Belgium, Edouard van Beneden (1846–1910) found that the egg and the sperm contribute equal numbers of chromosomes to the zygote. Around the same time, the young American physician Walter Sutton (1877–1916) demonstrated that each of the gametes contributes half of the normal set of chromosomes, leaving the zygote with the normal diploid set (Galun, 2003). After the rediscovery of Mendel's laws of inheritance (1900), Sutton, at the University of Columbia, and Boveri, at the University of Würzburg, developed the chromosome theory of heredity, in which chromosomes are carriers of the Mendelian factors (*Anlagen*) (Petronczki et al., 2003).

Gametes are produced by specialized organs—the ovaries and testicles, respectively. Egg cells are produced in the ovaries and sperm cells in the testicles. These gametes are haploid (from ancient Greek απλός (haploos), single) cells (i.e., cells containing only one of each pair of chromosomes and genes of the parent). Such gametes have only half the normal set of chromosomes and are known as a *haploid number of chromosomes* in order to stress the distinction from the diploid number of chromosomes that is characteristic of somatic cells. Thus, parents contribute equally to the development of the offspring.

The halving of the number of chromosomes in gametes is known as meiosis (from ancient Greek μείον (meion), less, and from the proto-Indo-European root **mei*, small).

Production of germ cells, egg cells, and sperm cells represent the coronation of the biphasic process of meiosis. Germ cells derive from a cluster of approximately 20 primordial germ cells (PGCs), which in mammals are specified by BLIMP1 (B-lymphocyte maturation-induced protein 1), early in the gastrulation process by the embryonic day 7. BLIMP1 suppresses the somatic differentiation of PGCs by repressing the premature expression of some germ cell-specific genes. Then PGCs migrate and on reaching the genital ridge by the embryonic day 12 (E12), they undergo epigenetic reprogramming, including the erasure of parental imprinting, as may be concluded from demethylation of many loci. Now PGCs activate the maternally inactivated X chromosome. At this early stage, when the sex of the embryo is determined, the oogonia and the primary oocytes establish their own epigenetic marks, especially DNA methylation.

Production of Egg Cells—Oogenesis

In vertebrates, PGCs migrate to reach the ovary, where they start dividing mitotically to form large numbers of oogonia. During the embryonic stage, oogonia enter meiotic divisions to arrest the cycle at prophase I until at the outset of sexual maturity. Inhibition of meiosis is primarily a result of the activation of gama-aminobutyric acid or dopamine central neurobiological brake, which happens soon after birth, that hold in check the secretion of GnRH by the GnRH pulse generator. Experimental release of the brake enables the induction of meiotic divisions.

The GnRH pulse generator is a network of 3000–4000 hypothalamic neurons. At the outset of puberty, coincident with substantial changes in the physiology and morphology of sexual organs, a decrease in neuropeptide Y (NPY) secretion releases the central neurobiological brake. This, as well as the secretion of the hormone kisspeptin by other hypothalamic neurons, triggers the activation of the GnRH pulse generator via the signal cascade.

Hypothalamic GnRH → pituitary FSH → ovarian E2.

This stimulates oogenesis and folliculogenesis. Development of ovarian follicles is stimulated by neurally induced pituitary gonadotropins (FSH and LH). FSH stimulates the development of the primary follicle into a secondary follicle, and the pituitary LH induces the transformation of the secondary follicle into the tertiary follicle. In the ovary, oogonia develop into primary oocytes, which arrest growth at the prophase I of the cell cycle. The first meiotic division results in the production of a large secondary oocyte and a small, first polar body (Figure 2.26).

The second meiotic division of the oocyte produces a larger haploid mature egg, and the second meiotic division of the polar body results in two polar bodies. The

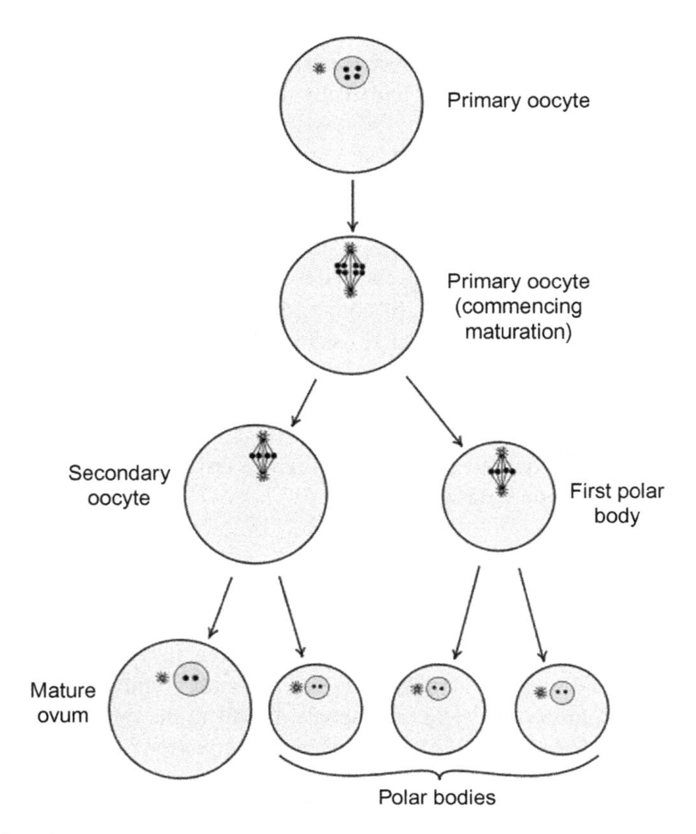

Figure 2.26 Diagram showing the reduction in the number of chromosomes during maturation of the ovum.
Source: From http://commons.wikimedia.org/wiki/File:Gray5.svg.

result of meiosis of the primary oocyte is the production of one mature egg and three polar bodies, which undergo apoptosis and are eventually reabsorbed by the organism. The oocyte grows into a mature egg cell and is released, ready for fertilization by a sperm cell.

Meiotic division during spermatogenesis is similar to oogenesis. The main difference is that while the meiosis of a primary oocyte results in the production of only one gamete and three polar bodies, the primary spermatocyte produces four sperm cells via meiotic divisions.

Hermaphroditism in Metazoans

Most metazoans have separate sexes, while most plants are hermaphroditic. The above discussion on sexual reproduction was about animals that have separate sexes. However, a considerable number of metazoans are hermaphroditic, displaying male and female breeding modes. There are two main forms of hermaphroditism. Simultaneous hermaphroditism is when the same organism has both the male and female sex organs and produces both types of gametes. Sequential hermaphroditism means that an organism switches from its inborn sex to the opposite sex, a development observed primarily in certain fish and gastropods. Many of these hermaphroditic species can reproduce asexually, through their own gametes, or sexually, when their eggs are fertilized by sperm cells from other conspecific individuals.

The most plausible hypothesis on the evolution of simultaneous hermaphroditism is the limited availability of mating partners. Accordingly, the low mobility and low population density favor the evolution of hermaphroditism, which offers selective advantages such as maximized fitness deriving from coupling male and female functions in a single organism, a higher likelihood of meeting a partner because all individuals are potential mates, and, in the absence of mating partners, the option of producing offspring by self-fertilization (Puurtinen and Kaitala, 2002). Statistical studies show that, besides low mobility and low population density, another important factor in driving the evolution of hermaphroditism is mate-search efficiency (Eppley and Jesson, 2008; Puurtinen and Kaitala, 2002). "When mate search is efficient, disruptive frequency-dependent selection on time allocation to mate search leads to the evolution of searching and nonsearching phenotypes and, ultimately, to the evolution of males and females." (Puurtinen and Kaitala, 2002). Returning from hermaphroditism to separate sexes is thought to be easier than the evolution of hermaphroditism because giving up a sex function might be easier than assuming one (Charnov, 1982).

So far, we have dealt with factors that may stimulate the evolution of hermaphroditism or separate sexes, but not with the mechanisms that produce this complex transformation. The reason is no secret; all the studies on hermaphroditism deal with the selection of the new trait instead of its evolution or emergence. However, there is compelling evidence of the mechanism of sequential hermaphroditism. It is about cases of sequential hermaphroditism described in at least 25 families of fish (Devlin and Nagahama, 2002). The fact that this incredible change occurs during an individual's life suggests that these fish have in place an inherited mechanism of switching to the opposite sex, which may be very relevant in understanding the mechanism behind simultaneous hermaphroditism, which still remains mysterious. Ample empirical evidence shows that sequential hermaphroditism occurs in many species that are known to have genetically determined sex, including birds, amphibians, and fish (Kuntz et al., 2004).

Sex chromosomes are identified only in 176, or about 10%, of 1700 fish species investigated cytogenetically (Devlin and Nagahama, 2002), but even these species can switch to the opposite of their genetic sex. Our knowledge of the genetic determination of sex in this group seems far from reliable when we read: "Male heterogamety (males are XY and females XX, as is generally the rule in mammals) and female heterogamety (females are WZ and males ZZ, the system at work in birds) are sometimes observed within the same fish genus, and even the same fish species. More complicated systems can involve multiple sex chromosomes and multiple gene loci (influence from autosomal loci on sex determination and polyfactorial sex determination)" (Volff, 2002).

The empirical evidence of sequential hermaphroditism in fish indicates that this is a function of an epigenetic mechanism (the switching to the opposite sex occurs within the life of an individual). As early as the mid-1970s, Tang et al. (1974) observed that injection of the mammalian pituitary LH in the females of the rice-field eel, *Monopterus albus* (Zuiew), induces the transformation of ovaries into male gonads displaying normal spermiogenesis.

Removal of a female fish from a protandrous (i.e., develop first as males) group or of a male from a protogynous group (i.e., develop first as females) induces the sex inversion of the largest individual into the sex of the removed fish. Such sex inversions are determined in the brain by the secretion of gonadotropins from hypothalamic neurons (Baroiller et al., 1999; Grober and Bass, 1991).

Unlike other vertebrate classes, the brain in fish is not fully sexualized, and hence, it is plastic enough to enable sex inversion in adult fish (McCarthy, 2009). The established role of social factors in sex inversion also supports the role of the brain in sex determination in fish. Effects of cortisol in sequential hermaphroditism in fish provide support to the idea that stress, a neurobilogical process, may be involved in environmental sex determination (Hattori et al., 2009).

Essential changes during sex inversion also occur in the number of neurons of the fish hypothalamus. Masculinization of fish is associated with an increase in the number of hypothalamic GnRH neurons in cases of socially determined sex inversion from female to male (Elofsson et al., 1997).

Sex Evolution and Sex Determination in Eumetazoans

There are many hypotheses on the evolution of sex. The oldest, and still prevalent among them, is that of the German biologist August Weisman (1834–1914), put forward by the end of the nineteenth century. According to Weisman, sexual reproduction generates variation upon which the natural selection acts. However, the opposite idea has also been expressed; sexual reproduction may counteract further evolution (Shcherbakov, 2010). Cavalier-Smith believes that recombinative variability is a consequence of crossing over, rather than the main selective advantage for the origin of sex (Mancebo Quintana and Mancebo Quintana, 2012). Yet another hypothesis is that obligate sexual reproduction is a result of multicellularity, based on the clear correlation between the two (Dacks and Roger, 1999). Other hypotheses posit that sexual reproduction evolved in response to the need for repairing DNA damages and so on.

However, these hypotheses have not been validated. None of them plausibly explains how presumed advantages of sexual reproduction could offset the sex-related disadvantages, such as the high costs of meiosis and production of gametes, the extra time required to find the corresponding gamete and/or the partner, the smaller number of offspring they produce (the twofold cost of sex), or the general cost of maintaining males along the reproductively sufficient females.

The fact that sexual reproduction offers evolutionary and some physiological advantages is unquestionably indicated by the ever-increasing frequency of the sexual mode of reproduction as we climb the evolutionary tree. Many sexual facultative species, such as the yeast *Sacchaaromyces cerevisiae*, aphid insects, and water fleas (*Daphnia* spp.) switch to sexual reproduction as soon as they find themselves in unfavorable conditions, indicating a significant advantage of the sexual reproduction when life conditions deteriorate.

The above-mentioned correlation between sexual reproduction and evolutionary progress is a major fact that seems to have been insufficiently considered in studies on the causes of the evolution and maintenance of sexual reproduction. It may hold clues about the selective pressures that might have driven that evolution.

The transition to multicellularity increased pressures for evolving mechanisms that would protect against genetic mutations. Stated as briefly as possible, these pressures were related to the following facts.

First, one of the consequences of the transition to multicellularity is the expansion of the multicellular genome related to the requirements of maintaining the homeostasis of a multicellular organism rather than a single-celled organism. The multiplication of the number of genes in multicellulars derives primarily from the increase in the number of nonhousekeeping genes. Based on a partial investigation, the human genome showed that about 84% of the investigated genes was nonhousekeeping genes (Hsiao et al., 2001). The multiplication of the number of genes, their regulatory sequences, and the genome size in general obviously lead to an increased probability of DNA mutation.

Second, the increased complexity of multicellulars led to the creation of complex gene regulatory networks (GRNs) and to participation of one nonhousekeeping gene in many GRNs so that mutation in any gene may affect the structure and function of many organs.

Diploidy, or the presence of two copies of each gene/chromosome, would be a good prevention against possible deleterious effects of DNA mutations. It may be argued, however, that diploidy alone would be equally efficient at thwarting the possible negative effects of genetic mutation, and sexual reproduction might not be necessary to evolve. Counter to this, remember that sexual reproduction wards off the chance of mutations affecting the function and structure of multicellulars by joining the haploid genomes of different individuals in each generation.

Epigenetic Determination of the Primary Sex in Vertebrates

Sex determination is considered a textbook example of the genetic determination of sex in metazoans. In mammals, including humans, sex is determined by combining two types of sex chromosomes, X and Y. Combination XX induces the development of female individuals and XY produces males. Females containing two identical sex chromosomes are homogametic, and males with two different sex chromosomes are heterogametic. The opposite is observed in birds, where males are homogametic (ZZ) and females heterogametic (ZW).

The prevailing opinion is that "sex genes" are first expressed in the gonads, and the secretion of sex hormones by gonads determines the differentiation of male and female phenotypes. Accordingly, the signal cascade of genetic determination of sex looks as follows (Crews, 1993):

> Zygote→gonad determining genes→gonad formation→hormones→sexual differentiation of phenotype.

However, in the past decade, evidence is accumulating to unambiguously demonstrate that differentiation of male and female phenotypes is epigenetically, and through gonads, determined in many species and higher taxa.

Epigenetic Determination of Secondary Sex in Fish

The temperature during early development causes changes in the normal sex ratio in many fish species (Blázquez et al., 2009). This is related to changes in the amount and type of aromatase the brain produces in response to the incubation temperature.

Contrary to the earlier belief that the sexual differentiation of gonads and secretion of gonadal hormones induced sexual differentiation in the brain in fish, ample evidence shows that sexual differentiation in the brain of fish embryos starts before gonadal differentiation, especially with respect to the expression of brain aromatase (Vizziano-Cantonnet et al., 2011); hence, it is more likely that brain signals induce sexual differentiation of gonads. The brain and gonads of the fish embryo express both types of aromatase, CYP19A and CYP19B, but their expression in the brain occurs before the sexual differentiation of gonads, and the expression of the CYP19B is higher in the brain (Matsuoka et al., 2006; Vizziano-Cantonnet et al., 2011).

The above evidence shows that, contrary to conventional wisdom, it is not the gonadal hormone secretion that determines the sexual differentiation of the brain. The reverse seems to be true: sexual differentiation of the brain determines the sexual differentiation of gonads and the animal phenotype in vertebrates. As Scott Gilbert puts it, "sex appears truly to reside in the brain" (Gilbert, 2005).

Epigenetic Determination of Sex in Reptiles

All reptiles (or at least reptile species that have no sex chromosomes) display temperature-dependent sex determination (TSD), rather than genetic sex determination. TSD is observed in many fish, all crocodilians, many turtles, and lizards. Sex in reptiles is determined by the temperature of the egg halfway through embryogenesis (Crews, 1993). When red-eared slider turtle (*Trachemys scripta elegans*) embryos are incubated at low temperatures, they develop into all-male clutches, and the transition from an all-male to all-female sex ratio takes place abruptly if the temperature increases by as little as 1°C (Crews et al., 2001).

In reptiles, the temperature influences sex via the secretion of aromatase in the brain, the enzyme responsible for the synthesis of estrogens via the aromatization of androgens. Aromatase is detected in the gonads only after the temperature-sensitive embryonic period, suggesting that the secretion of aromatase in the brain may induce sexual differentiation of gonads in reptiles. The only organ that secretes aromatase during the temperature-sensitive window in the embryos of the turtle *Malaclemys terrapin* in the southern regions of the United States is the embryonic brain (Jeyasuria and Place, 1998). Larvae of the genotypically female Spanish ribbed newt, *Pleurodeles waltl* (Michahelles, 1830), when reared at 32°C (before the formation of the genital ridge), developed into male newts (Dournon and Houillon, 1985).

Obviously, the temperature cannot directly affect the expression of sex genes—or any genes at all, for that matter. A correlation is observed to exist between sex inversing temperatures and the expression of the aromatase, which is responsible for the synthesis of estrogens and androgens. The fact that the temperature-sensitive period in turtles occurs before the full development of gonads proves that sex determination in turtles is not determined by gonads.

Curiously, brain (as opposed to gonadal) determination of sex is also observed in taxa that are known to have a clear genetically determined sex and sex chromosomes. For instance, sexual differentiation in chick embryos starts in the brain as early as the fourth embryonic day (Scholz et al., 2003) and the genetic sex of brain influences the sexual differentiation of the animal phenotype (Agate et al., 2003). A South Korean team found recently that a set of genes are expressed in the brains of male and female chick embryos on embryonic day 6, "before the influence of gonadal hormones is felt" (Lee et al., 2009).

A Single Genetic Toolkit, but Breathtaking Diversity of Forms

Looking at the diversity of forms, plants, animals, protozoa fungi, and bacteria, it is easy to be amazed by the immense diversity of living forms on Earth. Discussing the common origin of pigeon breeds in *On the Origin of Species*, Darwin wrote: "The diversity of the breeds is something astonishing" (Darwin, 1859), yet "all have descended from the rock-pigeon (*Columba livia*)" (Darwin, 1859, p. 23). Since then, the catalog of living forms has grown greatly. A recent study estimates the number of extant species on Earth at 8.7 million (±1.3 million), including 7.77 million species of animals, 298,000 plants, 36,000 protozoan species, and 611,000 fungi (Mora et al., 2011). In another estimate, the number of fungi species alone may be between 3.5 and 5.1 million. Such data led to the idea that the above figures may be an underestimation of the real number of extant living forms on Earth. When one takes into consideration the repeated mass extinctions of species during the last few hundred million years, that number increases dramatically.

In 1983, researchers in Walter Gehring's laboratory in Basel, Switzerland, discovered the *Hox* genes. *Hox* (from *Homeobox*) genes are a group of genes that have in common a DNA sequence (homeobox) that resides in one chromosome as a cluster. Of paramount importance to the development of multicellulars, they are TFs that bind to specific gene enhancers and function as switches for other genes. The expression patterns of these genes determine the establishment of the body axis and embryonic regions in eumetazoans. Several *Hox* genes seem to have existed in the first eumetazoans, the common ancestor of *Cnidaria* and *Bilateria* (Merabet et al., 2010). They are conserved among animal taxa with a clear trend of increasing numbers, ascending the evolutionary tree in eumetazoans.

The discovery of these "master" control genes and, more recently, the unexpected finding that no relationship exists between the number of genes and the evolutionary complexity or progress (a sheep has more genes than a human) contributed to the idea that these and other genes, such as those for cell membrane receptor proteins, cell adhesion proteins, and so on, represent the genetic or developmental-genetic toolkit. The differential patterns of activation of the toolkit genes, rather than changes in the number of genes or in DNA, determine the evolution of forms in the living world.

The concept of the developmental genetic toolkit contributed to further amplify scientists' amazement about the immense diversity of living forms on Earth. However, although the discovery of the *Hox* genes and other regulatory genes as the key to understanding mechanisms of animal development inspired great enthusiasm, it raises another problem. The high conservancy of *Hox* genes and many other developmentally important genes among animal taxa do not help much, if at all, to explain the tremendous diversification in morphology and functions within the animal kingdom. Gerhart and Kirschner (1997) have readily admitted that "where we most expect to find variation, we find conservation, a lack of change."

The fact that conservation of the genetic toolkit led to accelerated diversification in the multicellular world is a great paradox. The solution of this conundrum requires biologists to search for a "user" of the genetic toolkit.

A User of the Genetic Toolkit

If the genetic toolkit and other developmentally important genes are conserved, as biologists have found, the source of the enormous diversity of living forms cannot be related to these genes. Biologists now believe that "the animal form is not so much a matter of the genes an animal has, but how they are used during the process of development, of going from a single egg to the complete mature animal, the building of all of its body parts. So these genes are used in every animal, but the choreography is what differs" (Carroll, 2006).

We know that it is not genes, but the different patterns of their expression, that provide the living world with the enormous combinatorial potential embodied in the astonishing diversity of living forms. But choreography is an art, and the choreographer instructs. If one expands this metaphor, the question would be: What tells cells throughout the animal body when and where to express which genes?

This question is a formidable challenge to any biologist. The idea that a material entity, within an animal itself, may control the expression of thousands of genes in virtually an infinite number of cells of widely different types seems next to impossible. This is a question about a big "unknown"; hence, it would be wise to break it down into smaller pieces (or partial questions) by going stepwise from the known to the unknown.

We know about the classical Jacob-Monod model and a few other types of genetic feedback systems of regulation in unicellulars. We also know that a unicellular can adaptively respond to internal and external stimuli via epigenetic mechanisms (see the section "The Control System in Unicellulars" in Chapter 1) that regulate gene expression and DNA replication and determine cell structure and functions.

Transition to multicellularity raises the intricate issue of coordination of functions of billions to trillions of cells of different types, each type with a different function and structure. Even if one would agree with the simplicist Virchowian concept of the organism as a "republic of cells," a coordination of their activity at a supracellular and systemic level is indispensable to prevent the otherwise inescapable fall of the system into chaos.

This supracellular and systemic control of gene expression is now tangible in eumetazoans. Each cell in these organisms expresses a variable number of genes that are necessary for the cell's subsistence and normal functioning; they are called *housekeeping genes*. The rest (i.e., thousands of nonhousekeeping genes in cells of multicellular organisms) are expressed for the sake of the organism, to meet its actual requirements. They are fees that the cell pays for membership in the organismic community of cells.

Obviously, no cell knows what the organism needs at any moment in time. Even theoretically, cells in multicellulars need instructions on *when*, *where*, and *how long* they have to express each gene. These instructions have to reach the cell from extracellular sources in a genetically intelligible form via chemical signals; different signals must be sent to different cells at different times during their lives.

Theory aside, let us consider what we know about these nongenetic extracellular signals that the organism sends to all its cells. Most extracellular signals, such as hormones, growth factors, secreted proteins, and neurotransmitter neuromodulators, use signal transduction pathways (see the section "Epigenetic Information and Signal Cascades" in this chapter) to transmit their messages to genes to induce their expression. The first step to the activation of a specific signal transduction pathway is the binding of a ligand (e.g., hormone and growth factor) to its specific cell membrane receptor. The binding of the ligand to its receptor activates a particular signal transduction pathway by inducing a chain of phosphorylations, enabling the expression of

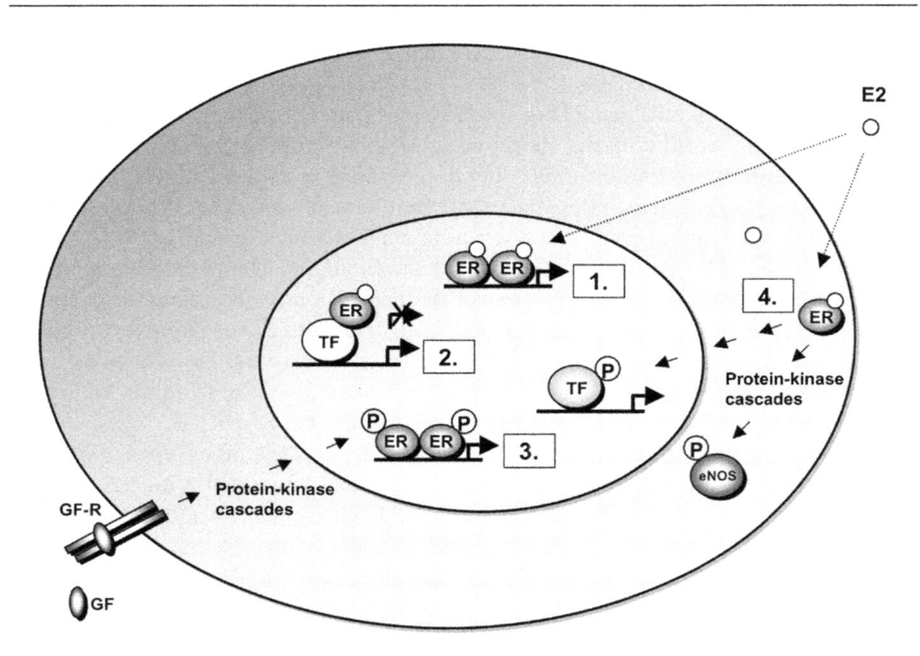

Figure 2.27 Schematic illustration of ER signaling mechanisms. (1) Classical mechanism of ER action. Nuclear E2–ERs bind directly to EREs in target gene promoters. (2) ERE-independent genomic actions. Nuclear E2–ER complexes are tethered through protein–protein interactions to a TF complex that contacts the target gene promoter. (3) Ligand-independent genomic actions. Growth factors (GFs) activate protein-kinase cascades, leading to phosphorylation (P) and activation of nuclear ERs at EREs. (4) Nongenomic actions. Membrane E2–ER complexes activate protein-kinase cascades, leading to altered functions of proteins in the cytoplasm, or regulate gene expression through phosphorylation (P) and activation of a TF (Björnström and Sjöberg, 2005).

a specific gene. A number of hormones and other inducers that have smaller molecular size are able to pass through the cell membrane and bind to their specific receptors in the cytoplasm or in the cell nucleus.

Let us illustrate the mechanisms of gene expression through cell membrane receptors and nuclear receptors with a simple example of the activation of estrogen-inducible genes by E2 and growth factors. The binding of E2 to its nuclear receptor ecdysone receptor (ER) in the cell nucleus causes a change in the receptor conformation, which facilitates the recruitment of coactivators (Figure 2.27). In this form, after dimerization, the complex binds an estrogen response element (ERE) in the regulatory sequence of E2-induced genes. E2 can also bind a TF complex before binding the ERE. In another mechanism, E2 binds its receptor in the cytoplasm and, from there, activates a signal cascade. A number of growth factors, via signal cascades, can also activate proteins that bind the ER in the nucleus and in this form bind ERE in estrogen-induced genes (Björnström and Sjöberg, 2005).

Hormones are secreted by endocrine glands, but growth factors are induced by hormones. So, for example, Wnt is induced by progesterone during the morphogenesis of the mammary gland; inhibin, a member of the TGF-beta family of growth factors, is induced by the pituitary FSH; expression of the *Igf*-1 (insulin growth factor-1) gene is induced by the GH and by the neurohormone VIP, and so on.

It is now common knowledge that in eumetazoans, the secretion of all hormones from the target endocrine glands (thyroid, pancereas, adrenals, ovaries, and testicles) is induced by secretion of specific "stimulating" hormones in the pituitary; hence, the designation of the pituitary as the "master gland." But soon the master gland turned out to be subordinate to another part of the brain, called the hypothalamus. Secretion of each of the pituitary hormones is induced by a specific "releasing" hormone produced by the hypothalamus.

Genetic marks, DNA methylation, and histone modifications also seem to be regulated by extracellular signal cascades. DNA methylation is the function of methyltransferases, but these enzymes are activated by hormones such as glucocorticoids and E2. In turn, these hormones are respectively regulated by the pituitary ACTH and FSH and, farther upstream, by the hypothalamic neurohormones CRH and GnRH.

Histone modification (e.g., acetylation, deacetylation, and methylation) are regulated via neuroendocrine cascades. So, for example, acetylation of histones in the ovarian granulosa cells induces the expression of FSH-responsive genes. Sometimes the nervous system is directly involved in histone modification. Such is the case with expression of AChR in the NMJ; axon terminals of the motor neuron release in the junction agrin, which, by activating a signal transduction pathway, enables the recruitment of acetylating enzymes, modification of the histone, and the exposure of genes for transcription. (For an expanded discussion of this issue, see the section "Neural Origin of Epigenetic Information for Epigenetic Structures" earlier in this chapter.)

By connecting the intracellular signal transduction pathways with the neuroendocrine cascades, we may see the expression of particular genes as the result of a general signal cascade that starts with brain signals. This seems to be a common mechanism of gene expression based on epigenetic information (chemical signals) released in the brain as an output of the processing of internal and external stimuli. This epigenetic information flows through neuroendocrine signal cascades, which via intracellular signal transduction pathways send the epigenetic message to one or a number of specific genes. In a generalized form, the neuroendocrine signal cascade for the expression of a gene is presented in Figure 5.16.

This neuroendocrine model, however, cannot explain the fact that these circulating chemicals induce gene expression in certain cells and regions of the animal body, while repressing it in other cells and regions. A neural mechanism of spatial restriction of gene expression in animal bodies will be described in Chapter 3.

The neural control of gene expression is a surprising revelation in modern biology. A further discussion on the nature of this control is beyond the scope of this book, but interested readers can find extensive information on this topic in my book *Epigenetic Principles of Evolution* (Cabej, 2012), especially in Chapter 2.

Epigenetic Programming of Gene Expression in the Egg

Procreation in animals is a monopoly of the gametes—egg and sperm—in dioecious animals and the egg in parthenogenetic animals. No other type of cell can develop into an adult organism. What makes gametes, or the egg cell in parthenogenetic animals, uniquely capable of producing adult organisms?

For a long time, it was taken for granted that the genome provides gametes the singular ability to reproduce. However, it has never been argued—let alone proved—that this is indeed the case.

There are at least three essential facts that unambiguously argue against a determining role of the genome in gametes' ability to start and proceed with individual development. First, that the other somatic cells in metazoans are incapable of developing into adult organisms, although they have the same species-specific genome and genes that gametes/zygotes have. Second, early individual development following the first cleavage division of the zygote depends not on zygotic genes but on the epigenetic information deposited in gametes mainly in the form of parental (maternal and paternal) cytoplasmic factors, primarily mRNA and also on other chemicals such as hormones and neurotransmitters. Third, cloning experiments require the oocyte cytoplasm exclusively, and no cytoplasm of any other cell can substitute for the oocyte cytoplasm.

Two important landmarks in the development of the knowledge of the role of cytoplasmic factors in development is the transformation of putative somatic cells into germ cells by transplanting germ cytoplasm by Illmensee and Mahowald (1974) and demonstration by Freeman and Lundelius (1982) that a maternal cytoplasmic factor in the egg cell is responsible for left/right coiling in a snail. We know that it is the epigenetic information that the egg cell is provided with maternally that is responsible for the unique procreational ability of eggs. This information is provided to the egg in the form of maternal factors that are deposited in an orderly fashion within the egg and determine the early embryonic development up to the phylotypic stage.

Accumulating evidence shows that cytoplasmic factors of the sperm cell are involved in the initial regulation of early development. The sperm cell plays a determining role in asymmetric cleavage divisions that lead to the formation of different types of cells. Here, we will deal only with the deposition of the maternal factors in the egg cell because it is the primary source of cytoplasmic factors (epigenetic information) in the zygote.

Deposition of Maternal Determinants in the Egg

Despite the crucial importance of the egg in development, studies of the ordered placement of maternal factors in the egg cytoplasm has never been central in biological investigation. It is generally taken for granted that concentration gradients of morphogenetic substances may be responsible in determining their location in the egg cell. This statement has not been tested, and it seems implausible that concentration gradients of numerous substances in the same nanospace might maintain the strict arrangement of the various morphogens in the egg; diffusion of the morphogen molecules within the egg cytoplasm tends to eliminate any concentration gradient, let alone allow the accumulation of morphogens to certain sites.

In insects, most cytoplasmic determinants are transported from nurse cells to oocytes via ring canals in a biphasic process of "slow transport" followed by "dumping" of the nurse cells' content after their apoptosis, which is regulated by ecdysone (Soller et al., 1999), whose synthesis is cerebrally regulated by a neurohormone called prothoracicotropic hormone (PTTH). In lower invertebrates, such as *Hydra* (class *Hydrozoa*), oocytes get their cytoplasmic determinants by actively phagocytizing apoptotized nurse cells (Miller et al., 2000).

Another way to deposit cytoplasmic determinants in insect oocytes is receptor-mediated endocytosis, which is responsible for vitellogenin uptake by oocytes. Handler and Postlethwait (1977) observed a "cephalic event" that, via ecdysone, controls the uptake by the egg cell of vitellogenin (Richard et al., 2000), as well as other maternal cytoplasmic factors (Chapman, 1998) of the hemolymph.

The deposition of cytoplasmic determinants, such as *bicoid* mRNA (Cha et al., 2001), *oskar* mRNA, vasa mRNA (Tomancak et al., 1998), and others, is based on their transport by motor molecules along microtubules and actin filaments of the oocyte cytoskeleton. Motors such as dynein, kinesin, and myosins carry out the transport of determinants in various regions of the oocyte cytoplasm.

If the transport of these determinants takes place along the microtubule rails, this indicates that the cytoskeleton might play a role in the final localization of determinants in the egg. This implies, in turn, that the egg or the organism can adaptively modify the structure of the cytoskeleton.

Regulation of the Length of Microtubules—Key to Transport of Maternal Determinants in the Oocyte

A number of hormones, neurohormones, and neurotransmitters are involved in the process of lengthening (polymerization) and shortening (depolymerization) cytoskeleton microtubules. Many of them perform their microtubule-modifying functions by acting on a special type of proteins known as *microtubule-associated proteins* (*MAPs*). By binding microtubules, MAPs induce their polymerization (cadherin) or depolymerization (Tournebize et al., 2000; Walczak et al., 1996).

Among neuroactive substances that modify the microtubule network by inducing changes in the structure of MAPs are the hypothalamic neuropeptide GnRH (Drouva et al., 1998), VIP, as well as the monoamine neurotransmitters dopamine and noradrenaline (Chneiweiss et al., 1992). The neuropeptide PTTH stimulates β-tubulin synthesis (Rybczynski and Gilbert, 1995), and it is believed that tubulin isotype composition may modulate microtubule polymerization dynamics in cells (Panda et al., 1994). The neurotransmitter acetylcholine (Ach) increases the Ca^{2+} in cells (Mäthger et al., 2003), which in turn induces changes in the structure and length of microtubules.

The ovaries and testicles possess an intrinsic catecholaminergic network of neurons (Mayerhofer et al., 1997) and it is plausible that Ach released by these neurons might modify the microtubule length, as illustrated later.

Recall that secretion of microtubule-modifying hormones is controlled by the CNS via neuroendocrine signal cascades. No theoretical explanation or hypothesis has been presented on the source of information necessary for the spatial patterns of the placement of cytoplasmic determinants in the egg cell. We are clearly in the midst of an explanatory conundrum, but the situation may not be as hopeless as it may seem.

Examination of the proximate mechanisms of modification of microtubules suggests that the involvement of not only neurohormones and neurotransmitters, but also of hormones of the target endocrine glands (which are secreted in response to neural signals), suggests that the nervous system may play an important role in the dynamics of microtubules in the oocyte, and, consequently, in the process of the ordered placement of maternal cytoplasmic factors in the oocyte.

Can the nervous system be a source of epigenetic information that determines the location of maternal factors in the oocyte? So far, we have no direct experimental evidence that neuroactive substances, neurohormones, and other hormones that affect the length and dynamics of the cytoskeleton microtubules also regulate the spatial patterns of deposition of maternal factors in the oocyte. Hence, I will restrict my efforts to finding any empirical evidence on the adaptive neural regulation of the length of microtubules. I chose to examine this possibility in studies on the adaptive coloration of fish and cephalopod species, an adaptive mechanism that in fish evolved more than 400 million years ago (Fujii, 2000).

The Nervous System Regulates "At Will" the Length of Microtubules

A number of fish species can, in seconds and seemingly at will, change their body color and pattern to match the perceived color of their background. Some can even mimic body colors and patterns of heterospecific fishes to avoid detection within the heterospecific crowd. This is the case with several fishes, such as the neon tetra (*Paracheirodon innesi*), blue damselfish (*Chrysiptera cyanea*), and the paradise whiptail (*Pentapodus paradiseus*), as well as several cephalopods.

The cells responsible for the adaptive change in body color in these fishes are iridophores, pigmentless light-reflecting cells of the skin that contain piles of thin transparent guanine platelets, which form multilayer reflectors. The distance between guanine plates is determined by the length of microtubules that connect plates to each other (Kasukawa et al., 1987; Mäthger et al., 2003). Any change in the length of the microtubules between plates will produce a change in the color reflected by iridophores, which is what the human eye perceives.

It was found that neurotransmitters like noradrenaline shift the light reflected by iridophores toward longer wavelengths, while Ach and noradrenaline antagonists such as adenosine shift it toward shorter wavelengths (Kasukawa, et al., 1987;

Mäthger et al., 2003). Neurotransmitter receptors are mediators of changes in the distance between platelets. The shifts in the light wavelengths reflected by guanine platelets is determined by corresponding changes (increases and decreases) in the length of microtubules and, consequently, of the distance between reflecting platelets (Kasukawa et al., 1987). Thus, the change in the length of interplate microtubules changes the wavelength and the color of light reflected by the skin. This is the "Venetian blind model" of adaptive color change (Yoshioka et al., 2011). The instantaneous change of body color is not instinctive in the classical meaning of the word, i.e., it does not result from any "fixed action pattern." Far from a standard indivisible response, it implies a practically infinite number of responses (i.e., the ability to change its body color in a multitude of ways so that it matches the perceived background or body color and patterns of other animals).

How does the animal translate the perceived color and pattern into its own body color and pattern? How does it verify whether the induced change matches the perception? While the neurobiological mechanism of processing of these stimuli and their "translation" in the brain is still unknown, we do know the neural pathway that leads to this form of astonishing adaptive pigmentless coloration.

The visual input from the retina is transmitted, via the optic nerve, to the CNS, where its processing results in an adaptive neural response that determines the adaptive coloration and pattern of the fish. The CNS selects the adaptive response, with the sympathetic nervous system acting as the mediator of the adaptive changes of skin color (Goda and Fujii, 1998). Severance of the sympathetic nerves prevents the fish from displaying the adaptive change of color and pattern, leading to the darkening of skin, typical of depression of the sympathetic activity, and the activation instead of the parasympathetic system in chromatophores (Fujii, 2000).

If the nervous system adaptively regulates the length of microtubules in the skin cells of fish to accomplish finely tuned adaptive coloration, is there any reason why it could not adaptively regulate the length of egg microtubules, and, consequently, the spatial arrangement of maternal factors in the egg?

References

Adcock, I.M., Ford, P., Ito, K., Barnes, P.J., 2006. Epigenetics and airways disease. Respir. Res. 7, 21. doi: 10.1186/1465-9921-7-21.

Agate, R.J., Grisham, W., Wade, J., Mann, S., Wingfield, J., Schanen, C., et al., 2003. Neural, not gonadal, origin of brain sex differences in a gynandromorphic finch. Proc. Natl. Acad. Sci. U.S.A. 100, 4873–4878.

Amberg, D.C., 1998. Three-dimensional imaging of the yeast actin cytoskeleton through the budding cell cycle. Mol. Biol. Cell 9, 3259–3262.

Aristotle, 1910. The History of Animals. Book V. Part 1 (D'Arcy Wentworth Thompson, Trans.). Clarendon Press, Oxford. Available: <http://ebooks.adelaide.edu.au/a/aristotle/history/book5.html>.

Aurich, C., Aurich, J.E., Parvizi, N., 2001. Opioidergic inhibition of luteinising hormone and prolactin release changes during pregnancy in pony mares. J. Endocrinol. 169, 511–518.

Balczon, R., Bao, L., Zimmer, W.E., Brown, K., Zinkowski, R.P., Brinkley, B.R.T., 1995. Dissociation of centrosome replication events from cycles of DNA synthesis and mitotic division in hydroxyurea-arrested Chinese hamster ovary cells. J. Cell Biol. 130, 105–115.

Baroiller, J.-F., Guiguen, Y., Fostier, A., 1999. Endocrine and environmental aspects of sex differentiation in fish. Cell. Mol. Life Sci. 55, 910–9131.

Beisson, J., 2008. Preformed cell structure and cell heredity. Prion 2, 1–8.

Beisson, J., Sonneborn, T.M., 1965. Cytoplasmic inheritance of the organization of the cell cortex in *Paramecium aurelia*. Proc. Natl. Acad. Sci. USA 53, 275–282.

Biswas, T., Ramana, C.V., Srinivasan, G., Boldogh, I., Hazra, T.K., Chen, Z., et al., 1999. Activation of human O6-methylguanine-DNA methyltransferase gene by glucocorticoid hormone. Oncogene 18, 525–532.

Björnström, L., Sjöberg, M., 2005. Mechanisms of estrogen receptor signaling: convergence of genomic and nongenomic actions on target genes. Mol. Endocrinol. 19, 833–842.

Blázquez, M., Navarro-Martín, L., Piferrer, F., 2009. Expression profiles of sex differentiation-related genes during ontogenesis in the European sea bass acclimated to two different temperatures. J. Exp. Zool. B Mol. Dev. Evol. 312, 686–700.

Brandan, E., Fuentes, M.E., Andrade, W., 1992. Decorin, a chondroitin/dermatan sulfate proteoglycan is under neural control in rat skeletal muscle. J. Neurosci. Res. 32, 51–59.

Brisken, C., Heineman, A., Chavarria, T., Elenbaas, B., Tan, J., Dey, S.K., et al., 2000. Essential function of Wnt-4 in mammary gland development downstream of progesterone signaling. Genes Dev. 14, 650–654.

Buckley, W.R., 2008. Computational ontogeny. Biol. Theory 3, 3–6.

Cabej, N.R., 2005. Neural Control of Development. Albanet, Dumont, NJ, pp. 43–45.

Cabej, N.R., 2012. Epigenetic Principles of Evolution. Elsevier Inc., London and Waltham, MA, pp. 393–397, 466–472.

Carroll, S., 2006. Interview on July 14, 2006 in NOVA Online. <http://www.pbs.org/wgbh/nova/body/genetic-factor.html/> (accessed 06.2012).

Catlett, N.L., Weisman, L.S., 2000. Divide and multiply: organelle partitioning in yeast. Curr. Opin. Cell Biol. 12, 509–516.

Cha, B.J., Kopetsch, B.S., Theurkauf, W.E., 2001. *In vivo* analysis of *Drosophila bicoid* mRNA localization reveals a novel microtubule-dependent axis specification pathway. Cell 106, 35–46.

Chant, J., 1999. Cell polarity in yeast. Annu. Rev. Cell Dev. Biol. 15, 365–391.

Chapman, R.F., 1998. The Insects—Structure and Function, fourth ed. Cambridge University Press, Cambridge, p. 306.

Charnov, E.L., 1982. The Theory of Sex Allocation. Princeton University Press, Princeton, NJ.

Chneiweiss, H., Cordier, J., Sobel, A., 1992. Stathmin phosphorylation is regulated in striatal neurons by vasoactive intestinal peptide and monoamines via multiple intracellular pathways. J. Neurochem. 58, 282–289.

Christophersen, N.S., Helin, K., 2010. Epigenetic control of embryonic stem cell fate. J. Exp. Med. 207, 2287–2295.

Cooper, G.M., 2000. The Cell: A Molecular Approach, second ed. Sinauer Associates, Sunderland, MA.

Crews, D., 1993. The organizational concept and vertebrates without sex chromosomes. Brain Behav. Evol. 42, 202–214.

Crews, D., Fleming, A., Willingham, E., Baldwin, R., Skipper, J.K., 2001. Role of steroidogenic factor 1 and aromatase in temperature-dependent sex determination in the red-eared slider turtle. J. Exp. Zool. 290, 597–606.

Dacks, J., Roger, A.J., 1999. The first sexual lineage and the relevance of facultative sex. J. Mol. Evol. 48, 779–783.

Darwin, C.R., 1859. On the Origin of Species by Means of Natural Selection, or the Preservation of Favoured Races in the Struggle for Life. John Murray, London, p. 21.

Devlin, R.H., Nagahama, Y., 2002. Sex determination and sex differentiation in fish: an overview of genetic, physiological, and environmental influences. Aquaculture 208, 191–364.

Doi, A., Park, I.H., Wen, B., Murakami, P., Aryee, M.J., Irizarry, R., et al., 2009. Differential methylation of tissue- and cancer-specific CpG island shores distinguishes human induced pluripotent stem cells, embryonic stem cells and fibroblasts. Nat. Genet. 41, 1350–1353.

Dournon, C., Houillon, C., 1985. Thermosensibilité de la différenciation sexuelle chez l'Amphibien Urodèle, *Pleurodeles waltlii* Michah. Conditions pour obtenir l'inversion du phénotype sexuel de toutes les femelles génétiques sous l'action de la température d'élevage. Reprod. Nutr. Dev. 25, 671–688.

Drouva, S.V., Poulin, B., Manceau, V., Sobel, A., 1998. Luteinizing hormone-releasing hormone-signal transduction and stathmin phosphorylation in the gonadotrope αT3–1 cell line. Endocrinology 139, 2235–2239.

Ebert, S.N., Balt, S.L., Hunter, J.P.B., Gashler, A., Sukhatme, V., Wong, D.L., 1994. Egr-1 activation of rat adrenal phenylethanolamine N-methyltransferase gene. J. Biol. Chem. 269, 20885–20898.

Elofsson, U., Winberg, S., Francis, R.C., 1997. Sex differences in number of preoptic GnRH-immunoreactive neurons in a protandrously hermaphroditic fish, the anemone fish *Amphiprion melanopus*. J. Comp. Physiol. A 181, 484–492.

Eppley, S.M., Jesson, L.K., 2008. Moving to mate: the evolution of separate and combined sexes in multicellular organisms. J. Evol. Biol. 21, 727–736.

Freeman, G., Lundelius, J., 1982. The developmental genetics of dextrality and sinistrality in the gastropod *Lymnaea peregra*. Roux's Arch. Dev. Biol. 191, 69–83.

Fujii, R., 2000. The regulation of motile activity in fish chromatophores. Pigm. Cell Res. 13, 300–319.

Galun, E., 2003. Transposable Elements. Kluwer Academic Publishers, Dordrecht, The Netherlands, p.13.

Gamble, M.J., Freedman, L.P., 2002. A coactivator code for transcription. Trends Biochem. Sci. 27, 165–167.

Gerhart, J., Kirschner, M., 1997. Cells, Embryos and Evolution. Blackwell Science, Malden, MA.

Gilbert, S.F., 2005. Mechanisms for the environmental regulation of gene expression: ecological aspects of animal development. J. Biosci. 30, 65–74.

Goda, M., Fujii, R., 1998. The blue coloration of the common surgeonfish, *Paracanthurus hepatus*—II. Color revelation and color changes. Zool. Sci. 15, 323–333.

Griffiths, P.E., 2000. Genetic information: a metaphor in search of a theory. Philos. Sci. 68, 394–412.

Griffiths, P.E., Gray, R.D., 1994. Developmental systems and evolutionary explanation. J. Philos. XCI, 277–304.

Grifiths, P.E., Gray, R.D., 1997. Replicator II—judgement day. Biol. Philos. 12, 471–492.

Grober, M.S., Bass, A.H., 1991. Neuronal correlates of sex/role change in labrid fishes: LHRH-like immunoreactivity. Brain Behav. Evol. 38, 302–312.

Gunderson, F.Q., Merkhofer, E.C., Johnson, T.L., 2011. Dynamic histone acetylation is critical for cotranscriptional spliceosome assembly and spliceosomal rearrangements. Proc. Natl. Acad. Sci. U.S.A. 108, 2004–2009.

Handler, A.M., Postlethwait, J.H., 1977. Endocrine control of vitellogenesis in *Drosophila melanogaster*: effects of the brain and *corpus allatum*. J. Exp. Zool. 202, 389–402.

Harrington, E.W., Trun, N.J., 1997. Unfolding of the bacterial nucleoid both *in vivo* and *in vitro* as a result of exposure to camphor. J. Bacteriol. 179, 2435–2439.

Hattori, R.S., Fernandino, J.I., Kishii, A., Kimura, H., Kinno, T., Oura, M., et al., 2009. Cortisol-induced masculinization: does thermal stress affect gonadal fate in pejerrey, a teleost fish, with temperature-dependent sex determination? PLoS ONE 4 (8), e6548. doi: 10.1371/journal.pone.0006548.

Hedges, S.B., Blair, J.E., Venturi, M.L., Shoe, J.L., 2004. A molecular timescale of eukaryote evolution and the rise of complex multicellular life. BMC Evol. Biol. 4, 2.

Hegstrom, C.D., Truman, J.W., 1996. Steroid control of muscle remodeling during metamorphosis in *Manduca sexta*. J. Neurobiol. 29, 535–550.

Herndon, C.A., Snell, J., Fromm, L., 2011. Chromatin modifications that support acetylcholine receptor gene activation are established during muscle cell determination and differentiation. Mol. Biol. Rep. 38, 1277–1285.

Hewitt, K.J., Shamis, Y., Hayman, R.B., Margvelashvili, M., Dong, S., Carlson, M.W., et al., 2011. Epigenetic and phenotypic profile of fibroblasts derived from induced pluripotent stem cells. PLoS ONE 6 (2), e17128. doi: 10.1371/journal.pone.0017128.

Hinojos, C.A.D., Sharp, Z.D., Mancini, M.A., 2005. Molecular dynamics and nuclear receptor function. Trends Endocrin. Met. 16, 12–18.

Hnilicová, J., Staněk, D., 2011. Where splicing joins chromatin. Nucleus 2, 182–188.

Hnilicová, J., Hozeifi, S., Dušková, E., Icha, J., Tománková, T., Staněk, D., 2011. Histone deacetylase activity modulates alternative splicing. PLoS ONE 6 (2), e16727. doi: 10.1371/journal.pone.0016727.

Holliday, R., Pugh, J.E., 1975. DNA modification mechanisms and gene activity during development. Science 187, 226–232.

Hsiao, L.-L., Dangond, F., Yoshida, T., Hong, R., Jensen, R.V., Misra, J., et al., 2001. A compendium of gene expression in normal human tissues. Physiol. Genomics 7, 97–104.

Illmensee, K., Mahowald, A.P., 1974. Transplantation of posterior polar plasm in *Drosophila*. Induction of germ cells at the anterior pole of the egg. Proc. Natl. Acad. Sci. U.S.A. 71, 1016–1020.

Jeyasuria, P., Place, A., 1998. Embryonic brain–gonadal axis in temperature-dependent sex determination of reptiles: a role for P450 aromatase (CYP19). J. Exp. Zool. 281, 428–449.

Juliandi, B., Abematsu, M., Nakashima, K., 2010. Epigenetic regulation in neural stem cell differentiation. Dev. Growth Differ. 52, 493–504.

Kasukawa, H., Oshima, N., Fujii, R., 1987. Mechanism of light-reflection in blue damselfish motile iridophore. Zool. Sci. 4, 243–257.

Kornblihtt, A.R., Schor, I.E., Allo, M., Blencowe, B.J., 2009. When chromatin meets splicing. Nat. Struct. Mol. Biol. 16, 902–903.

Kuntz, S., Chesnel, A., Flament, S., Chardard, D., 2004. Cerebral and gonadal aromatase expressions are differently affected during sex differentiation of *Pleurodeles waltl*. J. Mol. Endocrinol. 33, 717–727.

Laborie, C., Van Camp, G., Bemet, F., Montel, V., Dupouy, J.P., 2003. Metyrapone-induced glucocorticoid depletion modulates tyrosine hydroxylase and phenylethanolamine N-methyltransferase gene expression in the rat adrenal gland by a noncholinergic transsynaptic activation. J. Neuroendocrinol. 15, 15–23.

Lai, J.-C., Wu, J.-Y., Cheng, Y.-W., Yeh, K.-T., Wu, T.-C., Chen, C.-Y., et al., 2009. O6-methylguanine-DNA methyltransferase hypermethylation modulated by 17β-estradiol in lung cancer cells. Anticancer Res. 29, 2535–2540.

Lara, H.E., Hill, D.F., Katz, K.H., Ojeda, S.H., 1990. The gene encoding nerve growth factor is expressed in the immature rat ovary: effect of denervation and hormonal treatment. Endocrinology 126, 357–363.

Lee, S.I., Lee, W.K., Shin, J.H., Han, B.K., Moon, S., Cho, S., et al., 2009. Sexually dimorphic gene expression in the chick brain before gonadal differentiation. Poult. Sci. 88, 1003–1015.

Leong, W., Pawlik, J.R., 2010. Fragments or propagules? Reproductive tradeoffs among *Callyspongia* spp. from Florida coral reefs. Oikos 119, 1417–1422.

Loizou, J.I., Murr, R., Finkbeiner, M.G., Sawan, C., Wang, Z.Q., Herceg, Z., 2006. Epigenetic information in chromatin the code of entry for DNA repair. Cell Cycle 5, 696–701.

Lonard, D.M., O'Malley, B.W., 2007. Nuclear receptor coregulators: judges, juries, and executioners of cellular regulation. Mol. Cell 27, 691–700.

Lubin, F.D., Roth, T., Sweatt, J.D., 2008. Epigenetic regulation of bdnf gene transcription in the consolidation of fear memory. J. Neurosci. 28, 10576–10586.

Ma, D.K., Jang, M.-H., Guo, J.U., Kitabatake, Y., Chang, M.-l., Powanpongkul, N., 2009. Neuronal activity–induced gadd45b promotes epigenetic DNA demethylation and adult neurogenesis. Science 323, 1074–1077.

Maldonado, M., Uriz, M.J., 1999. Sexual propagation by sponge fragments. Nature 398, 476.

Mancebo Quintana, J.M., Mancebo Quintana, S., 2012. A short-term advantage for syngamy in the origin of eukaryotic sex: effects of cell fusion on cell cycle duration and other effects related to the duration of the cell cycle—relationship between cell growth curve and the optimal size of the species, and circadian cell cycle in photosynthetic unicellular organisms. Int. J. Evol. Biol. 2012, 1–25.

Maruyama, R., Choudhury, S., Kowalczyk, A., Bessarabova, M., Beresford-Smith, B., Conway, T., et al., 2011. Epigenetic regulation of cell type–specific expression patterns in the human mammary epithelium. PLoS Genet. 7 (4), e1001369. doi: 10.1371/journal.pgen.1001369.

Mäthger, L.M., Land, M.F., Siebeck, U.E., Marshall, N.J., 2003. Rapid colour changes in multilayer reflecting stripes in the paradise whiptail, *Pentapodus paradiseus*. J. Exp. Biol. 206, 3607–3613.

Matsuoka, M.P., van Nes, S., Andersen, O., Benfev, T.J., Reith, M., 2006. Real-time PCR analysis of ovary- and brain-type aromatase gene expression during Atlantic halibut (Hippoglossus hippoglossus) development. Comp. Biochem. Physiol. B Biochem. Mol. Biol. 144, 128–135.

Mayerhof, A., Dissen, G.A., Costa, M.F., Ojeda, S.R., 1997. A role for neurotransmitters in early follicular development: induction of functional follicle-stimulating hormone receptors in newly formed follicles of the rat ovary. Endocrinology 138, 3320–3329.

McCarthy, M.M., 2009. The two faces of estradiol: effects on the developing brain. Neuroscientist 15, 599–610.

McGuire, A.M., Pearson, M.D., Neafsey, D.E., Galagan, J.E., 2008. Cross-kingdom patterns of alternative splicing and splice recognition. Genome Biol. 9, R50.

Merabet, S., Sambrani, M., Pradel, J., Graba, Y., 2010. Regulation of hox activity In: Deutsch, J.S. (Ed.), Hox Genes Advances in Experimental Medicine and Biology, vol. 689 Springer Science + Business Media LLC, Landes Bioscience, New York, pp. 4.

Miller, M.A., Technau, U., Smith, K.M., Steele, R.E., 2000. Oocyte development in hydra involves selection from competent precursor cells. Dev. Biol. 224, 326–338.

Mora, C., Tittensor, D.P., Adl, S., Simpson, A.G.B., Worm, B., 2011. How many species are there on earth and in the ocean? PLoS Biol. 9 (8), e1001127.

Moseley, J.B., Goode, B.L., 2007. The yeast actin cytoskeleton: from cellular function to biochemical mechanism. Microbiol. Mol. Biol. Rev. 70, 605–645.

Nanney, D.L., 1966. Corticotype transmission in *Tetrahymena*. Genetics 54, 955–968.

Neiman, A.M., 2005. Ascospore formation in the yeast *Saccharomyces cerevisiae*. Microbiol. Mol. Biol. Rev. 69, 565–584.

Oesterreich, F.C., Bieberstein, N., Neugebauer, K.M., 2011. Pause locally, splice globally. Trends Cell Biol. 21, 328–335.

Otto, S., 2008. Sexual reproduction and the evolution of sex. Nat. Edu. 1, 1.

Pal, S., Gupta, R., Kim, H., Wickramasinghe, P., Baubet, V., Showe, L.C., et al., 2011. Alternative transcription exceeds alternative splicing in generating the transcriptome diversity of cerebellar development. Genome Res. 21, 1260–1272.

Panda, D., Miller, H.P., Banarjee, A., Ludueña, R.F., Wilson, L., 1994. Microtubule dynamics *in vitro* are regulated by the tubulin isotype composition. Proc. Natl. Acad. Sci. U.S.A. 91, 11358–11362.

Petronczki, M., Siomos, M.F., Nasmyth, K., 2003. Un Ménage à quatre: the molecular biology of chromosome segregation in meiosis (Review). Cell 112, 423–440.

Pruyne, D., Bretscher, A., 2000. Polarization of cell growth in yeast. II. The role of the cortical actin cytoskeleton. J. Cell Sci. 113, 571–585.

Puurtinen, M., Kaitala, V., 2002. Mate-search efficiency can determine the evolution of separate sexes and the stability of hermaphroditism in animals. Am. Nat. 160, 645–660.

Rachel Duan, W., Ito, M., Park, Y., Maizels, E.T., Hunzicker-Dunn, M., Jameson, J.L., 2002. GnRH regulates early growth response protein 1 transcription through multiple promoter elements. Mol. Endocrinol. 16, 221–233.

Rattner, J.B., Phillips, S.G., 1973. Independence of centriole formation and DNA synthesis. J. Cell Biol. 57, 359–372.

Ravel-Chapuis, A., Vandromme, M., Thomas, J.-L., Schaeffer, L., 2007. Postsynaptic chromatin is under neural control at the neuromuscular junction. EMBO J. 26, 1117–1128.

Riboni, L., Escamilla, C., Chavira, R., Dominguez, R., 1998. Effects of peripheral sympathetic denervation induced by guanethidine administration on the mechanisms regulating puberty in the female guinea pig. J. Endocrinol. 156, 91–98.

Richard, D.S., et al., 2000. Hormonal regulation of yolk protein uptake by vitellogenic oocytes in diapausing and mutant female *Drosophila melanogaster*. International Conference in Honour of Professor David Saunders. Edinburgh, 20–24 March 2000. (Abstract).

Riggs, A.D., 1975. X inactivation, differentiation, and DNA methylation. Cytogenet. Cell Genet. 14, 9–25.

Robinson, G.W., Hennighausen, L., Johnson, P.F., 2000. Side-branching in the mammary gland: the progesterone–Wnt connection. Genes Dev. 14, 889–894.

Rybczynski, R., Gilbert, L.I., 1995. Prothoracicotropic hormone elicits a rapid, developmentally specific synthesis of beta tubulin in an insect endocrine gland. Dev. Biol. 169, 15–28.

Salvador, L.M., Park, Y., Cottom, J., Maizels, E.T., Jones, J.C., Schillace, R.V., et al., 2001. Follicle-stimulating hormone stimulates protein kinase A-mediated histone H3 phosphorylation and acetylation leading to select gene activation in ovarian granulosa cells. J. Biol. Chem. 276, 40146–40155.

Scholz, B., Kultima, K., Mattsson, A., Axelsson, J., Brunström, J., Halldin, K., 2006. Sex-dependent gene expression in early brain development of chicken embryos. BMC Neurosci. 7, 12.

Severing, E.I., van Dijk, A.D.J., Stiekema, W.J., van Ham, R.C.H.J., 2009. Comparative analysis indicates that alternative splicing in plants has a limited role in functional expansion of the proteome. BMC Genomics 10, 154.

Shannon, C.E., 1993. The lattice theory of information. In: Sloane, N.J.A., Wyner, A.D. (Eds.), Collected Papers. IEEE Press, New York, NY, pp. 180.

Sharma, S., Kohlstaedt, L.A., Damianov, A., Rio, D.C., Black, D.L., 2008. Polypyrimidine tract binding protein controls the transition from exon definition to an intron defined spliceosome. Nat. Struct. Mol. Biol. 15, 183–191.

Shcherbakov, V.P., 2010. Biological species is the only possible form of existence for higher organisms: the evolutionary meaning of sexual reproduction. Biol. Direct 5, 14.
Shea, N., 2007. Representation in the genome and in other inheritance systems. Biol. Philos. 22, 313–331.
Shea, N., 2011. Developmental systems theory formulated as a claim about inherited representations. Philos. Sci. 78, 60–82.
Shea, N., 2012. Inherited representations are read in development. Br. J. Philos. Sci. 0, 1–31.
Soaje, M., Deis, R.P., 1997. Opioidergic regulation of prolactin secretion during pregnancy: role of ovarian hormones. J. Endocrinol. 155, 99–106.
Soaje, M., de Di Nasso, E.G., Deis, R.P., 2002. Regulation by endogenous opioids of suckling-induced prolactin secretion in pregnant and lactating rats: role of ovarian steroids. J Endocrinol. 172, 255–261.
Soller, M., Bownes, M., Kubli, E., 1999. Control of oocyte maturation in sexually mature *drosophila* females. Dev. Biol. 208, 337–351.
Sonneborn, T.M., 1970. Gene action in development. Proc. R. Soc. London B Ser. Biol. Sci. 176, 347–366.
Sonoda, J., Pei, L., Evans, R.M., 2008. Nuclear receptors: decoding metabolic disease. FEBS Lett. 582, 2–9.
Szyf, M., Weaver, I., Meaney, M., 2007. Maternal care, the epigenome and phenotypic differences in behavior. Reprod. Toxicol. 24, 9–19.
Takahashi, K., Yamanaka, S., 2006. Induction of pluripotent stem cells from mouse embryonic and adult fibroblast cultures by defined factors. Cell 126, 663–676.
Tang, F., Chan, S.T.H., Lofts, B., 1974. Effect of mammalian luteinizing hormone on the natural sex reversal of the rice-fieldeel, *Monopterus albus* (Zuiew). Gen. Comp. Endocrinol. 24, 242–248.
Tomancak, P., Guichet, A., Zavorsky, P., Ephrussi, A., 1998. Oocyte polarity depends on regulation of *gurken* by Vasa. Development 125, 1723–1733.
Tournebize, R., et al., 2000. Control of microtubule dynamics by the antagonistic activities of XMAP215 and XKCM1 in *Xenopus* egg extracts. Nat. Cell Biol. 2, 13–19.
Tress, M.L., Martelli, P.L., Frankish, A., Reeves, G.A., Wesselink, J.J., Yeats, C., et al., 2007. The implications of alternative splicing in the ENCODE protein complement. Proc. Natl. Acad. Sci. U.S.A. 104, 5495–5500.
Vandegehuchte, M.B., Lemière, F., Vanhaecke, L., Vanden Berghe, W., Janssen, C.R., 2010. Direct and transgenerational impact on *Daphnia magna* of chemicals with a known effect on DNA methylation. Comp. Biochem. Phys. C 151, 278–285.
Vedral, V., 2010. Decoding Reality. Oxford University Press Inc., New York, NY.
Vizziano-Cantonnet, D., Anglade, I., Pellegrini, E., Gueguen, M.M., Fostier, A., Guiguen, Y., et al., 2011. Sexual dimorphism in the brain aromatase expression and activity, and in the central expression of other steroidogenic enzymes during the period of sex differentiation in monosex rainbow trout populations. Gen. Comp. Endocrinol. 170, 346–355.
Volff, J.-N., 2002. Sex determination in fish. Genome Biol. 3 reports0052.
von Neumann, J., 1966. The Theory of Self-Reproducing Automata. Edited and completed by Burks, A.W. University of Illinois Press, Urbana and London.
Walczak, C.E., Mitchison, T.J., Desai, A., 1996. XKCM1: a *Xenopus* kinesin-related protein that regulates microtubule dynamics during mitotic spindle assembly. Cell 84, 37–47.
Xiong, J., Lu, X., Zhou, Z., Chang, Y., Yuan, D., Tian, M., et al., 2012. Transcriptome analysis of the model protozoan, *Tetrahymena thermophila*, using deep RNA sequencing. PLoS One 7 (2), e30630. doi: 10.1371/journal.pone.0030630.

Yoshioka, S., Matsuhana, B., Tanaka, S., Inouye, Y., Oshima, N., Kinoshita, S., 2011. Mechanism of variable structural colour in the neon tetra: quantitative evaluation of the venetian blind model. J. R. Soc. Interface 8, 56.

Zhou, Y., Lu, Y., Tian, W., 2012. Epigenetic features are significantly associated with alternative splicing. BMC Genomics 13, 123.

3 Epigenetic Control of Animal Development

In Chapter 2 (in Section "Epigenetic Programming of Gene Expression in the Egg"), it was pointed out that the zygote's monopoly over procreation is the result of the epigenetic information provided to gametes in the form of cytoplasmic factors and epigenetic marks, rather than the genetic information in DNA, which is identical in all animal cells. Using sophisticated epigenetic contrivances at the cell (cytoskeleton) and organismic (neural and neuroendocrine) levels, parents provided gametes with the irreplaceable dowry of epigenetic information necessary for the zygote to build up its incipient economy until the embryo reaches a stage when it becomes informationally self-sufficient for erecting its complete adult structure.

Cloning experiments have shown that the epigenetic information in the egg cytoplasm, rather than the DNA or the cell nucleus, is what primarily makes cloning possible. This is why in all cloning experiments, biologists always use eggs (and eggs of the same species, for that matter) alone. The transfer of a sheep somatic nucleus into an ewe egg may produce a cloned sheep, but it is barely imaginable that it would produce any viable organism if it were transferred into a somatic sheep cell or a cow's egg. In cloning, there is no substitute for the epigenetic information contained in the egg cytoplasm; it has the copyright on development.

Fertilization—Fusion of the Egg and Sperm Cell

Fertilization is an epigenetic process in which the genome is dormant. The development of multicellular organisms from single-celled gametes occurs via zygotes (eggs in parthenogenetic organisms), but since most animals are dioecious, herein I will briefly deal with development via zygotes that form by the union of eggs and sperm cells. Both are morphologically and physiologically prepared for the union.

No matter where the union of gametes takes place, whether in the environment (external fertilization) or within the female organism (internal fertilization), the process is generally conserved among the animal taxa and is similar in principle.

In the evolutionary division of labor, sperm is destined to seek and find the egg. In its search for the egg, it is epigenetically guided by the gradient concentration of chemoattractants released by the egg. In 1991, it was discovered that capacitated spermatozoa are attracted to the egg by factors "released from the egg and its surrounding cells" without identifying the factors (Ralt et al., 1991). Later it was reported that progesterone secreted by cumulus oophorus, a group of the ovarian follicle cells surrounding the oocyte, serves as a chemoattractant, guiding sperm toward

Building the Most Complex Structure on Earth. DOI: http://dx.doi.org/10.1016/B978-0-12-401667-5.00003-1

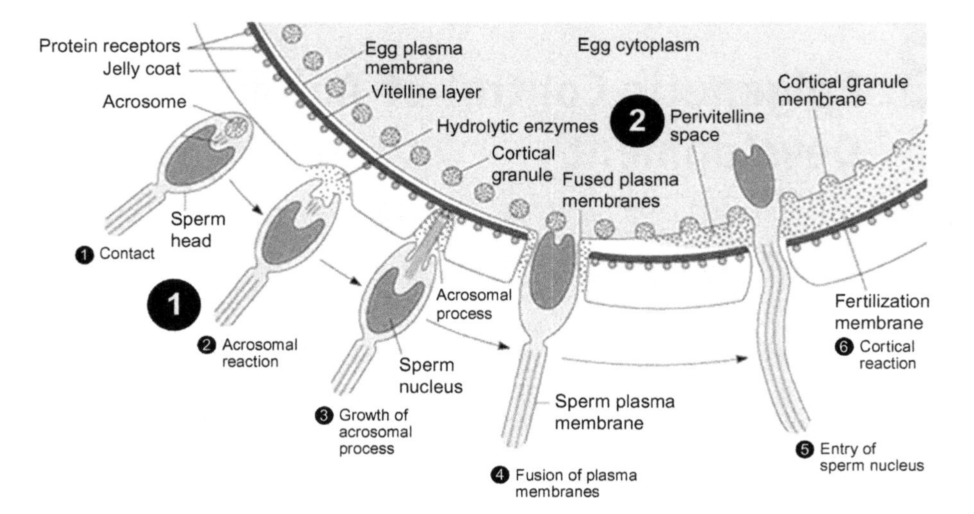

Figure 3.1 Staged penetration and fusion of the sperm nucleus and the centriole in the egg in the process of fertilization.
Source: From http://www.bio.davidson.edu/courses/molbio/molstudents/spring2005/dresser/review%20paper.htm

the egg (Guidobaldi et al., 2008). However, some biologists think that the sperm guidance mechanism still remains elusive.

The sperm first penetrates the cumulus oophorus layer of cells. Then, when it comes in contact with zona pellucida, the acrosome releases enzymes and begins an acrosome reaction that brings the sperm head in contact with the sperm receptors on the vitelline layer (Figure 3.1). Reaching the plasma membrane, the acrosome membrane fuses with it, thus releasing the acrosome content, which contains hydrolytic and proteolytic enzymes that break the egg coat and allow penetration of the sperm nucleus and the centrosome into the egg cytoplasm.

Egg Activation

The entry of sperm in the egg induces Ca^{2+} release, which triggers egg activation and exocytosis of the cortical granules, secretory organelles containing enzymes, and glycosylated components that fill the previtelline space, thus preventing other sperm cells from entering the egg cytoplasm (Gilbert, 2000; Liu, 2011). The sperm thus "lights the fuse" for a Ca^{2+} fertilization wave, inducing the egg nucleus to resume meiotic division, which results in the production of a polar body that is extruded outside the egg, leaving the egg in a haploid state. The Ca^{2+} waves from internal calcium stores are induced by activation of two main pathways, inositol 1,4,5-trisphosphate (IP3), and nicotinic acid adenine dinucleotide phosphate (NAADP) (Stricker, 1999; Whitaker, 2008) (Figure 3.2). Although

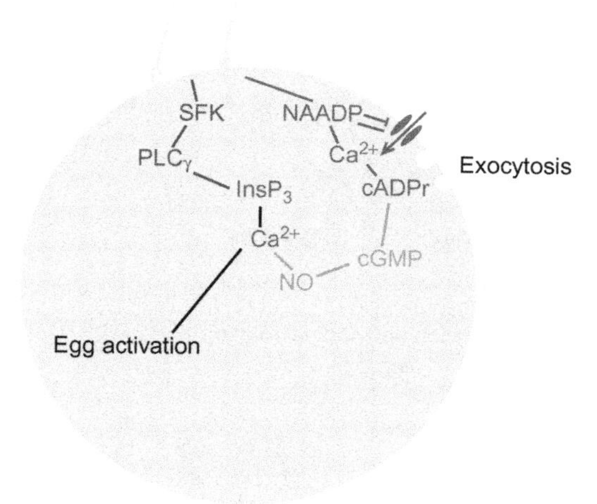

Figure 3.2 Src-family kinases (SFK) activate PLCγ to produce $InsP_3$ and trigger the calcium waves (blue pathway); in sea urchins. In mammals, sperm–egg fusion introduces PLCζ into the egg cytoplasm, producing $InsP_3$ (yellow pathway). Sperm–egg fusion may also introduce NAADP in echinoderms; NAADP activates plasma membrane calcium channels (red pathway). In ascidians, NAADP deactivates plasma membrane channels, while cADPr triggers local calcium release to trigger cortical granule exocytosis (red pathway). In sea urchins, calcium activates nitric oxide production, which generates cADPr via cGMP (green pathway). *Abbreviations*: PLC, phosphoinositide phospholipase C; cADPr, cyclic ADP-ribose; cGMP, cyclic guanosine monophosphate; $InsP_3$, inositol trisphosphate. (For interpretation of the references to color in this figure legend, the reader is referred to the web version of this book.) *Source*: From Whitaker (2008).

fertilization-induced calcium signals are widely shared throughout the animal kingdom, in some groups the egg experiences a single fertilization Ca^{2+} wave, while others experience more than one.

In *Xenopus*, while the genome is still quiescent, fertilization of the egg induces rotation of the egg cortex relative to the cytoplasm, pushing the dorsalizing maternal cytoplasmic factors to the part of the egg cortex that is opposite to the sperm entry point (Figure 3.3). A maternal transcript, Fatvg, plays the key role in the process of cortical rotation (Chan et al., 2007). Just before gastrulation, the epigenetically driven cortical rotation leads to formation of the Spemann organizer. The cortical rotation performs the so-called slow transport of the dorsalizing factors. The slow transport is complemented by the fast dorsalward transport of maternal factors along the polymerizing microtubules. A special role is played by the maternal β-catenin, a component of the Wnt pathway (Moon and Kimelman, 1998): it induces the expression of a number of downstream genes, including its own gene, leading to formation of the dorsoventral and anteroposterior axes of the embryo (Weaver and Kimelman, 2004).

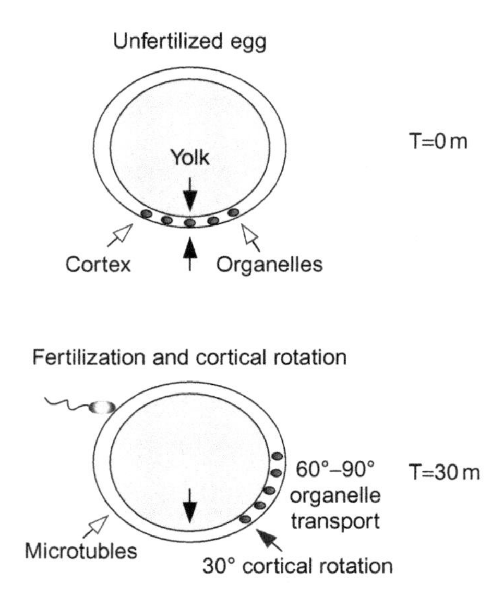

Figure 3.3 In the unfertilized egg, there is no evident dorsoventral polarity, and at the vegetal pole, there are small organelles between the cortex and the inner yolk. A rotation of the cytoplasm relative to the cortex is initiated 30 min after fertilization (note how the cortex, represented by the upward black arrow, has shifted 30° relative to the inner yolk, represented by the downward black arrow) There is a 60–90° translocation of the organelles along the parallel microtubule array in the vegetal hemisphere.
Source: From Moon and Kimelman (1998).

Genomic Restoration in the Zygote

Both the sperm and egg pronuclei are encased in a nuclear membrane. The male pronucleus is associated with the sperm centrosome, whose centrioles organize an aster (a starlike microtubular structure), which slowly drives the male pronucleus from its cortical position toward the egg pronucleus until they meet and fuse around the middle of the egg (Figure 3.4). However, the female pronucleus is not a passive partner; it also goes the "extra mile" to meet the male partner by moving along microtubules of the sperm aster (Reinsch and Gönczy, 1998). In species in which the egg centrosome disintegrates (Schatten, 1994) and is absorbed, the male centrosome serves as the only organizing center of the egg for building the cytoskeleton, which drives the sperm pronucleus in the direction of the egg pronucleus toward their fusion. On their way to the middle of the egg, both pronuclei replicate their decondensed DNA, which again is condensed into chromosomes when pronuclei fuse with each other. Now that each pronucleus has restored the diploid set of chromosomes, the zygotic nucleus has doubled the number of both paternal and maternal chromosomes and is ready to start the first cleavage division, producing two diploid cells.

Cleavage divisions may be equal when it produces two cells that are similar both genetically and phenotypically, but often this division is unequal and leads to

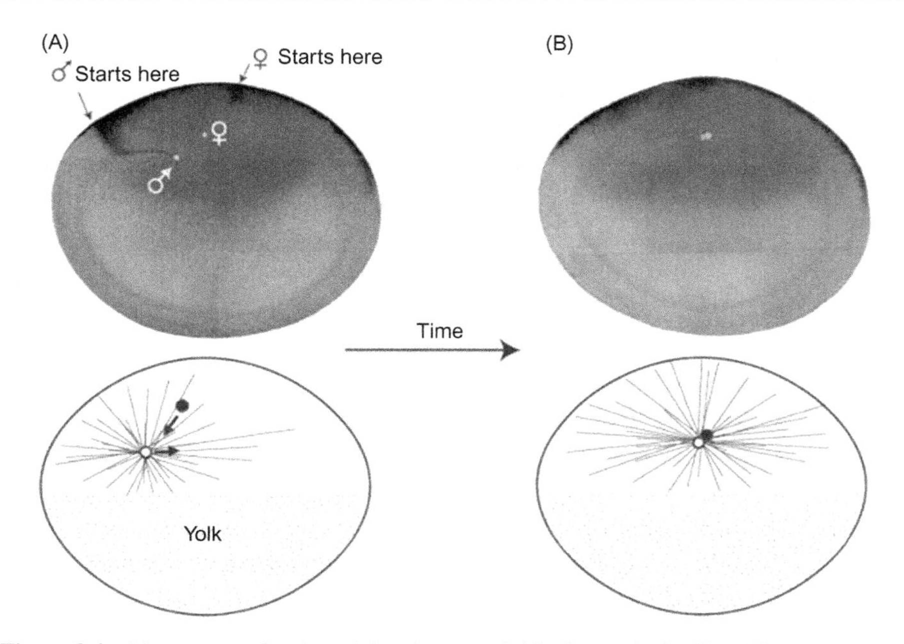

Figure 3.4 Movements of male and female pronuclei in the newly fertilized *Xenopus* egg. The positions of the male and female pronuclei are depicted 30 min (A) and 40 min (B) after fertilization. Sections of *Xenopus* eggs stained with Azure B are shown on top, and corresponding schematic representations below. Thin black lines represent microtubules; white and black disks represent male and female pronuclei. The male pronucleus is associated with the centrosome, which nucleates microtubules to form the sperm aster. The growth of the sperm aster drives the male pronucleus from the cell cortex toward the center of the egg. Organelle-like motility propels the female pronucleus along microtubules from the cell cortex toward the centrosome located in the center of the sperm aster. After pronuclear meeting, the aster and the associated male and female pronuclei continue to migrate until the center of the egg is reached.
Source: From Reinsch and Gönczy (1998).

the formation of two cells, which are genetically identical but epigenetically (and hence phenotypically) different. Thus, in many metazoans, differentiation of genetically identical cells may begin as early as the first cleavage division, suggesting that the zygote is epigenetically programmed to produce two different cells of the same genotype.

The DNA in the sperm nucleus is packed densely with protamines, which prevent transcription, although it retains part of the histones (Miller et al., 2010) that it possessed before transforming into a mature sperm cell. After entering the egg cell, it has to be epigenetically reprogrammed in order to replicate its DNA. This, obviously, requires establishment of a new, "zygotic" epigenetic state in the epigenetically condensed sperm genome. In mammals, the male pronucleus begins DNA demethylation and the replacement of protamines by histones (Oswald et al., 2000;

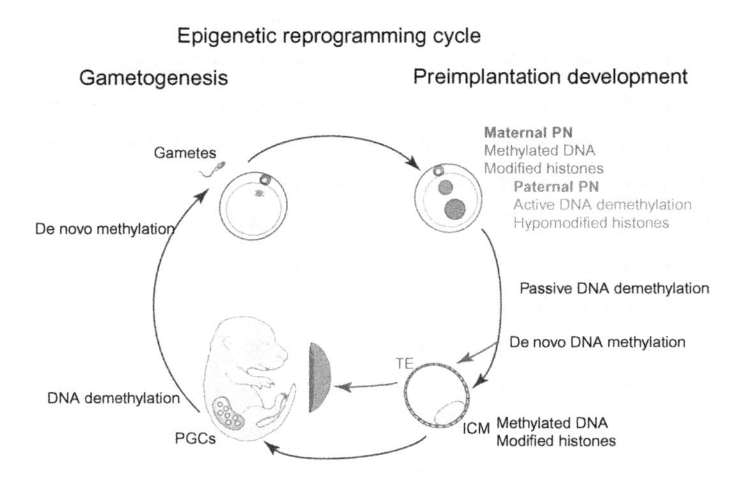

Figure 3.5 Epigenetic reprogramming cycle. Epigenetic modifications undergo reprogramming during the life cycle in two phases: during gametogenesis and preimplantation development. Primordial germ cells (PGCs) arise from somatic tissue and develop into mature gametes over an extended period of time. Their genome undergoes DNA demethylation in the embryo between E11.5 and E12.5, including at imprinted genes. Following demethylation, the genomes of the gametes are *de novo* methylated and acquire imprints; this process continues up to E18.5 in males and in maturing oocytes before ovulation in females. Fertilization signals the second round of reprogramming during preimplantation development. The paternal genome is actively demethylated, and its histones initially lack some modifications present in the maternal pronucleus (PN). The embryo's genome is passively DNA-demethylated during early cell cycles before blastulation. Despite this methylation loss, imprinted genes maintain their methylation through this preimplantation reprogramming. *De novo* methylation roughly coincides with the differentiation of the first two lineages of the blastocyst stage, and the inner cell mass (ICM) is hypermethylated in comparison to the trophectoderm (TE). These early lineages set up the DNA methylation status of their somatic and placental derivatives. Histone modifications may also reflect this DNA methylation asymmetry. Particular classes of sequences may not conform to the general genomic pattern of reprogramming shown here.
Source: From Morgan et al., 2005.

Santos et al., 2002). An enzyme called Tet3 oxidase may be responsible for this demethylation. Demethylation of the paternal genome is independent of the DNA replication and represents a general feature of fertilization in almost all examined mammals (Lepikhov et al., 2008). Before the first cleavage division in the male genome, processes of histone modification also occur.

In contrast to this early dynamic behavior of the male pronucleus, the processes of the DNA demethylation and histone modification/chromatin remodeling in the female genome are delayed. Demethylation of the egg DNA after fertilization is a passive process determined by the lack of methylation enzymes during the replication of the egg DNA (Figure 3.5). Imprinting of parental genes is protected against this epigenetic reprogramming of the genome (Morgan et al., 2005).

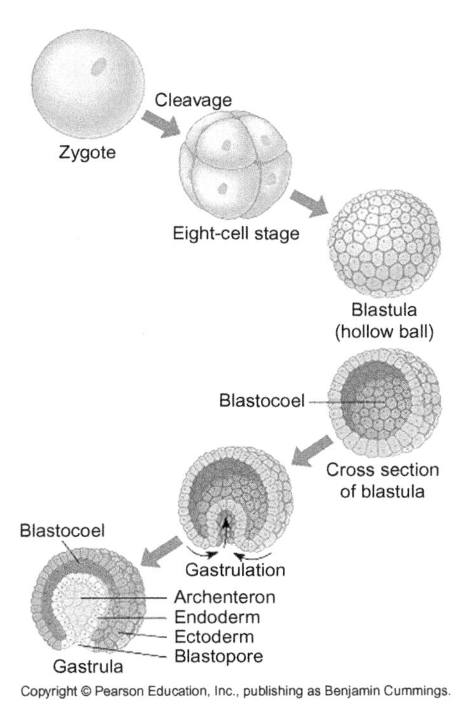

Figure 3.6 Generalized representation of the early development in animals from zygote to gastrulation stage.
Source: From Pearson Education Inc. (http://www.devbio.biology.gatech.edu/?page_id=752).

Characteristically, cleavage divisions are not preceded by the normal growth of the dividing cells. What happens instead is that the zygote (and, depending on species, several of its subsequent cell generations) are continually subdivided in the absence of growth, in a process where daughter cells continue to get half of the cytoplasm of the mother. Generally, this process leads to the formation of an embryonic spherical structure (which is named *morula* due to its resemblance to a mulberry) consisting of a spherical layer of cells surrounding a fluid-filled cavity called the *blastula* (Figure 3.6), which has the same size as the zygote. As mentioned earlier, cleavage divisions may be equal or unequal. During unequal division, daughter cells get different quantities of cytoplasm and different qualities of the parentally provided epigenetic information, leading to early phenotypic differentiation of cells that are genetically identical. The causes and mechanisms of this epigenetic differentiation of early embryonic cells are not fully known and understood. However, some of the basic mechanisms of cell differentiation are known.

Epigenetic Regulation of Cell Differentiation

Cell differentiation is a typical epigenetic process where, from a single cell and genotype, tens to hundreds of different cell types and phenotypes develop. The most obvious and the better-known changes that lead to different types of cells are changes in patterns of gene expression during the development of an individual. From the viewpoint of genetics and the genetic program, it is paradoxical that from a single cell with a single genotype, many different cell phenotypes arise during development. If the zygote (the egg in parthenogenetic organisms) has a genetic program, as many believe it does, differentiation in a higher-vertebrate organism of hundreds of cell types, which differ from each other widely both in function and morphology, would suggest that the embryo has to develop 100 different genetic programs—one for each cell type.

Some biologists are still reluctant or hesitant to acknowledge that cell differentiation is determined by other than genetic mechanisms. This sounds weird when one remembers that more than 70 years ago, in 1941, Sewall Wright (1889–1988), the great American geneticist and one of the founders of population genetics, admitted, "The usual and most probable view is that cellular differentiation is cytoplasmic and must therefore persist and be transmitted to daughter cells by cytoplasmic heredity" (Wright, 1941).

Epigenetic Modes of Cell Differentiation

The first mode of cell differentiation observed during development is determined by the *asymmetric distribution* of cytoplasmic factors (epigenetic information) in the zygote. Even when the cleavage divisions are equal, cells resulting from the division of the parental cell will have different fates, and these fates, obviously, are epigenetically determined by the differential distribution of parental factors in the zygotic cytoplasm.

During animal development, cell differentiation is often induced by extracellular signals, which determine their transformation into different cell types. For example, Wnt and sonic hedgehog (Shh) signals from the neural tube in somites lead to the expression of skeletal muscle genes and differentiation of muscle cells (Schmidt et al., 2000); signals from the brain and spinal cord, primarily vascular endothelial growth factors (VEGFs), induce the formation of angioblasts in somites (Hogan et al., 2004), etc.

Another mechanism of cell differentiation is *asymmetric division* of cells. Asymmetric division of cells results in differential allocation of the cytoplasm in daughter cells, leading to the emergence of different phenotypes in cells of the same genotype. The mechanism of the asymmetric division that leads to cell differentiation is an epigenetic mechanism ultimately based on the properties or behavior of centrioles (microtubule organizing centers, more generally), which organize spindle poles. In many species, the asymmetric divisions that determine cell fates start since the early cleavage divisions. Neither genes nor chromosomes (nor DNA in general) is involved in the asymmetric division of cells. This seems to be an exclusive function of the cytoskeleton, microtubules, and actin filaments. One mechanism is based on differences in the pushing/pulling forces of microtubules of the spindle (Kaltschmidt and Brand, 2002) (Figure 3.7).

The unequal cell division in the *Drosophila* neural larval stem cells results from the furrow forming closer to one of the spindle poles. The two centrosomes resulting from centrosome duplication migrate to the apical cortex, where they form a single

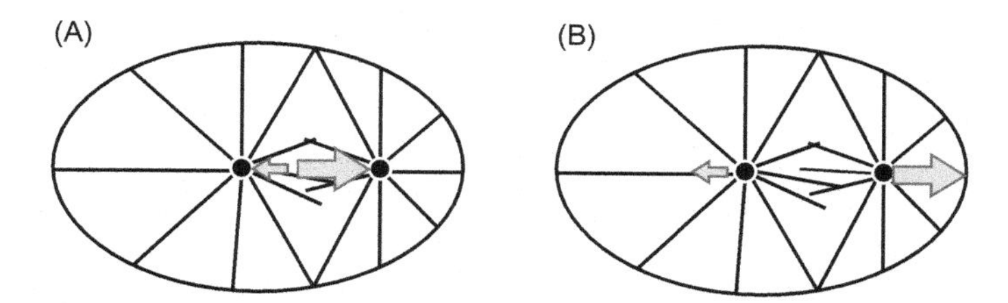

Figure 3.7 Possible models by which an unequal pulling force could be generated. (A) The force generated by the pushing apart of the overlapping midzone microtubules is unequal. (B) The force of the pulling astral microtubules is unequal.
Source: From Kaltschmidt and Brand (2002).

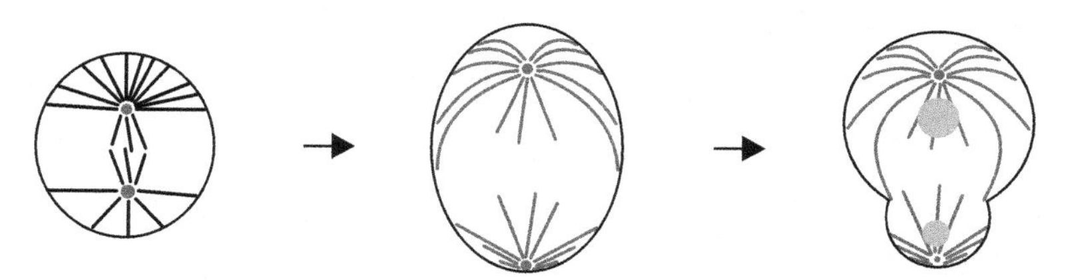

Figure 3.8 Generation of an eccentrically placed cleavage plane in the *Drosophila* neuroblast. The spindle forms symmetrically between the two spindle poles. Apical is up and basal is down. At the onset of anaphase, the microtubules appear to shorten on the basal side of the cell and elongate/enrich on the apical side. At telophase (gray ovals represent telophase DNA), the centrosome of the basal aster is smaller than that of the apical aster.
Source: From Kaltschmidt and Brand (2002).

major aster. Then one of them loses the pericentriolar material and migrates to the basal cortex, where it acquires pericentriolar material and forms the second aster (Figure 3.8). Microtubules elongate on the apical side and shorten on the basal side, leading to the formation of the furrow closer to one of the spindle poles and to the asymmetric division of cells (Kaltschmidt and Brand, 2002; Rebollo et al., 2007).

The fact that "[t]he direction of division and the ability of a cell to divide symmetrically or asymmetrically in size is brought about by rearrangement of the cytoskeleton" (Kaltschmidt and Brand, 2002) unambiguously indicates the epigenetic nature of the cell differentiation in animals. Messenger RNAs (mRNAs) distributed throughout the cytoplasm are localized, in a microtubule-dependent mode, in the pericentriolar material, as proved by the fact that experimental depolymerization of microtubules prevents accumulation of these mRNAs in the centrosome. Then, during the cleavage division, they dissociate from the centrosome and, via actin filaments, are transported to particular sites of the presumptive daughter cell cortex, leading to asymmetric division of mRNAs in daughter cells (Lambert and Nagy, 2002) (Figure 3.9). Asymmetric behavior of centrosomes seems to play an important role in the asymmetric division of cells (Yamashita and Fuller, 2008).

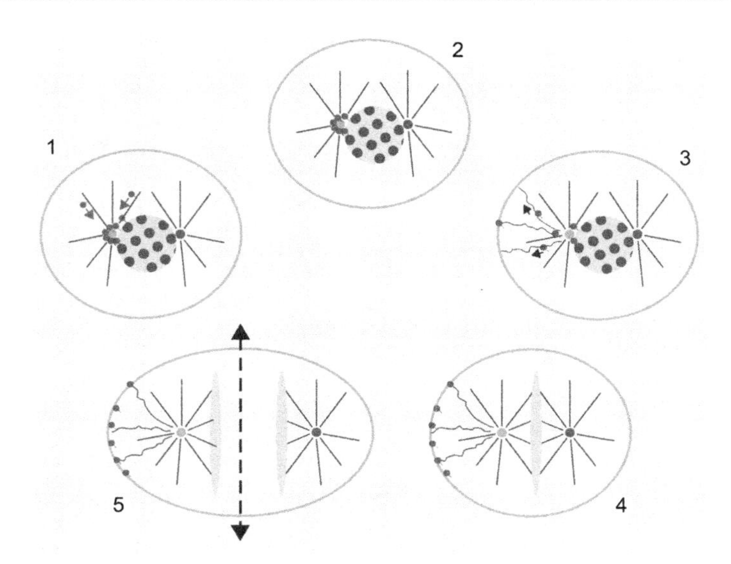

Figure 3.9 Asymmetric division of mRNAs. 1 and 2. During prophase, certain mRNAs bound to protein factors (blue) are transported to only one of the chromosomes (yellow circle) via microtubules (red). The centrosome (green circle) lacks the competence to trap (bind or catch) mRNAs. In the absence of microtubules, mRNA localization persists around the pericentrosomal material. 3. Then mRNA-protein complexes, via the actin cytoskeleton, are transported to a certain region of the cell cortex. 4. Localized correctly, mRNAs are bound independently of actin. 5. During cell division, mRNAs go to only one daughter cell. As a result, each daughter cell will have a different fate. (For interpretation of the references to color in this figure legend, the reader is referred to the web version of this book.)
Source: From Giet and Prigent (2003).

During telophase of a cell type in *C. elegans*, in response to extracellular Wnt signals, the spindle microtubule structure is restructured, thus inducing the asymmetric localization of WRM-1 (*C. elegans*' β-catenin homologue) in the cell nucleus and consequent asymmetric cell division (Sugioka et al., 2011).

The reorganization of cytoskeleton microtubules is responsible for determining the form and certain structural features of multinucleate myotubes during embryonic muscle formation; instead of the radial array of myoblast microtubules originating in the microtubule organizing center (MTOC), the fusion of myoblasts is associated with the formation of microtubule longitudinal linear arrays originating from the already diffuse MTOC and/or myotube extracentrosomal sites. An enzyme, LKB1, destabilizes microtubules and facilitates their reorganization in the process of the myotube development (Mian et al., 2012) (Figure 3.10). During muscle fiber differentiation, several centrosomal proteins are redistributed to microtubule-nucleating sites along nuclear membrane and through cytoplasm (Bugnard et al., 2005).

A typical case of asymmetric division is observed in *Caenorhabditis elegans* embryos; the sperm of this 1-mm-long roundworm provides the fertilizing egg with

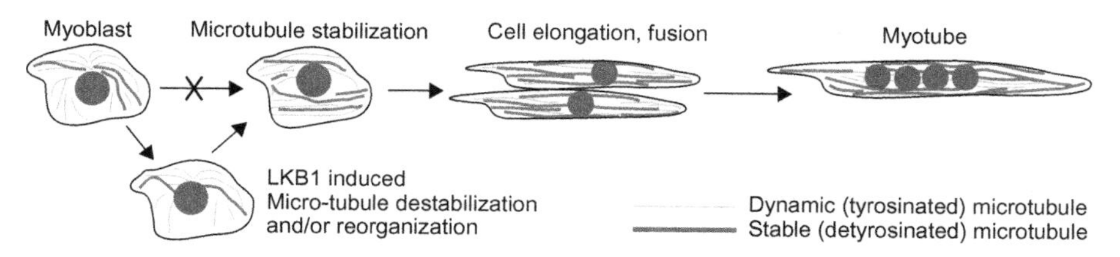

Figure 3.10 Myoblasts contain a radial microtubule array that is a mixture of dynamic microtubules and stabilized microtubules. Fully differentiated myotubes show a linear microtubule array consisting of abundant detyrosinated/stable microtubules. Simple microtubule stabilization blocks the formation of myotubes, and a transient decrease in microtubule stabilization precedes cell elongation and fusion into myotubes. LKB1 plays a role in this microtubule destabilization and/or reorganization, which accounts for its role in the differentiation process.
Source: From Mian et al. (2012).

a substance that induces mechanical loosening of the actin cytoskeleton, leading to its translocation to the opposite side of the sperm entry (presumptive anterior pole) (Mammoto and Ingber, 2010).

The differential distribution of mRNAs determines the different fates of daughter cells of common origin. It is epigenetically determined, and the visible source of epigenetic information necessary for the specific asymmetric distribution of mRNAs seems to be the cell centrosome. The fact that different cells distribute mRNAs differently in their daughter cells suggests that the widely held opinion that centrosomes are informationally equivalent may be incorrect.

Epigenetic Determination of Early Development—The Embryonic Genome Is Still Dormant

Epigenetic Determination of Cleavage Divisions

A number of maternal cytoplasmic factors are necessary to establish the dorsoventral axis. In *Xenopus laevis*, since the first cleavage divisions, this axis is dertermined by translating the maternal *Wnt*11 mRNA that is deposited into the vegetal pole of the egg (Ku and Melton, 1993), and in the absence of *Wnt*11, these embryos do not develop dorsal structures (White and Heasman, 2008). Wnt is primarily localized in Spemann's organizer. As a dorsalizing maternal factor, it is involved in the process of neurulation, and the neural tube becomes the main source of Wnt in early development.

In zebrafish, another maternal transcript involved in specifying the dorsoventral axis is the Nodal-related morphogen, *Squint* (*Sqt*), which localizes asymmetrically in two blastomeres as early as the four-cell stage, thus determining the formation of the dorsal embryonic structures. Early removal of cells containing *sqt* transcripts leads

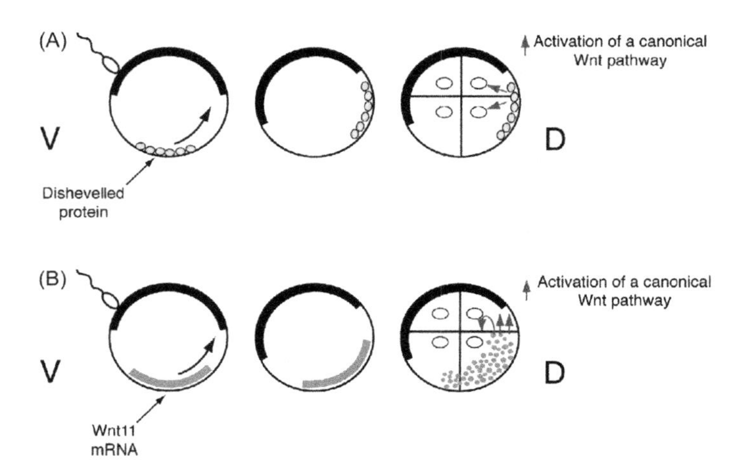

Figure 3.11 A model of the initiation of axis formation by an extracellular Wnt11 signal. (A) The prevailing model of initiation of the canonical Wnt signaling pathway in *Xenopus* axis formation by an intracellular signal. Vesicles (shown in blue) containing disheveled protein are transferred from a vegetal to a dorsal location by cytoplasmic movements. They stabilize β-catenin and activate the signaling cascade intracellularly. (B) A model of the initiation of axis formation by an extracellular Wnt11 signal secreted by dorsal vegetal cells. Purple represents cortically localized *Wnt*11 mRNA. Speckled purple represents the dispersal of *Wnt*11 mRNA from the cortex during cleavage stages. (For interpretation of the references to color in this figure legend, the reader is referred to the web version of this book.)
Source: From Tao et al. (2005).

to ventralization or prevention of the development of dorsal structures (Gore et al., 2005).

Two models are presented to explain the formation of dorsal structures in *Xenopus*. According to the prevailing model, an intracellular maternal protein, Dsh (disheveled), involved in the Wnt dorsalizing pathway, moves along the egg cortex from the vegetal to the dorsal side of the egg, thus determining the asymmetric dorsal localization of this Dsh (Figure 3.11A). According to a newer model, the dorsalizing pathway is activated by the maternal transcripts of *Wnt*11 with the involvement of the maternal FRL1 and HSPG (heparin/heparan sulfate proteoglycans) (Tao et al., 2005) (Figure 3.11B).

The Wnt-β-catenin pathway has key roles in pattern formation, cell-fate determination, and cell polarity (Hsieh, 2004). The components of the canonical Wnt pathway are well conserved across species, and the key in the pathway is the regulation of the β-catenin (Willert and Nusse, 1998). The pathway is active during embryonic development but is generally dormant in adult organisms (Katanaev, 2010). Binding of Wnts to the receptor Frizzled on the cell surface activates the intracellular protein Dsh, which in turn inactivates the trimeric GSK3beta/Axin/APC (adenomatous polyposis coli) complex. Inactivation of the trimeric β-catenin degradation complex by Wnt signaling allows free β-catenin to accumulate in the cytoplasm and enter the nucleus (hence the term *Wnt/β-catenin pathway*). Nuclear β-catenin in turn interacts

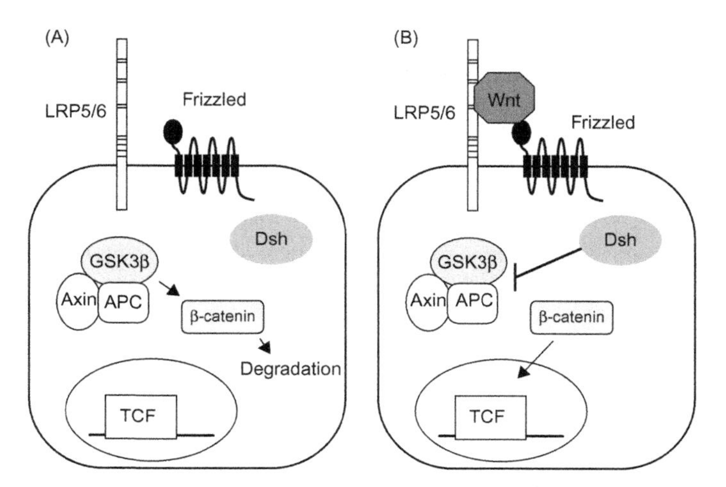

Figure 3.12 The canonical Wnt signaling pathway. (A) In the absence of Wnt, little β-catenin is present in the cytoplasm because it is degraded as a result of phosphorylation by the Axin/APC/GSK3beta complex; (B) Interaction of Wnt with Frizzled and LRP5/6 activates Dsh, which in turn inactivates the Axin/APC/GSK3beta complex, allowing β-catenin to accumulate and enter the nucleus. Upon entering the nucleus, β-catenin interacts with the transcription factor of the TCF family to activate gene transcription (Hsieh, 2004).

with members of the T-cell factor/lymphoid enhancer factor (LEF/TCF) transcription factor family to promote the activation of downstream target genes involved in the dorsal axis specification. (Figure 3.12B). In the absence of Wnt, the trimer induces degradation of β-catenin (Hsieh, 2004; Soriano et al., 2001) (Figure 3.12A).

Another important embryonic pathway is the Wnt-calcium pathway, which seems to regulate the formation of the early embryonic organizer through antagonism with the Wnt-β-catenin pathway (Slusarski and Pelegri, 2007).

Epigenetic Control of Early Development

Theodor Boveri, based on his embryological investigations, 110 years ago came to the now generally accepted conclusion that early development from the zygote to the blastula stage is controlled by extranuclear (*epigenetic*, as it is known today) factors.

I would like to ascribe to the cytoplasm of the sea urchin egg only the initial and simplest properties responsible for differentiation ... it provides the most general basic form, the framework within which all the specific details are filled in by the nucleus (Boveri, 1902).

The term *early embryonic development* will be used here to describe the entire period spanning from egg fertilization to the phylotypic stage.

In brooding oviparous animals, the egg genome is dormant and the egg fertilization is triggered by epigenetic factors activated by the entry of the sperm. As mentioned earlier, the first cleavage division and a varying number of subsequent divisions are exclusively controlled and regulated by epigenetic factors parentally (there is evidence that the paternal cytoplasmic factors are also involved in early embryonic development along the maternal factors) provided to gametes.

Depending on the species, zygotic genome activation (ZGA) occurs sometime between the two-cell stage and the gastrulation. It is interesting to note that in *Drosophila*, ZGA starts with the expression of intronless genes (De Renzis et al., 2007), which is likely unavoidable given the fact that the mechanisms for eliminating introns through gene splicing are not established yet. Some maternal products are supplied in insufficient quantities, and this is the reason for starting the expression of relevant zygotic genes (Wieschaus, 1996).

Live-bearing, viviparous animals also deposit cytoplasmic factors in their gametes, but their role in development is less important and lasts for shorter periods than oviparous animals. This is evolutionarily plausible: the mother can, and does, control the development by transplacentally supplying epigenetic factors to the embryo for the whole *in utero* life (embryonic and fetal). In mice, for example, 90% of maternal transcripts are degraded by the two-cell stage. A weak, "minor," ZGA may occur before the cleavage, primarily in the sperm pronucleus (Aoki et al., 1997), but the relatively small number of transcripts are translated only in the two-cell stage (Hamatani et al., 2004) when the "major" ZGA begins (Latham et al., 1991). Although the embryonic genome in a mammal like the mouse is activated as early as the first cleavage, the maternally provided mRNAs and proteins persist until the eight-cell stage (Hamatani et al., 2004), suggesting that they may play some role in early development until that point in time (and possibly even later). Recall that even the activation of the zygotic genome and initial patterns of gene expression are determined by maternal transcriptional factors.

In *Drosophila*, the almost exclusive epigenetic control of development via cytoplasmic factors persists until the 13th division cycle (i.e., with formation of the syncitial blastoderm and the end of cleavage divisions). Repression of zygotic genes is associated with condensed zygotic chromatin, as well as hypoacetylation and methylation of the histone H3 (Meehan et al., 2005). Zygotic genes are generally repressed until the 13th cell cycle, and are associated with condensed, hypoacetylated, and H3-methylated chromatin (Meehan et al., 2005).

At the blastoderm stage, the *Drosophila* embryo has 8192 cells. It is estimated that the majority of the approximately 20,000 transcriptional units in *Drosophila* must be provided maternally (Wieschaus, 1996). However, by the 14th division cycle in *Drosophila*, 33% of maternal factors go down significantly, and part of the zygotic transcription is intended to compensate for the loss of maternal mRNAs/proteins (De Renzis et al., 2007). Such facts led biologists to the conclusion that "[g]enes that must be supplied zygotically represent only a small fraction of the *Drosophila* genome" (Wieschaus, 1996).

The Mid-Blastula or Maternal–Embryonic Transition

In many oviparous animals, after formation of morula, which occurs sometime during the blastula stage, the embryo begins a large-scale expression of zygotic genes. Individual development increasingly depends on the zygotic gene expression, although most of the maternal factors persist for varying periods of time in the embryo until the phylotypic stage. So, for example, after gastrulation, formation of the complete germ band (the phylotypic stage of insects) in the leafhopper *Euscelis plebejusis* is under the control of three maternal cytoplasmic factors (Vogel, 1983).

This stage in the activation of the embryonic genome is known as *mid-blastula transition* (MBT). The term was introduced by Kane and Kimmel (1993) to describe development's passage from maternal to embryonic control. From here, embryonic cells divide asynchronously and show a level of motility that is necessary for gastrulation and cell division. The asynchronic division of cells results from variations in their nucleocytoplasmic ratios (Kane and Kimmel, 1993).

In *Xenopus*, the MBT begins by the 12th cell division, when the transcription of the zygotic genome is detected for the first time (Lund et al., 2009); and in zebrafish, the MBT ends after the 13th division cycle (Kane and Kimmel, 1993). The MBT coincides with the silencing and degradation of maternal mRNAs (Thatcher et al., 2007; Giraldez, 2010).

The transition is associated with maternal and zygotic enzymatic degradation of maternal mRNAs. Among maternal factors engaged in the clearing of maternal mRNAs are a few maternal microRNAs (miRNAs). Initially, destruction of maternal mRNAs is a suicidal phenomenon (maternal proteins eliminate maternal mRNAs), and later zygotic transcription leads to the production of miRNAs and proteins that enhance the efficiency of maternal mRNA degradation (Tadros and Lipshitz, 2009). Among them is the maternal miR-430, which is activated at the MBT and induces degradation of maternal mRNAs by facilitating their deadenylation (Giraldez, 2010; Giraldez et al., 2006). In mammals as well, the mouse egg and zygote contain maternal miRNAs, of which the most abundant are miRNAs of the let-7 family. Compared to the zygote, the two-cell stage embryo has lost 60% of maternal miRNAs, but in the four- and eight-cell stages, the amount of miRNAs increases more than 10-fold as a result of the activation of zygotic genes for miRNAs.

That maternal miRNAs are essential for the early development in mammals is demonstrated by the fact that mouse eggs fail to undergo the first cleavage division when depleted of all maternal miRNAs (Tang et al., 2007). Recently, a number of RNA-binding proteins are identified that seem to determine the direction of maternal degradation machinery to mRNAs targeted for degradation (Walser and Lipshitz, 2011).

After the MBTs, in *Xenopus*, formation of a preorganizer is induced by the maternal β-catenin in the β-catenin-rich dorsal side of the mid-blastula embryo. Soon, on the vegetal dorsal side of the embryo, appears the Nieuwkoop center, which is induced by maternal signals of β-catenin, VegT, and Vg1 in the presence of Nodal-related (Xnr) proteins, as well as a maternal Wnt component (Vonica and Gumbiner, 2007) (Figure 3.13). On the dorsal animal-marginal region of the amphibian embryo forms the blastula chordin and noggin-expressing (BCNE) center (Kuroda et al., 2004), from which develops Spemann's organizer and much of the anterior central nervous system (CNS) during gastrulation. Establishment and activation of the BCNE center requires xNorrin, another maternal cytoplasmic factor that is essential

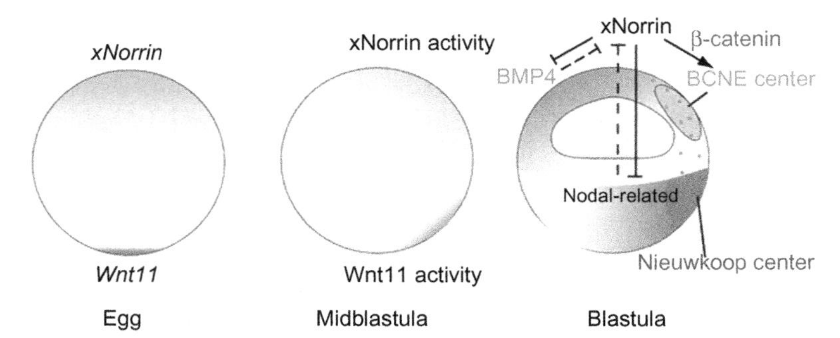

Figure 3.13 A model of dorsal specification in *Xenopus*. During oogenesis, maternal *xNorrin* and *Wnt*11 are localized to the animal and vegetal poles, respectively. After fertilization, both mRNAs are enriched on the dorsal side, leading to two localized activity domains: the BCNE center and the Nieuwkoop center. xNorrin in the dorsal animal cells helps to specify neuroectoderm fate by activating a Wnt/β-catenin signaling domain, the BCNE center, and also participates in antagonizing the Nodal-related signal from the vegetal half and the bone morphogenetic protein (BMP) signal from the ventral side. Wnt11 in the dorsal vegetal domain is required for β-catenin activation in all dorsal regions, including in the BCNE center. Yellow: xNorrin and BCNE center. Purple: Wnt11. Red dots indicate stabilized β-catenin. Green: Nieuwkoop center. (For interpretation of the references to color in this figure legend, the reader is referred to the web version of this book.)
Source: From Xu et al. (2012).

for early neuroectoderm specification and for expression of the zygotic β-catenin (Xu et al., 2012).

Gastrulation—Formation of Embryonic Layers

The generalized spherical blastula soon develops into a new embryonic structure known as the *gastrula*. The term was coined by Ernst Haeckel (1834–1919) due to the gastrula's resemblance to a pouch (from ancient Greek *gastros*, stomach). The zygotic genome is now functional, although, depending on species, various proportions of maternal factors are still present and active. Many cells begin to differentiate and move around in specific patterns. Their differentiation is correlated to changes in the patterns of gene and mRNA expression. Despite the activation of the zygotic genome, many maternal factors are still necessary for gastrulation to proceed. Among these maternal factors are the maternally provided serotonin (Colas et al., 1999) and maternal *Xsmad1RNA*, whose translation is necessary for early gastrulation, as suggested by the failure of *Xenopus* eggs depleted of maternal X*smad*1*RNA* to gastrulate (Miyanaga et al., 2002). The maternal interferon regulatory Factor 6 mRNA (*irf6*) recently was added to the number of maternal factors necessary for gastrulation in *Xenopus* and the zebrafish *Danio rerio*. Depletion of this maternal factor in *Xenopus* and the expression of a dominant negative variant of Irf6 in

D. rerio lead to the failure in the development of the primary superficial epithelium and gastrulation defects (Sabel et al., 2009).

During gastrulation, formation of two or three embryonic germ layers occurs, respectively in diploblastic (lower invertebrates) and triploblastic (higher invertebrates and all vertebrates) organisms. Formation of these layers results from directed cell movements and proliferation of cells. A better-known mechanism of directed cell movement is that in which migrating cells have membrane receptors that, on their way to target sites, detect and bind specific ligands (primarily growth factors and hormones) released by other cells, which at this stage should be maternal by origin. Not much is known about the forces driving the movement and the shape of migrating cells. The most likely candidates are contractile forces generated by the actin–myosin cytoskeleton (Martin, 2010).

As a result of invagination, the dorsal side of the embryo now consists of three germ layers. Later, the endoderm and mesoderm will produce the majority of the organ systems, including musculature and skeleton as well as ectoderm, which will give rise to the nervous system and epidermal derivatives, such as various forms of integument, hairs, nails, etc.

Cell division is now driven by the nucleocytoplasmic ratio instead of maternal cyclins, and cell differentiation is determined by extracellular signals instead of asymmetric distribution of maternal factors in cells. Production of changes in cells under the influence of factors released by other cells, which is observed during gastrulation and during the whole development of an individual, was termed *induction* in a famous article published in 1924 by Hans Spemann and Hilde Mangold (Spemann and Mangold, 1924). As a result of the interaction between the preorganizer and the Nieuwkoop center, Spemann's organizer, which is responsible for the patterning of embryonic germ layers during gastrulation, develops. Embryonic structures homologous to the amphibian Spemann's organizer in birds and mammals were given the name *nodes* or *Knoten* (German for *nodes*), in 1876 by the physiologist Viktor Hensen (1835–1924). Hensen used the term to describe a mass of cells in the 7-day rabbit embryos in which the germ layers were indistinguishable (Blum et al., 2007).

Embryonic Induction

Hans Spemann (1869–1941), one of the founders of experimental developmental biology, is credited with the discovery of the embryonic organizer. He entered biology unconventionally; beginning as an apprentice in his father's book business, he studied preclinical medicine, finally settling in the field of zoology. His research culminated with the discovery of the organizer, made public in 1924 in an article coauthored with Hilde (Proescholdt) Mangold (1898–1924), a young biologist working on her thesis in his laboratory at Freiburg University, Germany. They implanted cells of the dorsal lip of a salamander species embryo to the opposite side of another salamander species embryo and observed that the latter developed a new notochord, CNS, somites, etc., giving rise to a Siamese twin organism (Figure 3.14). Thus, they

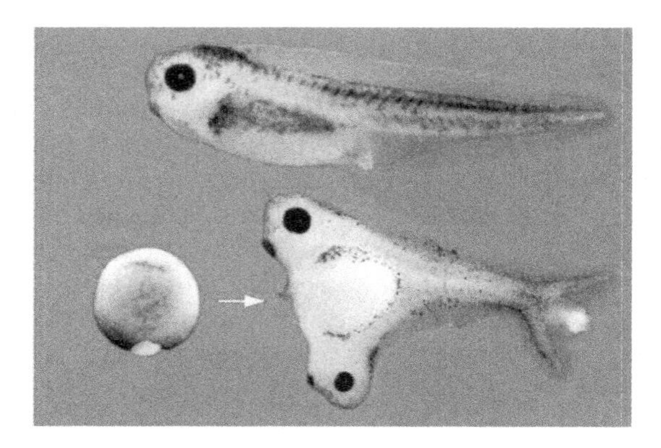

Figure 3.14 The Spemann–Mangold organizer experiment in 1924.
Source: From De Robertis (2006).

discovered that amphibian embryos have an "Organisator Region," whose transplantation to other parts of another embryo induced the formation of a secondary embryonic axis. They coined the term *induction* to describe such signaling from a group of cells and its morphogenetic effects on another group of cells. The work won him (his young collaborator died accidentally just before the publication of the article) the 1935 Nobel Prize for Medicine and Physiology. He also developed the first nuclear transfer method actually used in cloning experiments: he transferred the nucleus of an early (16-cell) embryo to an enucleated salamander embryo cell, which developed into an adult salamander. His successful transplantation of the cell nucleus to an egg heralded the advent of experimental cloning by the nuclear transfer method.

This is a classic experiment because of its elegance and importance in experimental and theoretical developmental biology, and it still remains, in De Robertis and Wessely's expression, the "Holy Grail of experimental embryology" (De Robertis and Wessely, 2004).

Spemann's organizer synthesizes a number of neuralizing factors and antagonists of Wnts, especially BMPs secreted by the ventral center, thus enabling the dorsal side to adopt the default neural fate, but Spemann's organizer itself is an epigenetically induced structure:

> *The organizer … is formed in the blastula stage of the amphibian embryo by cells that have responded to two maternal agents: a general meso-endoderm inducer (involving the TGF-beta signaling pathway) and a dorsal modifier (probably involving the Wnt signaling pathway).*
>
> *Harland and Gerhart (1997)*

Formation of Embryonic Germ Layers

In *Xenopus*, gastrulation leads to the formation of three embryonic layers and the establishment of definitive anteroposterior and dorsoventral axes (Birsoy et al., 2006).

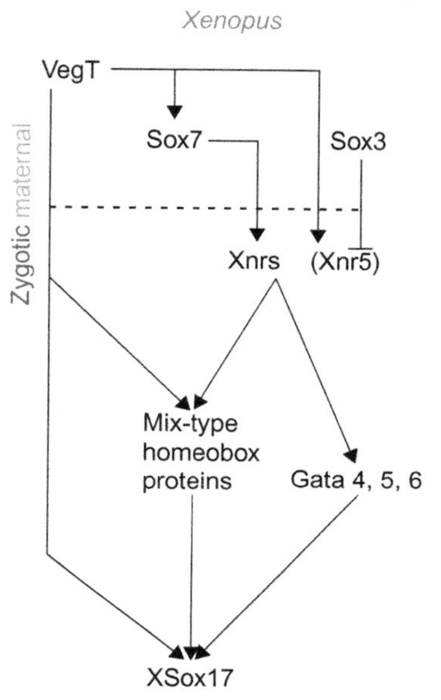

Figure 3.15 Molecular pathway leading to endoderm in the zebrafish and *Xenopus*. In contrast, Xnrs are Xnr1, Xnr2, Xnr4, Xnr5, and Xnr6. Mix-type homeobox proteins are Mixer, Mix.1, Mix.2, Bix.1, Bix.2, Bix.3, and Bix.4.
Source: From Fukuda and Kikuchi (2005).

Formation of Endoderm

The maternal transcription factor VegT controls the formation of endoderm in *Xenopus* and zebrafish (Figure 3.15) and is also necessary for mesoderm formation (Fukuda and Kikuchi, 2005; Zhang et al., 1998). Two maternal factors for *Sox*7 and *Sox*3 are translated downstream of the maternal VegT. *Sox*7 is necessary for inducing *Xnrs* (*Xenopus* Nodal-related genes) and the downstream gene regulatory network, which induces other zygotic genes to converge to the expression of *XSox*17 (Fukuda and Kikuchi, 2005; Zhang et al., 2005). Certainly, the diagram in the figure is very simplified and it is estimated that only about 10% of more than 300 genes involved in endoderm formation are regulated by the transcription factors presented in the figure (Sinner et al., 2006). Another maternal protein necessary for proper endoderm and mesoderm formation is Vg1, localized in the vegetal pole of the *Xenopus* egg (Birsoy et al., 2006).

Formation of Mesoderm

Mesoderm formation is controlled and regulated epigenetically by maternal factors. Key in mesoderm induction is the group of transcription factors VegT, transcribed from maternal mRNAs in the vegetal pole. Diffusing dorsally, the transcription factor VegT, along the maternal β-catenin, activates *Xenopus* Nodal-related genes (*Xnr*), which then initiate the formation of the mesoderm by activating fibroblast growth factor (FGF)–, Wnt-, and BMP pathways. The two first pathways are responsible for maintaining the mesoderm state and the latter one is responsible for patterning it

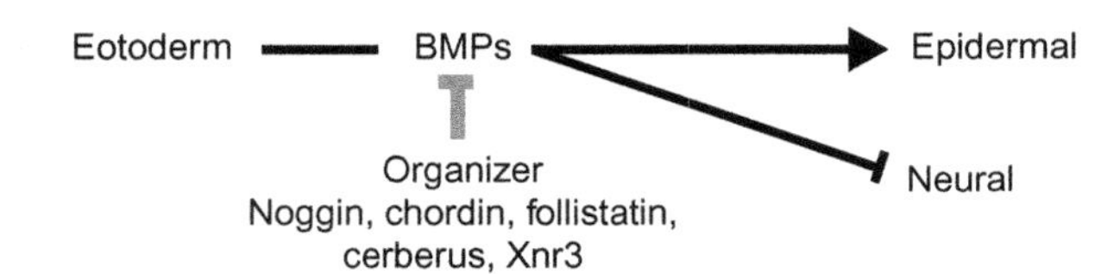

Figure 3.16 BMP signaling and the specification of ectodermal fate (default model) in *Xenopus*. Proposed model for the role of BMP signaling in the specification of the ectodermal fate in frog embryos.
Source: From Muñoz-Sanjuán and Brivanlou (2002).

(Kimelman, 2006). Lack or elimination of VegT prevents formation of the mesoderm (Kimelman and Bjornson, 2004; Kofron et al., 1999). Maternal VegT is responsible for 90% of the mesodermal tissues (Kofron et al., 1999).

Formation of the Neural Tube

Neural Induction

Neural induction follows gastrulation and leads to the formation of the neural tube and the CNS. In terms of the commitment to neural fate, it begins before the gastrula stage and ends when gastrulation finishes (Stern, 2004, 2005), and according to Müller (1996), the cascade of signals leading to formation of the CNS starts in the uncleaved egg.

During neural induction, many maternal factors are still active in the embryo, along the zygotic expression of genes. Given the maternal activation of the zygotic gene expression and that neural induction and specification starts earlier during the blastula stage (Streit et al., 2000; Wilson and Edlund, 2001), neural induction is an epigenetic process:

> *The development of the nervous system is a largely epigenetic phenomenon in which events induce subsequent events in what is usually a highly orderly sequence.*
> *Tierney, 1996*

According to the default model for the neural induction, ectodermal cells in the absence of BMPs and the presence of the BMP antagonists, noggin, chordin, and follistatin, released by Spemann's organizer, most of the dorsal region of the embryo adopts a neural fate (Muñoz-Sanjuán and Brivanlou, 2002; Weinstein and Hemmati-Brivanlou, 1999) (Figure 3.16), and as much as 50% of the ectoderm is included in the neural plate (Gilbert, 2000, http://www.ncbi.nlm.nih.gov/books/NBK10080/). Inhibition of BMPs is sufficient for neural induction, and in the absence of BMPs, the CNS forms even without Spemann's organizer (Reversade et al., 2005).

It is now understood that BMP inhibition is not sufficient for the neural induction of ectoderm. In *Xenopus*, neural induction requires both weak FGF4 or embryonic

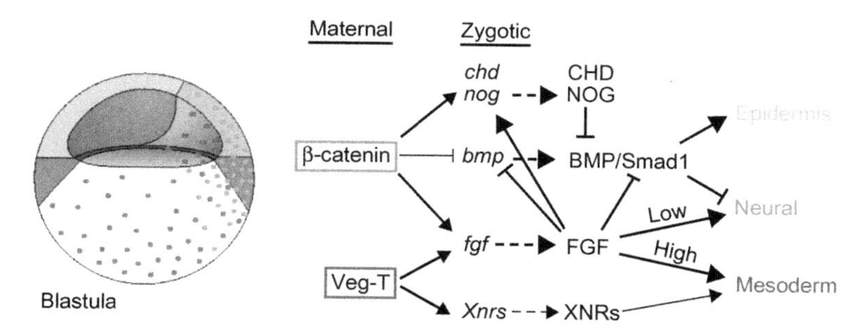

Figure 3.17 A model for neural induction in *Xenopus*. Colored domains indicate prospective tissues. Orange and purple dots indicate β-catenin and VegT protein distribution in blastula nuclei, respectively. (Arrows are not intended to represent direct regulation. The goal of this figure is to represent how the major molecular pathways known to affect early embryonic patterning are integrated. Regulatory arrows shown in black are inferred in part or totally from our work, while those shown in gray originate from other studies.) Neural specification requires concomitant BMP inhibition and low FGF signaling acting in a BMP-independent manner. (For interpretation of the references to color in this figure legend, the reader is referred to the web version of this book.)
Source: From Delaune et al. (2005).

fibroblast growth factor (eFGF) and BMP inhibition. eFGF serves as both a neural inducer and as a BMP inhibitor and seems to be a conserved initiator of neural specification among chordates (Delaune et al., 2005). This finding led to the development of the combinatorial model of neural induction (Figure 3.17).

The fact that transplants from Spemann's organizer and Hensen's node can induce the formation of neural tubes in other species' embryos indicates that the neural induction is highly conserved in vertebrates.

The Primary Neurulation

After the maternal-zygotic transition and at the beginning of gastrulation, 12 neural fate stabilizing (NFS) transcription factors, products of a number of maternal and zygotic mRNAs and neurogenic mRNAs expressed in response to anti-BMP signals released by the BCNE center and the organizer, are detected in the dorsal ectoderm. These maternal and zygotic NFSs interact in a gene regulatory network (Figure 3.18), initiate the formation and patterning of the neural plate and the neural differentiation. Among these genes, $foxD_5$ plays a crucial role as an upstream signal upregulating the function of the neuroectodermal genes *gem*, *sox*11, *geminin*, and other genes. In turn, $foxD_5$ expression is induced by FGF signals (Rogers et al., 2011). Several of the genes upregulated by $foxD_5$, in turn, feed back to downregulate its expression, thus contributing to the medial–lateral patterning of the neural plate for maintaining a progenitor neural cell population and preventing the neural differentiation of the neuroctoderm (Rogers et al., 2009; Yan et al., 2009).

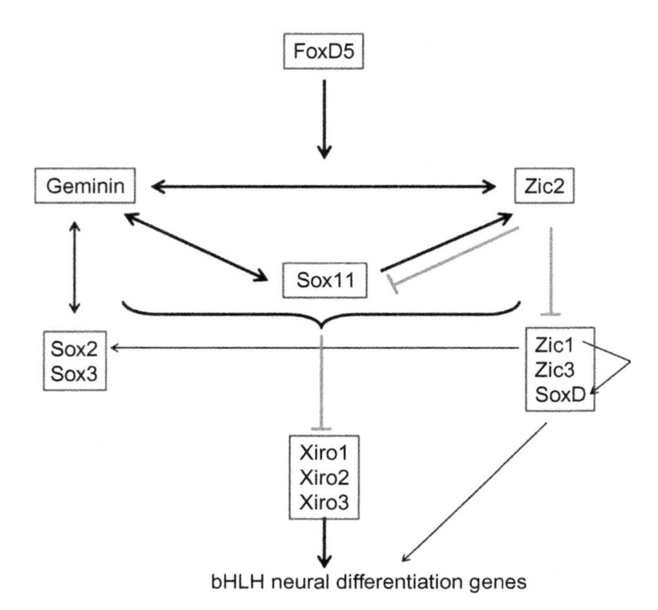

Figure 3.18 A gene regulatory network results in the stabilization of neural fate. After these 12 NFS genes are induced, FoxD5 upregulates *Geminin, Sox*11, and *Zic*2 expression. These three genes regulate each other, and together (bracket), they differentially affect the expression of the other eight NFS genes. *Geminin* and *Sox2/Sox*3 directly regulate each other, and increased *Zic*1 upregulates *Sox*2 and *SoxD*. *FoxD*5, *Geminin, Sox*11, and *Zic*2 maintain neural ectodermal cells in an immature, stemlike state, whereas *Zic*1, *Zic*3, *SoxD*, and *Xiro*1-3 promote the onset of neural differentiation.
Source: From Rogers et al. (2009).

Both the notochord and the prechordal plate, at the anterior end of the notochord, contribute to the induction of the neural plate. Their formation is also epigenetically determined and regulated by maternal factors. In ascidians, for example, primary notochord cells derive from four blastomeres in the anterior marginal zone of the vegetal hemisphere of the 64-cell embryo (Nishida, 2002). It is noteworthy that a maternal factor *mys* (misty somite) continues to be active during the formation of the neural tube as a determinant of the boundaries of somites, transient embryonic structures from which axial skeleton and associating muscles develop in vertebrates (Kotani and Kawakami, 2008).

In humans, the neural plate starts in the form of a flat sheet of about 100,000 cells, and it is believed that all human neurons and glial cells originate from the neural plate (Dowling, 2007). Once the neural plate forms, its edges elevate dorsally until they close at the midline to form the roof plate and the neural tube.

This mechanism of formation of the neural tube is known as *primary neurulation*. The secondary neurulation is another mechanism of the formation of the neural tube. In this case, the neural plate forms a medullary cord, which hollows out to form the neural tube.

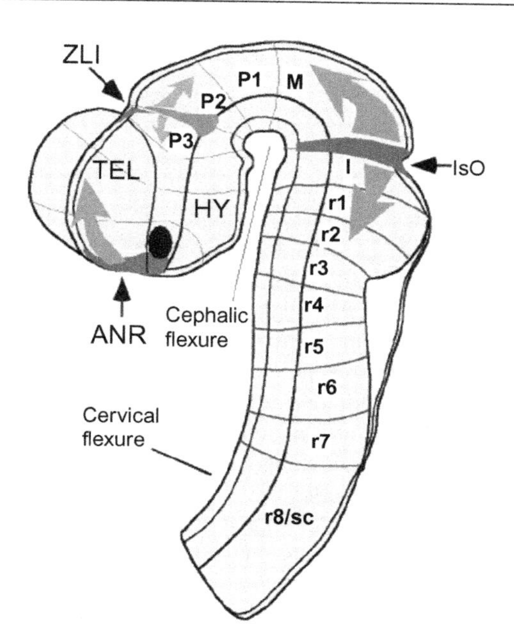

Figure 3.19 Representation of a lateral view of an embryonic day-10.5 mouse neural tube showing the main neuronal regions and the transverse segments of the neural tube in relation to the secondary organizers. *Abbreviations*: ANR, anterior neural ridge; AVE, anterior ventral endoderm; AM, axial mesoderm; DI, diencephalon; M, midbrain; HY, hypothalamus; I, isthmus; IsO, isthmic organizer; RH, rhombencephalon; TEL, telencephalon; ZLI, zona limitans intrathalamica; p1–p3, prosomeres; r1–r8, rhombomeres; sc, spinal cord. *Source*: From Vieira et al. (2010).

In some species, like chick embryos, primary neurulation occurs in the anterior parts of the trunk and secondary neurulation in the caudal parts (Osorio et al., 2009).

Just prior to the closure of the neural folds, the neural crest cells (NCCs) delaminate from the neural tube and begin their epigenetically programmed migration. The anterior part of the neural tube then forms three bulges, the prosencephalon (forebrain), mesencephalon (midbrain), and rhombencephalon (hindbrain). The prosencephalon differentiates into the anterior secondary prosencephalon (telencephalon and hypothalamus) and the more caudal diencephalons (Pombero and Martinez, 2009), from which two secondary bulges extend laterally to form the eye Anlagen or optic vesicles. At the boundary between the hindbrain and midbrain, in the isthmus region, a secondary organizing center referred to as an *isthmic organizer* develops at this stage.

The prosencephalon and the rhombencephalon consist of segmental units respectively called *prosomeres* and *rhombomeres* (Figure 3.19).

At the beginning of the *Xenopus* late tailbud (phylotypic) stage (stages 29–44), the maternal serotonin is almost exhausted. But immediately after stage 32, the level of serotonin increases dramatically. This coincides with the formation of axons by serotonergic neurons, differentiated earlier at stage 25 in the embryonic rostral brain stem (van Mier et al., 1986).

The Phylotypic Stage

In 1828, Karl Ernst Ritter von Baer (1792–1876) published the first volume of the influential *On the Development of Animals: Observation and Reflection*. In this work, von Baer pointed out that at his time, the prevailing opinion of biologists was "that the embryo of higher animals goes through adult forms of lower animals" (von Baer, 1828). He observed that different animals display morphological similarities during various periods of their development: "The further we go back in the development, the more we find also in very different animals such similarities" (von Baer, 1828, p. 223).[1] According to him, all the typical vertebrate characters appear very early during development, so that "[t]he vertebrate embryo from the beginning is a vertebrate and in no time it matches with that of an invertebrate animal" (von Baer, 1828, p. 220).[2]

There, von Baer criticized the idea that higher animals go through adult forms of lower animals, but this concept was revived and popularized by Ernst Haeckel (1834–1919) in his book *General Morphology* (1866), an attempt to support the Darwinian idea of evolution and of common ancestry Haeckel (1866). His concept of "ontogeny as a brief and rapid recapitulation of phylogeny" is flawed, and supporting drawings that he published to popularize the idea received deserved criticism. That said, the theory had a positive influence on the study of evolution, in the development of the concept of homology, and in the modern concept of the phylotypic stage, a central concept of modern evo-devo. Haeckel doctored his drawings to substantiate his biogenetic law to give Darwinism supporting evidence from embryology (Richardson and Keuck, 2002).

He was criticized during his lifetime for exaggerating the similarities between embryonic forms of higher animals and those of the adult forms of lower ones. Now, a century later, this criticism about "inaccurate and misleading" drawings resumed, and these inaccurate drawings are used to deny the existence of a phylotypic stage in the animal world. While admitting that "Haeckel recognized the evolutionary diversity in early embryonic stages, in line with modern thinking," critics have identified some "potential sources for several of the drawings, and find some evidence of doctoring," yet they fairly admit that "Haeckel's much-criticized embryo drawings are important as phylogenetic hypotheses, teaching aids, and evidence for evolution. While some criticisms of the drawings are legitimate, others are more tendentious" (Richardson and Keuck, 2002).

Richardson's main argument against a conserved phylotypic stage is the variation in timing; that is, that different species and other taxa show considerable temporal

[1] Je weiter wir also in der Entwickelung zurückgehen, um desto mehr finden wir auch in sehr verschiedenen Thieren eine Uebereinstimmung.

[2] Der Embryo des Wirbelthiers ist schon anfangs ein Wirbelthier, und hat zu keiner Zeit Übereinstimmung mit einem wirbellosen Thiere.

variation at the onset of the phylotypic stage (Bininda-Emonds et al., 2003). If this turned out to be true, the phylotypic stage would be about a common Bauplan of organisms of a phylum/subphylum, and it does not necessarily imply that the stage will follow the same tempo across the taxa.

Recall that scientific knowledge is always relative rather than final, and in hindsight, it can be criticized as imperfect. Such a viewpoint, also emphasizes his scientific contribution: "Haeckel sensed a commonality in early embryos. Although that sense could have biased his depiction of embryos, Haeckel was more correct than the data of his day allowed." (Elinson and Kezmoh, 2010). The commonality was rediscovered only almost half a century after his death, when Friedrich Seidel (1897–1992) demonstrated that the third law was not valid for the earlier stages of development and that many organisms tend to converge to a morphologically similar stage of the basic body pattern (*Stadium der Körpergrundgestalt*), which is now known as the *phylotypic stage*. The phylotypic stage is considered to be "the central event in early embryonic development" (Sauer et al., 1996). However, the fact that vertebrate embryos starting from different forms of early development converge later at a common morphology at the phylotypic stage (O'Farrell et al., 2004) to develop more divergent morphologies later is incontestable (Figure 3.20).

In humans, the phylotypic stage is between 25 and 30 days after fertilization; in zebrafish, it occurs at 24–48 h; and in mouse embryos, at the E (embryonic day) 8.0–8.5. In *Xenopus*, the phylotypic stage is known as the *tailbud stage*. It corresponds to the Nieuwkoop and Faber stages 20–44, which is divided into the early tailbud stage (20–29) and the late tailbud stage (29–44). The modern hypotheses based on von Baer's and Haeckel's concepts on patterns of embryonic development, are presented in Figure 3.21.

In vertebrates, generally, the phylotypic stage coincides with the onset of neurulation and ends with the formation of the full set of somites (Galis and Metz, 2001). At the phylotypic stage, different chordates are similar in size, and this is valid for both a mouse and a whale, although the latter is a millionfold heavier (Kirschner and Gerhart, 2005).

The phylotypic stage as the most conserved developmental stage in vertebrates and the hourglass model are also supported by gene expression data. At the gene expression level, it coincides with the most conserved expression pattern of the *Hox* and other developmental genes, with the onset of intricate networks of both global and local inductive signals for organogenesis, which make most changes lethal, making the phylotypic stage highly conserved (Irie and Sehara-Fujisawa, 2007).

Formation of the Neural Crest

The NCC is a transient vertebrate structure that evolved around 450 Mya. Developmental and phylogenetic evidence suggests it evolved from the epidermal nerve plexus of protochordates (Gans and Northcutt, 1983; see Figure 3.22). It is involved in the development of almost all vertebrate organs and structures. It may also be the most important event related to the evolution of vertebrates.

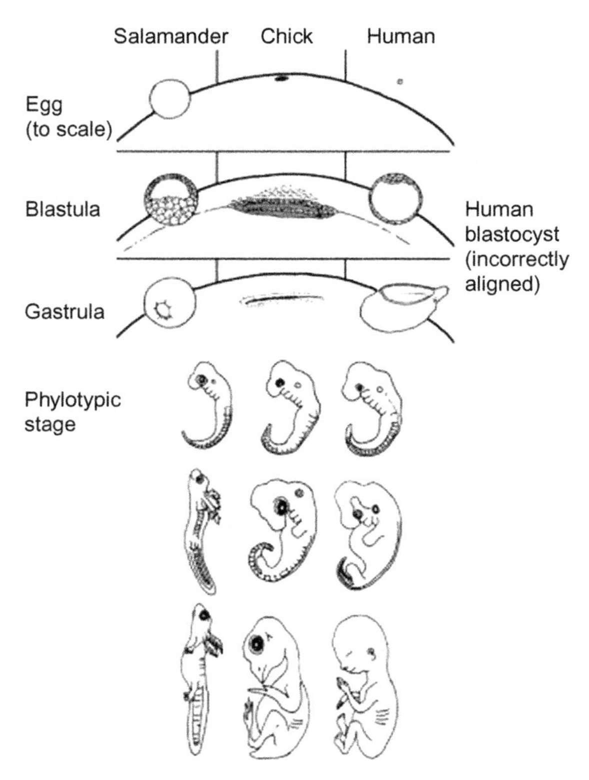

Figure 3.20 Phylotypic stages of chordates and arthropods. The morphology of chordate embryos, represented by salamander, chick, and human embryos, initially converges to look more similar at the so-called phylotypic stage (or pharyngula stage) and subsequently diverges. The high degree of similarity between organisms allows an unambiguous alignment of stages near the phylotypic stage, but ambiguities in alignment can occur at other stages. We argue that the illustrated and generally accepted alignment of the human blastocyst with the blastula stages of amphibian and chick is incorrect.
Source: From O'Farrell et al. (2004).

NCCs and neurons derive from the same precursor cells. NCCs differentiate in the neural plate; that is, in the dorsal portion of the neural tube (Figure 3.23). Prior to leaving the neuroepithelium, they undergo the transition from epithelial to mesenchymal (Barembaum and Bronner-Fraser, 2005), a transformation that is also typical for many neuronal cells (Baker and Bronner-Fraser, 1997). In the neural plate, NCCs are hardly distinguishable from neuroepithelial cells, and when a part of the neural tube is removed, the adjacent neural tube compensates for the NCCs of the ablated portion Schwerson et al. (1993).

The neural induction is determined by maternal cytoplasmic factors. Then, signals from the neural plate stimulate the expression of the "neural crest specifiers" (*Slug/Snail, AP*-2, *FoxD*3, *Sox*10, *Sox*9, and *Msl/*2). These genes enable expression of the

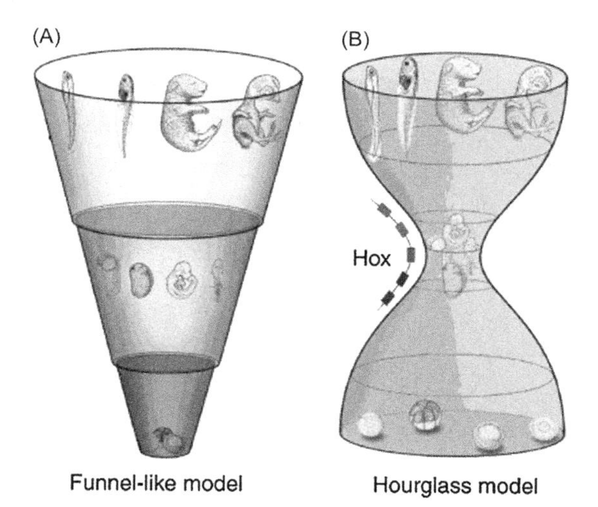

Figure 3.21 The two major hypotheses about how developmental processes are conserved against evolutionary changes. In both models, embryogenesis proceeds from bottom to top, and the width represents the phylogenetic diversity of developmental processes, which are deduced from morphological similarities. (A) The funnel-like model predicts conservation at the earliest embryonic stage. During embryogenesis, diversity increases additively and progressively. (B) The hourglass model predicts the conservation of the organogenesis stage. During this stage, a highly intricate signaling network is established consisting of inductive signals, including the Hox 11 genes, which leads to conservation of the animal body plan. *Source*: From Irie and Kuratani (2011), slightly simplified.

"neural crest effector genes" (Muelemans and Bronner-Fraser, 2004), which determine the formation of mature NCCs in the neural plate.

NCCs from each region of the neural tube follow distinct itineraries of migration to their target organs/tissues throughout the animal body. Before leaving the neural tube, they are provided with epigenetic information on the following:

- Where to go and which are the relevant cues to follow?
- What to do in the specific organs/tssues they settle in?

On their journey, NCCs take their cues from certain receptors that express cells that happen to be on the way to the target sites. Based on their origin along the neural tube and to the site of destination in the animal body, different NCCs have to follow different cues. It follows that different NCCs must be provided with different information on the cues they must follow. It is clear that "cues" that an NCC follows on its path to the target site are not cues *per se*, in the sense that they are meaningless signs to something other than cues. It is the NCC, and even a particular NCC type, that choose what chemical to take as a cue or a landmark on its way to the target site. The destination site is the lodestar of the migration of NCCs, not the chemicals that it may encounter on its travels through the labyrinthine pathways of the animal body.

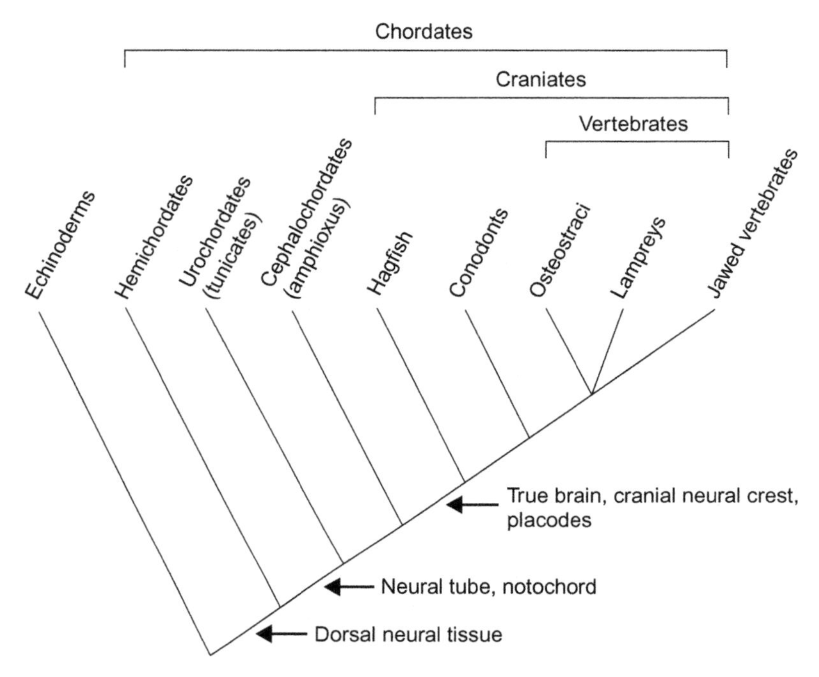

Figure 3.22 Features associated with the evolution of the chordate/vertebrate dorsal neural tube are shown in this phylogenetic tree. NCCs and placodes are vertebrate features that follow the evolution of a neural tube and notochord.
Source: From Hall (2009), modified from Holland and Graham (1995).

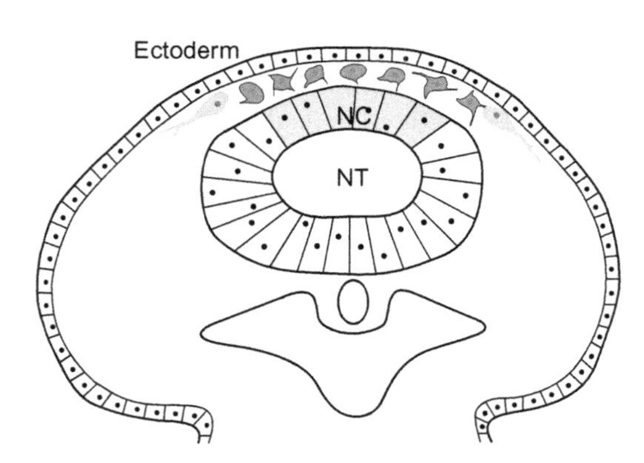

Figure 3.23 Cranial NCC migration starts around Hamburger and Hamilton (HH) Stage 11 in chick. The lead cranial NCCs emerge from the dorsal neural tube (beige) within a short distance from the dorsal neural tube, they acquire directionality (yellow NCCs). (For interpretation of the references to color in this figure legend, the reader is referred to the web version of this book.)
Source: From Kulesa et al. (2010).

Eventually NCCs settle in the destination site, and like zealous missionaries "leading by example," start instructing the local uncommitted population of cells on what type of cell they need to become. Adequate evidence suggests that this conversion is induced by NCC newcomers. Experiments of transplantation of duck neural crest for beaks in quails and vice versa provide elegant proof: in both cases, the chimera inherits the beak of the donor (i.e., duck beaks in quails and quail beaks in ducks). According to a study by Schneider and Helms (2003):

> *Neural crest cells provide patterning information for beak morphology. Not only do NCCs direct their own morphogenesis, they also pattern nonneural crest beak tissues in a manner characteristic of donor species.*

Epigenetic Control of Postphylotypic Development in Animals

Parental Epigenetic Information Is Exhausted at the Phylotypic Stage—Where Does the Information for Further Development Come from?

Although certain maternal factors are active during the formation of the nervous system and even during the phylotypic stage in relation to the formation of somites (Kotani and Kawakami, 2008), as we have seen, the parental information that drives the early embryonic development until the formation of the Bauplan at the phylotypic stage is practically exhausted.

The weaning of the embryo from the source of the parental information on which the early development and the Bauplan relied on is a turning point from an informational view. Exhaustion of the parental epigenetic information, far from an informational catastrophe, comes as an informational change of the guard: the CNS, which is already operational, takes over the postphylotypic development up to adulthood, including the formation of gametes provided with the same epigenetic information that drove the development of the embryo to the phylotypic stage. The CNS, thus, repays the epigenetic information that the parent or parents invested for its development not to the "lender," but to the next generation in a fair-relay informational scheme.

To reiterate, the epigenetic information spent to erect the Bauplan was provided to the embryo via maternal gametic and paternal cytoplasmic factors arranged in specific patterns in the gametes and the zygote. That information consists of many thousands of mRNAs, proteins, hormones, neurotransmitters, and other elements, which, as we have shown, determines the development of the zygote up to the formation of the Bauplan at the phylotypic stage.

The parents neither copy themselves nor provide the full design of the adult organism. Instead, at the phylotypic stage, they provide the embryo with a young "designer" that, in the process of development, figures out the structure of the next stage from the structure of the actual embryonic stage and produces information to fashion it. Sufficient empirical evidence shows that the nervous system is responsible. For the details of how this happens, we must wait until the nature of the computing in the nervous system is fully comprehended, which we hope will come eventually.

The role of the nervous system in animal communication with their environment and in controlling and regulating their physiology is generally recognized and represents objects of various disciplines of biology. This epigenetic theory of development and evolution recognizes an additional function for the nervous system: the control and regulation of animal morphology. The role of the nervous system as a source of epigenetic information for animal development is a new idea and already has been covered in various forms in my previous works. Although the developmental role of the nervous system has almost never been a special object of biological investigation, the evidence to substantiate it, surprisingly or expectedly, is more than adequate.

Now, I will present succinctly the essential evidence on the role of the nervous system in the postphylotypic development of vertebrates.

Apoptosis in Invertebrates

Apoptosis as a form of programmed cell death (PCD) was described first by Carl Vogt in 1842 (Seipp et al., 2001). During metamorphosis, animals have to rid themselves of cells, tissues, and organs that are no longer necessary for postlarval stages. PCD is a well-coordinated process that includes morphological, physiological, and behavioral changes. The initiation of apoptisis requires an acute sensibility of structural changes and the function of the larvae in order to determine when to stop growing, eliminate old structures, and develop new structures. There is no other way to seek out the organs responsible for sensing those changes and determining the metamorphic transformations than to examine the molecular cascades that initiate metamorphosis. We have a clear picture today of the signal cascades that control and regulate metamorphosis in invertebrates. The causal chain in all of them starts with signals from the CNS, which trigger the activation of signal cascades that cause the apoptosis of the obsolete organs and formation of postlarval structures. The facts that the sensing organ is the CNS and that it starts signal cascades for metamorphosis are not surprising. The nervous system is the only system that pervades thoughout the animal body up to a cell level. The nervous system is well known for its ability to "direct the destruction of obsolete larval tissues and their replacements by tissues and structures that form the adult fly ... via the precise stage- and tissue-specific regulation of key death effector genes" (Draizen et al., 1999).

In insects, the signal cascade begins in the brain. Responding to afferent input on body size and other physiological changes, at a determined point in time, the insect's brain releases the neuropeptide prothoracicotropic hormone (PTTH), which activates secretion by the prothoracic glands of ecdysone, which in turn activates a downstream gene regulatory network, leading to the PCD (apoptosis) of the salivary gland, as represented in a simplified form in Figure 3.24.

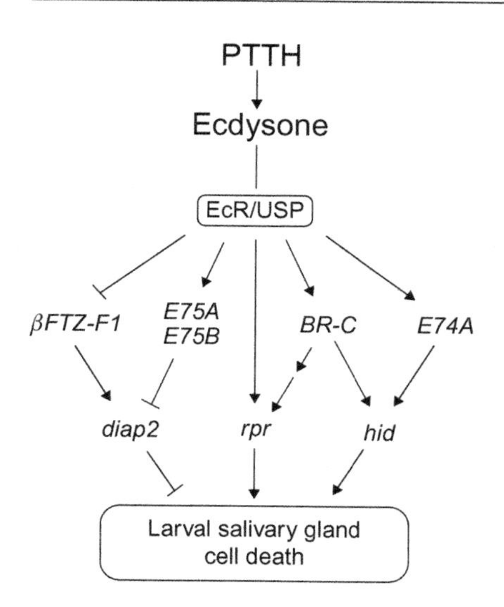

Figure 3.24 A model for the steroid regulation of salivary gland programmed cell death. b*FTZ-F*1 induces the *diap*2 death inhibitor in the salivary glands of late prepupae. The prepupal pulse of ecdysone then represses b*FTZ-F*1 and induces *E75A* and *E75B*, which function redundantly to repress *diap*2. The prepupal ecdysone pulse also directly induces *rpr* transcription, as well as the transcription factors encoded by *BR-C* and *E74A*. *BR-C* is required for both *rpr* and *hid* transcription, and *E74A* is required for maximal levels of *hid* expression. *Source*: From Jiang et al. (2000).

Left–Right Asymmetry

Around the beginning of the twentieth century, Spemann and his coworkers observed that partial or complete ligature of embryos of two newt species caused the normal asymmetric placement of inner organs to invert (*situs inversus*) in 50% of the embryos. Shortly thereafter, in 1921, Hilde Wilhelmi, based on her experimental results with ligature experiments, noticed that laterality defects were correlated with ablations on the left side of the embryo at mid-neurula stage and concluded that "situs inversion in general was explained by the fact that the left side of the germ has something that the right half does not have" (Blum et al., 2009). Seven decades later, Levin et al. (1995) discovered what "the left side of the germ has." They found that a molecular cascade responsible for the establishment of the L–R asymmetry in respect to AP and DV axes, was activated on the left side of chick embryos. In their model, the left–right specification in chick embryos results from an interaction of the products of genes for activin, Shh, and NR1 (nodal-related 1).

Considerable progress has been made in recent years in understanding the ultimate source of the signals that trigger the cascade of gene activation to establish the laterality of various organs in the vertebrate body. It was proposed, in 2001, that cilia "on the node, or organizer, of the gastrulation-stage mouse embryo" cause a leftward movement of the fluid and induce the left–right asymmetry (Brueckner, 2001; McGrath and Brueckner (2003)) during neurulation pIn *Xenopus* as well, the source of the forces moving the fluid leftward is generated by the cilia of the gastrocoel roof plate (GRP) or its equivalents, the Kupfer vesicle in fish and the ventral node in mices. The flow transports a morphogen to the left side of the ciliated epithelium and, from there, to the left lateral plate mesoderm (LPM), creating an asymmetric

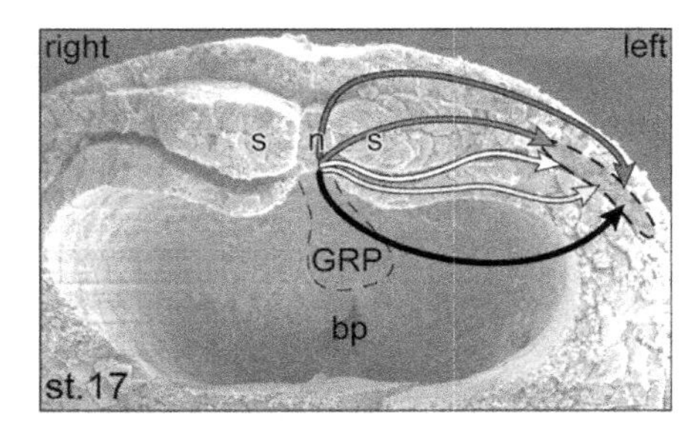

Figure 3.25 Possible routes of signal transfer from the GRP to the left LPM. Signal(s) generated by leftward flow at the GRP could travel through the endoderm (yellow), mesoderm (red), notochord and ectoderm (blue), or archenteron (black), or between endoderm and mesoderm (white) to reach the left LPM (green, outlined by dashed line). *Abbreviations*: bp, blastoporus; n, notochord; s, somite. (For interpretation of the references to color in this figure legend, the reader is referred to the web version of this book.)
Source: From Blum et al. (2009).

distribution of the morphogen and expression of a "leftward cascade" (Blum et al., 2009; see Figure 3.25).

The nodal flow model of the left–right asymmetry is conserved across vertebrates at both the molecular and organ levels (Oteíza et al., 2008), and it was a feature in the common ancestor of the vertebrates (Schweickert et al., 2006).

According to another model, laterality is determined by maternal *ATP*4 mRNA, which enables the passage of the maternal and/or embryonic serotonin to particular blastomeres before gastrulation. However, their asymmetric distribution in the zygote and embryo remains controversial (Walentek et al., 2012). The *Xenopus* fertilized egg contains considerable amounts of maternal serotonin, which declines during early development, only to rise again as a result of neurally derived serotonin secretion after the formation of the neural tube. In *Xenopus*, downregulation of serotonin expression (Beyer et al., 2012) prevents the flow and asymmetry, and the blockage of serotonin receptors (the mediators of serotonin function) alters the normal left-sided expression of *Xnr*1 (*Xenopus* nodal-related 1) and the downstream *X-lefty* (Fukumoto et al., 2007). ATP4a and serotonin converge to activate the Wnt cascade, the former by inducing expression of the Foxjl and the latter by inducing formation of superficial mesoderm and ciliated GRP (Beyer et al., 2012; Shook et al., 2004). Both models of the establishment of laterality rely on epigenetic factors: on the asymmetric distribution of maternal serotonin and ATP4a and on the directed leftward movement of cilia of the GRP cells.

We close this brief review with an interesting example of the neural regulation of the laterality of organs in the big-clawed snapping shrimp (*Alpheus heterochelis*). This shrimp has a pair of chelae (claws) that initially develop symmetrically, with similar size and morphology, but by the sixth juvenile stage, they start differentiating into a large snapper claw for defense and display on one side, and a much smaller claw for feeding and burrowing on the other. Denervation of limbs leads to the development of identical symmetrical snappers (Mellon, 1999). This result, as well as experiments of the removal of one claw and/or denervation (Read and Govind, 1997), have shown the following:

> *[T]he normal bilateral differentiation in these animals is maintained as a result of the imbalance in the nature and amount of the sensory traffic returning to the central ganglia from the snapper and pincer claws.*
>
> *Young et al. (1994)*

Myogenesis in Invertebrates

Myogenesis in insects is epigenetically determined. In *Drosophila*, local innervation is required for the proliferation and distribution of myoblasts; hence, it is essential for muscle development (Currie and Bate, 1991). The indirect flight muscles (IFMs) in this insect consist of the dorsal longitudinal muscles (DLMs) and dorsoventral muscles (DVMs). Their development occurs in two stages. In the first stage, they form the myoblast pool in a nerve-dependent mode, and in the second stage, which is also nerve-dependent, motoneurons establish the critical threshold of the pool and regulate myoblast patterning and muscle formation (Figure 3.26). According to Roy and VijayRaghavan:

> *[I]nductive instructions from the metamorphosing motor nerves mediate splitting (of larval muscles—N.C.) and are essential for the process to proceed. Roy and VijayRaghavan (1998).*

Denervation of DVMs causes a decline in the proliferation rate of myoblasts and prevents myoblast patterning and formation of the muscle Anlagen. In contrast, DLMs develop even after denervation. The reason for the difference is that pupal DLMs develop from persisting larval muscles, which serve as template for myoblast and muscle patterning, whereas DVMs are eliminated during metamorphosis and have to develop *de novo* (Fernandes and Keshishian, 2005). In the tobacco hawkmoth (*Manduca sexta*), neurectomy of the larval leg nerve prevents proliferation and accumulation of myoblasts in the DLM Anlage and the development of muscle fiber bundles for muscle development (Bayline et al., 2001; Consoulas and Levine, 1997).

Another paradigmatic example of the crucial role of the innervation in muscle development is the development of the dorsal oblique 1 (DEO1) in *Manduca*. Larval DEO1 consists of five muscle fibers; all of them but one are lost during metamorphosis, with the leftover fiber serving as an Anlage for the development of the insect's adult muscle. Biologists now know why this fiber alone is privileged to survive and lead to the adult DEO1: it is the only one in which a particular ecdysone receptor, known as EcR-B1, is expressed. Binding of ecdysone to the receptor induces local myoblast proliferation. The fiber is a beneficiary of an extrinsic savvy, rather than its own. The credit goes to the motoneuron innervating larval muscle; in the process of metamorphosis, its axonal arbor recedes from the four other larval

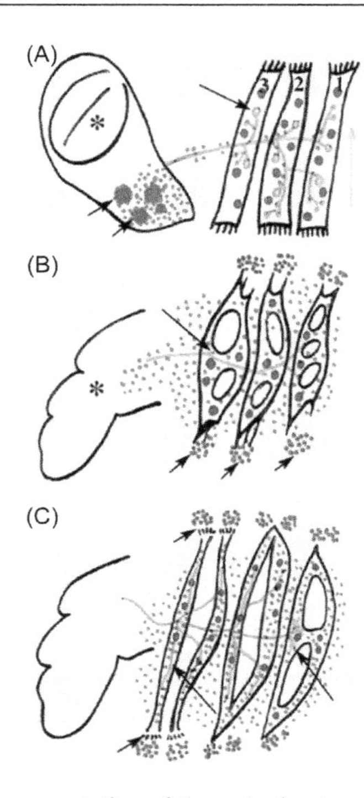

Figure 3.26 Diagrammatic representation of the early developmental events in the patterning of DLMs. (A) During larval life, myoblasts (red dots) that give rise to the DLMs and other dorsal mesothoracic muscles remain sequestered in the wing disc (asterisk) and associated with motor nerves (green) innervating the larval muscles, dorsal oblique 1, 2, and 3. The attachment sites for the IFMs are prefigured on the notal epithelium of the wing disc by expression of stripe in groups of cells (short arrows). *Blue*, larval muscle nuclei; *long arrow*, a synaptic bouton on the larval muscle. *Top*, anterior; *vertical green arrow*, dorsal midline. Orientation of the subsequent panels is similar. (B) During early pupal development (6–10 h APF), the myoblasts migrate out from the everting wing disk (asterisk) and swarm over the three remnant larval fibers, which unlike other larval muscles, escape histolysis. The larval muscles appear vacuolated, and there is a regression of the synaptic terminals of the motor neurons (long arrow). The stripe expressing epidermal cells (purple spots) that will become the attachment sites for the DLMs position themselves adjacent to the larval muscles (short arrows, posterior attachment sites). (C) By 16 h APF, the larval fibers have begun to split longitudinally to form the templates for DLM development, and the motor neurons send out fresh arborizations over the splitting muscles (long arrows). Filopodial extensions from the ends of the templates anchor them to the attachment sites on the epidermis (short arrows). (For interpretation of the references to color in this figure legend, the reader is referred to the web version of this book.)
Source: From Roy and VijayRaghavan (1998).

muscles and remains connected to only one of them. The motoneuron releases a "diffusible" agent that induces the muscle fiber to produce the specific receptor EcR-B1. The remaining uninnervated larval fibers produce only the isoform EcRA of the ecdysone receptor that, in a suicidal process, induces the PCD and elimination of the larval muscle fibers. The fact that the innervation by the motoneuron is responsible for the expression of EcR-B1 is also demonstrated by the experimental denervation of muscle fibers that leads to the dramatic reduction and full repression of EcR-B1 expression (Hegstrom et al., 1998).

It is true that in insects, muscle development usually requires ecdysone regulation along the innervation, but as is well known, ecdysone secretion is itself under strict regulation by the CNS (primarily by the neuropeptide PTTH).

Myogenesis in Vertebrates

Somites are transient structures from which the skeleton and muscles develop. The formation of muscles is one of the earliest results of the neural tube inductive actions in somites. In *Xenopus laevis*, the development of myotome and muscles from somites requires Wnt signals from the dorsal neural tube and Shh and signals from the ventral neural tube and the notochord (Cossu and Borello, 1999; see Figure 3.27).

The mediator of Wnt and Shh action in myogenesis is *Pax*3, which induces the expression of two myogenic genes, *MyoD* and *Myf*5. Myogenesis in many animals is primarily determined by the expression of transcription factors from the *MyoD* family (*Myf*5, *MyoD*, and myogenin) after somite formation (Piran et al., 2009). A simplified gene regulatory network for the initiation of myogenesis and chondrogenesis is presented in Figure 3.28.

As noted above, Shh and Wnt signals for the expression of MyoD genes originate in the neural tube, as demonstrated by the fact that somites do not express *MyoD* genes when separated from the neural tube (Alves et al., 2003) and experimental misexpression of MyoD and Myf5 in the chick neural tube induces the development of ectopic skeletal muscles (Delfini and Duprez, 2004). Signals from the dorsal neural tube, and to a lesser extent from the ventral neural tube, are basic inducers of myogenesis and the ablation of the neural tube prevents the formation of muscles in somites (Stern, 2005). In chick embryos, the neural tube regulates the directed elongation of myocytes; removal of half of the neural tube at the presomitic mesoderm level, before differentiation of myocytes, causes disorganization of myocytes in comparison with the control side after 24 h (Gros et al., 2009).

Denervation leads to the upregulation and downregulation of 32 genes and to muscular atrophy. The altered expression of five of these genes indicates that Wnt signaling may be reduced after denervation (Magnusson et al., 2005). Denervation leads to the increased expression of the myogenin gene (Voytik et al., 1993), suggesting that the nervous system may also act as a brake in muscle development, probably by determining the growth rate and/or the end of muscle growth.

Examples of the direct influence of the CNS and local innervation on muscle development in vertebrates are plentiful. In electrical fish, innnervation is necessary for the transformation of muscle fibers into electrocytes, and the interruption of neural input causes the dedifferentiation of electrocytes back into muscle fibers

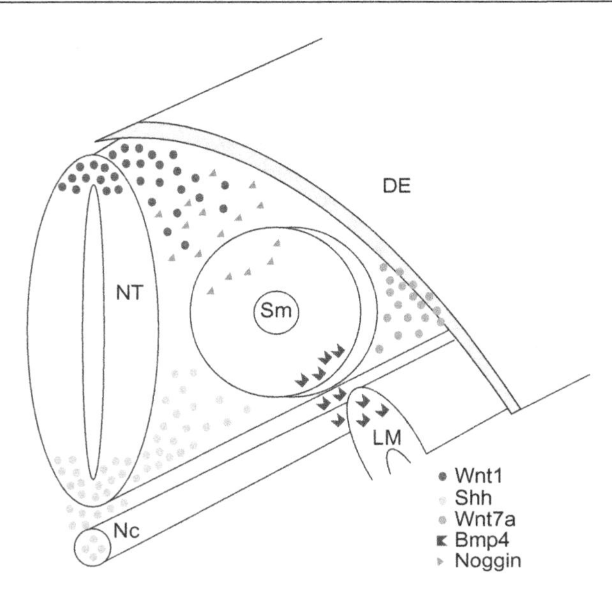

Figure 3.27 A simplified scheme of signaling molecules in newly formed epithelial somite. Shh (ochre dots), produced by notochord (Nc) and floor plate, acts on the ventral domain of newly formed epithelial somites, inducing sclerotome, and also on the dorso-medial domain, inducing medial dermomyotome. Wnt1 (red dots), produced by dorsal neural tube (NT), acts (with Shh) on the dorso-medial domain of newly formed somites (Sm), where *Myf5* expression is observed soon after and epaxial progenitors are specified. Wnt7a (blue dots), produced by dorsal ectoderm (DE), acts on the dorso-lateral domain, where hypaxial progenitors are specified. BMP4 (brown polygons), produced by lateral mesoderm (LM), prevents *MyoD* activation and early differentiation in the lateral domain of somites. Its action is counteracted by direct binding of Noggin (green triangles) produced by dorsal neural tube. *Source*: From Cossu and Borello (1999).

(Unguez and Zakon, 2002). In juvenile *Xenopus* frogs, the laryngeal muscle, *musculus dilator laryngis*, is monomorphic (i.e., identical in both sexes). After metamorphosis, the muscle becomes sexually dimorphic. The male muscle grows six to seven times larger than that of the female. The difference is explained by the high levels of circulating androgen in males at puberty (Sassoon and Kelley, 1986). Although it is well known that puberty and the associating rise of androgen levels in males is determined and regulated by neural mechanisms, further studies show that the nervous system induces the "masculinization" of the muscle, not only via the long-range neurohormonal pathway, but also directly via muscle innervation. The manifold increase in muscle fibers in males depends on both the circulating androgen and muscle innervation. The latter also affects the fiber growth. Denervation of the muscle causes its atrophy (Tobias et al., 1993).

In chick embryos, motor neurons, upon entering the limb Anlagen, release the retinoic acid-synthesizing enzyme (RALDH-2), inducing muscle cell differentiation and muscle formation. Recently, evidence has been presented that in chick

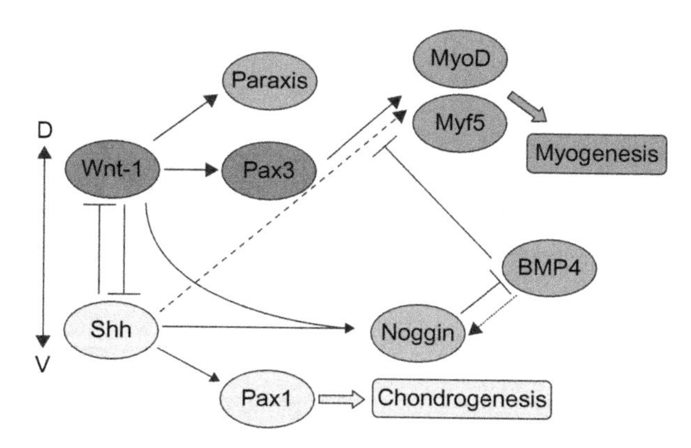

Figure 3.28 Schematic presentation of signaling molecules affecting somitic myogenesis. The relationships between signaling molecules and important transcription factors that participate in compartmentalization and differentiation of the somite. The dotted black line indicates that Bmp4 presumably creates the ability to express noggin. The dashed gray line indicates a possible Shh signal upregulating early myotomal markers. *Abbreviations*: D, dorsal; V, ventral.
Source: From Piran et al. (2009).

hindlimbs, RALDH-2 precedes the arrival of motoneurons (Wang and Scott, 2008). This, and the fact that denervation performed before the nerves enter the limb impairs muscle growth, may suggest that the innervation is necessary for survival and proper growth of muscles, rather than for initial muscle differentiation (Butler et al., 1982).

Firing patterns of motor neurons that innervate chick muscles determine muscle fiber developmental type by differentially regulating the expression of genes for myosin-heavy chains in fast and slow muscle fibers, which differ in morophology and contractility (Chin et al., 1998; Jordan et al., 2004).

Development of the Heart

The neural tube negatively controls heart development. Despite the presence of heart-inducing BMP in the mesoderm, the anti-BMP signals (Wnt-3 and Wnt-8) released by the neural tube inhibit the formation of the heart Anlage in the mesoderm in the vicinity of the neural tube (Tzahor and Lassar, 2001; see Figure 3.29). The inhibitory action of the neural tube in heart development is overcome by the secretion of the anti-Wnt signals (dkk-1 and crescent) from the Spemann organizer and the ectopic expression of *dkk*-1 or *crescent* induces the expression of cardiogenic genes in the noncardiogenic mesoderm (Schneider and Mercola, 2001).

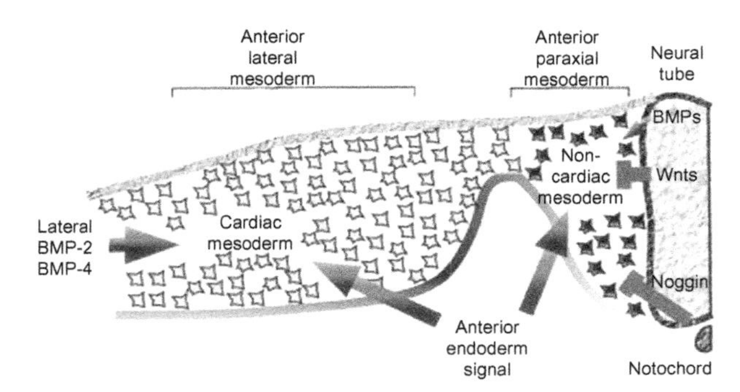

Figure 3.29 Heart formation is cued by a combination of positive and negative signals from surrounding tissues. Although a signal or signals from the anterior endoderm work to promote heart formation in concert with BMP signals in the anterior lateral mesoderm (blue arrows), signals from the axial tissues (red) repress heart formation in the more dorsomedial anterior paraxial mesoderm. Inhibitory signals that block heart formation in anterior paraxial mesoderm include Wnt family members expressed in the dorsal neural tube (Wnt-1 and Wnt-3a) and anti-BMPs expressed in the axial tissues (i.e., noggin in the notochord). We suggest that the sum of these positive and negative signals determine the medial-lateral borders of the heart-forming region. (For interpretation of the references to color in this figure legend, the reader is referred to the web version of this book.)
Source: From Tzahor and Lassar (2001).

Along the signal cascade starting from Spemann's organizer and the local cardiogenic induction, NCCs migrating to the heart Anlage from the cardial neural crest are crucial to heart development (Figure 3.30). These NCCs originate in rhombomeres 6, 7, and 8 of the hindbrain, where they are differentiated into various mesenchymal cells and migrate to regulate formation of the heart outflow tract and smooth muscles of the aortic arches (Hutson and Kirby, 2007; Phillips et al., 1987). They induce the septation of the common outflow into separate aortic and pulmonary arteries and influence ventral myocardial function and/or cardiomyocyte proliferation (Snider et al., 2007). Recruitment of the NCCs in the development of the cardiovascular system was crucial in the evolution of vertebrate cardiovascular system as a "new" heart (Fishman and Chien, 1997). In contrast to other vertebrates, in zebrafish, the cardiac neural crest invades the myocardium in all segments of the heart, besides the outflow tract, atrium, atrioventricular junction, and ventricle (Sato and Yost, 2003).

Removal of the cardiac neural crest in quail chick chimeras leads to anomalies of all outflow and inflow heart structures in whose formation the NCC is involved (Miyagawa-Tomita et al., 1991). Some recent evidence reducing the role of the cardiac neural crest in the cardiovascular system of amphibians may result from the smaller portion of the experimentally removed cardiac neural crest, but may also reflect a loss in this structure's ability to migrate deep in the heart tube during the

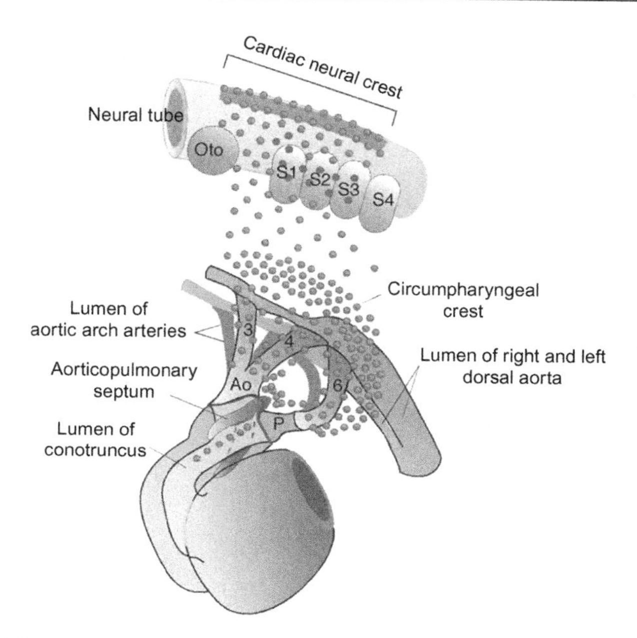

Figure 3.30 Migration and distribution of cardiac NCCs from their origin to the caudal pharynx and from there into the outflow tract.
Source: From Hutson and Kirby (2007).

evolutionary transition from fish into amphibians and the evolutionary reversion of that ability in amniotes (reptiles, birds, and mammals) (Lee and Saint-Jeannet, 2011).

Vasculogenesis and Angiogenesis

It is a well-established observation in anatomy that blood vessels do not form randomly and are closely associated with peripheral nerves. Only recently have biologists begun to figure out a possible causal relationship between the two. Beginning in the mid-1990s, a group of five VEGFs was reported to play a role in vasculogenesis (Carmeliet et al., 1996), proliferation of vasculoendothelial cells (Ash and Overbeek, 2000), and directed migration of endothelial cells (Esser et al., 1998). Mukoyama et al. (2002) presented direct experimental evidence on the role of the nervous system in angiogenesis: mice lacking sensory nerves in limbs develop large vessels that branch directly into small vessels rather than in normal intermediate vessels. It turned out that sensory neurons during development secrete VEGF, regulating both the arterial differentiation and branching on nerve templates (James and Mukouyama, 2011; Mukoyama et al., 2002; see Figure 3.31). This vasculogenetic function of the peripheral nerves seems to explain the causal basis between the association of nerves and blood vessels.

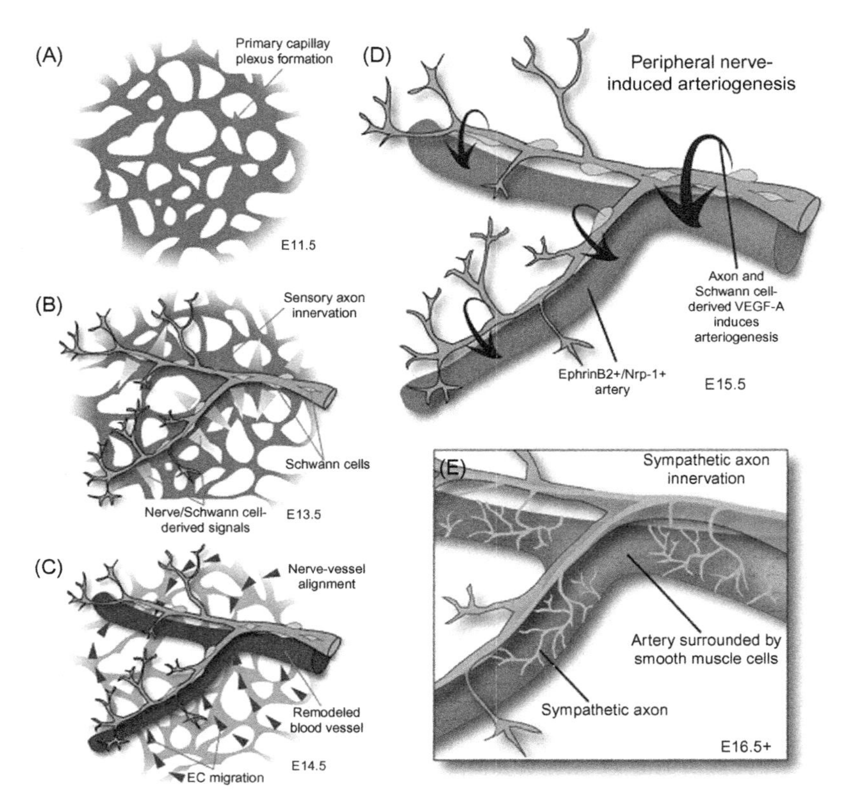

Figure 3.31 Model of blood vessel patterning by peripheral nerves in developing limb skin. (A) Blood vessels coalesce into a primary capillary plexus and begin to undergo remodeling by E11.5 before peripheral axon innervation develops in the limb. (B) Sensory axons and associated Schwann cells innervate the limb by E13.5 and secrete signals to pattern the primary vascular plexus. (C) In response to nerve and Schwann cell-derived signals, endothelial cells migrate toward and begin to align with peripheral nerves. (D) Blood vessels that have aligned with peripheral nerves undergo arteriogenesis in response to nerve and Schwann cell-derived VEGF-A. Aligned blood vessels upregulate arterial markers such as ephrinB2 and Nrp-1. Upregulation of Nrp-1 is thought to increase endothelial cell response to VEGF-A signaling. (E) After sensory nerve-induced arteriogenesis is complete, sympathetic axons innervate the limb, utilizing the blood vessels and sensory nerves as a template for migrations. Sympathetic axons invade the arterial smooth muscle layer to regulate local control of vascular tone.
Source: From James and Mukouyama (2011).

Mukoyama's model explains the vasculogenetic function of the peripheral nervous system. Another model is presented to explain the patterning of the perineural vessel plexus (PNVP) and the intraneural vessel plexus in the neural tube by the neural tube itself and the formation of the blood–brain barrier (James and Mukouyama,

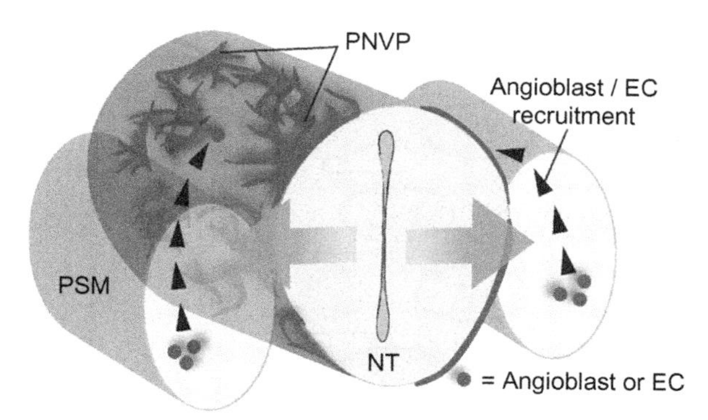

Figure 3.32 Model of PNVP recruitment in mice. At E8.5, angioblasts and ECs from the PSM respond to positive vessel patterning signals secreted from neural cells (green arrows) by differentiating, proliferating, and migrating to the surface of the neural tube. They surround the neural tube, forming a blood vessel plexus. *Abbreviations*: EC, endothelial cells; NT, neural tube; PNVP, perineural vessel plexus; PSM, presomitic mesoderm; E, embryonic day. (For interpretation of the references to color in this figure legend, the reader is referred to the web version of this book.)
Source: From James and Mukouyama (2011).

2011; James et al., 2009). The development of the vasculature around and within the neural tube is itself regulated by the endogenous release of VEGF-A in the neural tube (James et al., 2009). VEGF-A acts as a substitute for the neural tube in experiments (Hogan et al., 2004). The role of the neural tube as a vascular signaling center during the formation of the PNVP from the presomitic mesoderm is illustrated in Figure 3.32.

Formation of the Gastrointestinal Tract

The control and regulation of development of the gastrointestinal tract is complex and related not only to the tract itself, but also to the development of other organs, such as the liver, pancreas, and lungs. I will restrict this discussion to a brief outline of the development of the digestive tract of the worm *C. elegans*. This is not just because the development of the intestinal tract of this worm is better known than that of any other organism, although that is true. Most important, however, this worm's case illustrates best the early epigenetic determination of the gastrointestinal tract by maternal cytoplasmic factors.

The sperm's entry into the worm oocyte induces the polarization of the zygote. This and the high asymmetry of the first division lead to the formation of two cells that are very different in size and content of maternal cytoplasmic factors. These are the anterior AB-cell and the posterior P1 cell (Figure 3.33). The latter's division

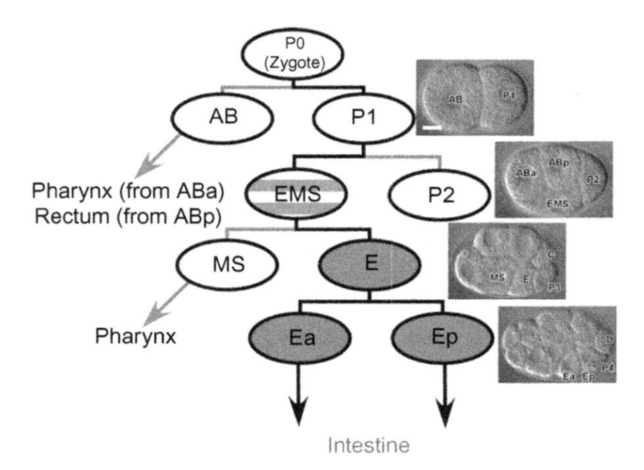

Figure 3.33 Early cell lineage of *C. elegans*.
Source: From Bossinger and Hoffmann, (2012).

leads to the production of the P2 and EMS (endoderm–mesoderm–stomodeum) cell in the four-cell stage (Bossinger and Hoffmann, 2012).

The EMS blastomere contains a maternal *skn*-1 mRNA that is translated into the transcription factor SKN-1. In turn, SKN-1 induces the expression of the *med*-1 and *med*-2 genes, which produce GATA-like transcription factors, necessary for the specification of the MS blastomere, but are less important for E blastomere specification. Two other genes, *end*-1 and *end*-3, are also induced by SKN-1. They produce two GATA factors, END-1 and END-3, which are primarily responsible for specifying the E blastomere. The asymmetric division of the ESM results in the production of the anterior MS blastomere and the posterior E blastomere. All 20 intestine cells and 38 of the pharynx cells of the worm derive from these two blastomeres (Kormish et al., 2010). Intestine cells in *C. elegans* derive from a single cell, known as the E blastomere, which present in the eight-cell embryo of the worm. One of the GATA factors, ELT-2, regulates about 80% (McGhee et al., 2009) or possibly all (Kormish et al., 2010) of the genes expressed in the *C. elegans* intestine during the late embryonic, larval, and adult stages.

Another epigenetic aspect of gastrointestinal development in vertebrates relates to the enteric nervous system (ENS). The ENS forms from NCCs, which during early development migrate primarily from the hindbrain and sacral regions of the neural crest to colonize the presumptive gastrointestinal tract, where they differentiate into a dozen neuron phenotypes and glia cells (Figure 3.34).

The role of the ENS in development and cell differentiation in the intestines is demonstrated in denervation experiments. Denervation has a restraining influence on enterocyte (Vespúcio et al., 2008) and goblet cell proliferation (Hernandes et al., 2003) of gut epithelium. Chemical denervation may induce a hyperplasia of endocrine cells in enteric villi and crypts (Santos et al., 2000), and the loss of mucosal

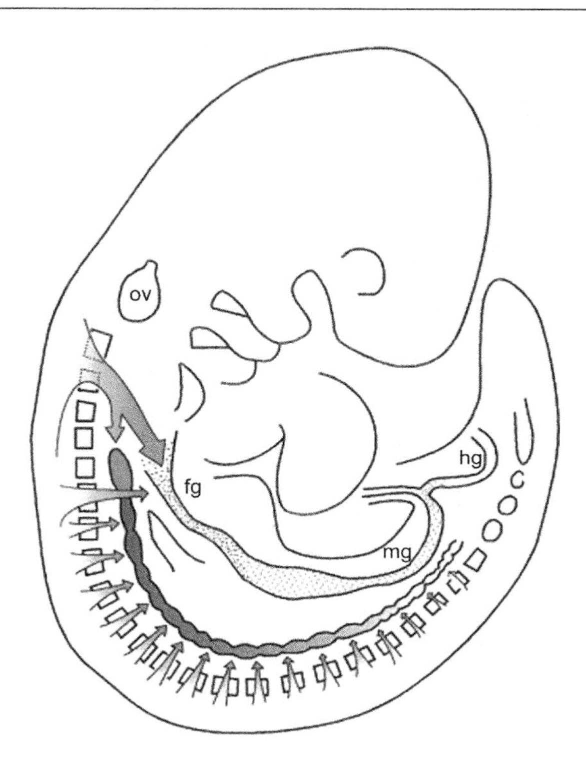

Figure 3.34 Schematic presentation of the derivatives of the sympathoadrenal (SA) and the sympathoenteric (SE) lineages. The SE lineage (shown in red) originates in the vagal neural crest of the hindbrain and migrates ventrally to populate the entire gut, as well as the superior cervical ganglion (SCG). The SE lineage (shown in blue), originates in the trunk neural crest (posterior to somite 6) and populates the foregut, as well as the ganglia of the sympathetic chain posterior to the SCG. The foregut is therefore populated by both SE and SA derivatives. *Abbreviations*: fg, foregut, hg, hindgut; mg, midgut; ot, otic vesicle. (For interpretation of the references to color in this figure legend, the reader is referred to the web version of this book.) *Source*: From Durbec et al. (1996).

innervation could result in the stimulation of the stem cell region by the substance P (Delbro, 2012).

There is evidence that intestinal stem/progenitor cells also are under nervous control (Lundgren et al., 2011) and parasympathetic innervation maintains the epithelial progenitor cell population in an undifferentiated state (Knox et al., 2010).

Pneumogenesis

In human embryos, the lung Anlage emerges during the fourth week, from an invagination of the ventral wall of the primitive foregut in the presumptive respiratory

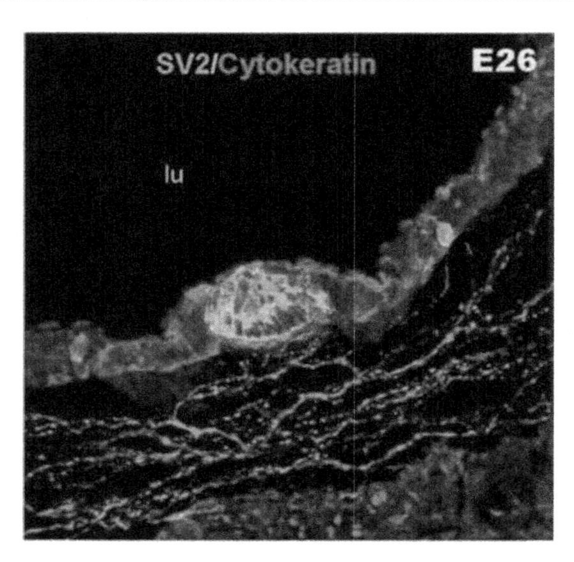

Figure 3.35 Airway epithelium in E26 lung, visualized using antibody against cytokeratin (red). Corpuscular NEB and two single PNECs (green) localized within epithelial lining. A rich network of submucosal nerve plexus (green) sends branches toward NEB base. No apparent neural contacts are seen with single PNEC in this section plane. *Abbreviation*: lu, airway lumen. (For interpretation of the references to color in this figure legend, the reader is referred to the web version of this book.)
Source: From Pan et al. (2004).

territory and develops into two endodermal bronchial buds. The Anlagen bifurcate and evaginate, concomitant with migration of the hindbrain NCCs that colonize the lung buds, where they differentiate into neurons and glia cells (Freem et al., 2012). On their way to the lung buds, they apparently follow the glial cell line-derived neurotrophic factor (GDNF) as an attractant. The incipient bronchi and epithelial tubules are encircled by a mesenchyme-derived smooth airway muscle (ASM) and are closely related to NCCs (Burns et al., 2008).

The first lung epithelial cells to differentiate are the pulmonary neuroendocrine cells (PNECs) at 8 weeks (Delgadoa et al., 2000; Emanuel et al., 1999). Shortly thereafter, clusters of PNECs, known as neuroepithelial bodies (NEBs), appear (Figure 3.35).

These neuroendocrine cells are innervated very soon after their differentiation. They release the neurotransmitter serotonin and bombesin-like peptides (BLPs), thus inducing the differentiation of various epithelial cell types (Emanuel et al., 1999). Lung vascularization and angiogenesis are also involved in lung development (Warburton et al., 2000), but it is well known that these processes are regulated by the local innervation that forms from NCCs. The serotonin released by lung neurons is believed to play an important role in lung growth, lung cell proliferation, and differentiation (Pan et al., 2006), and the presence of NEBs (predominantly in the branching points of bronchi) is believed to reflect its role in lung branching.

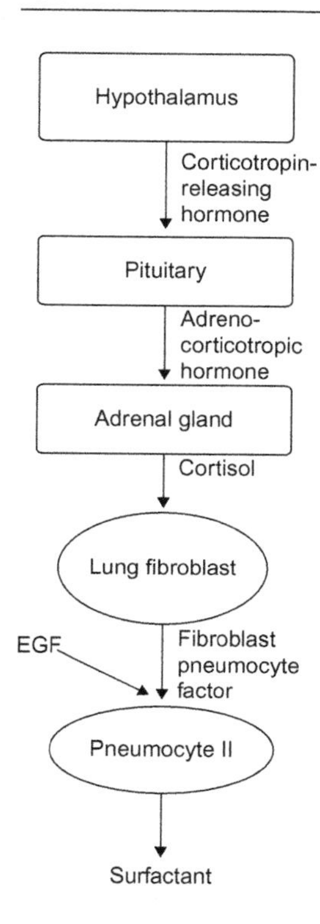

Figure 3.36 The signal cascade for production of surfactant starts in the brain. In the secretion of the surfactant, the following are also involved: EGF, via its receptor (EGF-R) (Dammann and Nielsen, 1998); TGFbeta-R; (dihydrotestosterone (DHT); Damman et al., 2000), as well as thyroid hormones (DiFiore and Wilson, 1994), paracrine hormones parathyroid hormone-related protein (PTHrP), and leptin (Torday et al., 2002).
Source: From Cabej (2012).

Mechanical strain produced by secreted fluids and diaphragm contractions induces the release of serotonin, which stimulates the proliferation of lung cells and formation of alveoli. Neural control in lung development is especially evident in the induction of surfactant by the alveolar cells of the lung that is essential for survival at the time of birth (Figure 3.36).

Nephrogenesis

By the middle of the twentieth century, Clifford Grobstein (1916–1998), used the transfilter technique (separation of two interacting tissues by filters) to demonstrate that the embryonic brain and spinal cord are potent inducers of the differentiation of the metanephric mesenchyme into epithelial cells and the formation of kidney tubuli (Grobstein, 1956). The brain and the spinal cord maintain this ability only for a small period during early development. Later, it was found that a secreted glycoprotein, Wnt-1, released by the CNS (i.e., the spinal cord) was capable of inducing the nephrogenic transformation of fibroblasts, and it was suggested that nephrogenesis was induced by the secretion of Wnt-1 by the spinal cord (Herzlinger et al., 1994) (Figure 3.37). Other tissues, such as the ureter bud and salivary gland, display a nephrogenic effect in direct contact with the mesenchymal cells, but only

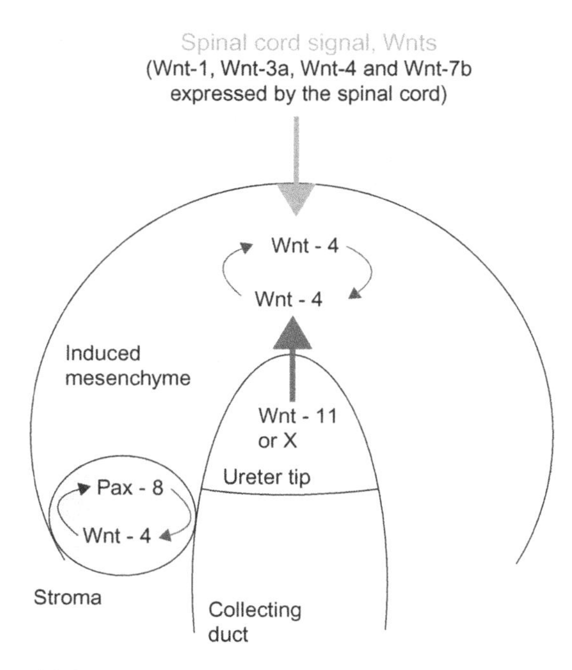

Figure 3.37 A model showing how Wnt-4 operates as a second tubule inductive signal. A ureter-derived signal, in the form of Wnt-11 or X, induces the expression of Wnt-4, which autoregulates itself and triggers tubule morphogenesis. Wnt-4 signaling involves *Pax*8. The spinal cord acts as an inducer tissue as it expresses a panel of Wnts, including Wnt-4, that triggers the autoregulated Wnt-4 gene expression to induce tubules.
Source: From Vainio et al. (1999).

the embryonic brain and spinal cord can induce their nephrogenic differentiation (Sariola et al., 1989).

Embryonic neurons, whose axons establish contacts with mesenchymal cells, are the most effective inducers of differentiation of mesenchymal cells *in vitro*, and maybe *in vivo* as well (Sariola et al., 1989).

Neural Control of Primary Sex Determination

For more than a century, it has been established that sex in animals is determined by sex chromosomes. The idea originates in an observation made in 1891 by the zoologist Hermann Henking (1858–1942) that a chromosome is present only in a part of the sperm cells of the European bug, *Pyrrhocoris apterus L.* He named the chromosome the "X-Faktor" and expressed the idea that it could play a role in insect sex determination. Later, the American biologist Clarence Erwin McClung (1870–1946) expanded the research, identifying the "X-element" as a possible factor of sex determination in other insects and animals; he also observed that only half of the sperms derived from grasshopper spermatogonia had the mysterious chromosome that he called the "accessory chromosome" (it turned out to be identical to the X chromosome). He related that the "accessory chromosome" was present in only half of sperm cells with sex determination. But it was the American geneticist Nettie Stevens (1861–1912) who, in 1905, identified the Y chromosome and posited that the gender of insects and worms depends on the presence of the Y chromosome.

The chromosomal determination of sex soon became almost a truism in biology. And the discovery of the temperature-dependent sex determination took biologists by surprise. Some amphibians have temperature-dependent sex determination, although they have "sexual chromosomes." Genotypically, female larvae of the salamander *Hynobius retardatus* reared at a high temperature (28°C) during the thermosensitive period (15–30 days after hatching) develop into female adults. Thus, a nongenetic factor, such as the temperature during the thermosensitive period, overrides the chromosomal sex determination.

A paracrine effect of a neuroendocrine factor would be to act as morphogen in the undifferentiated gonad to induce differentiation into the ovaries or testicles. Salame-Mendez et al. (1998).

The role of gonads and the brain in primary sex development and the development of sexual dimorphism is the source of some controversy. According to the prevailing view, the development and functioning of male and female sexual organs causes the masculinization/femininization of the brain via the secretion of sex hormones. Recent evidence, however, challenges the above view and may indicate that the reverse is true.

The embryonic mouse brain starts displaying sex differences in its expression of at least seven genes "before any gonadal hormone influence" (Dewing et al., 2003). In chicks, sex-specific expression of genes in the brain begins on the fourth embryonic day; that is, 1 day before it occurs in any other embryonic structure, including gonads (Scholz et al., 2006).

In marine turtles, and in many other reptiles, sex is also determined by the incubation temperature rather than "sex chromosomes." Generally, low temperatures induce the production of males and higher temperatures females, regardless of the genetic sex.

Sexual differentiation in the sea turtle (*Lepidochelis olivacea*) starts in the brain (diencephalon) rather than gonads. Here, the temperature is assessed and estradiol expression first rises. Salame-Mendez et al. (1998) observed that nerve fibers from the spinal cord pervade the gonads of the *L. olivacea* embryos and suggested that "[t]he spinal cord and the innervation derivating from it could play a role in driving or modulating the process of the temperature-dependent gonadal sex determination and/or differentiation."

The study was challenged by others (Pieau and Dorizzi, 2004; Davies and Wilkinson, 2006; Kuntz et al., 2004), but the most recent evidence seems to suggest central neural control of the gonadal differentiation. It is reported that sensory nerves innervating the genital ridge before gonadal differentiation are located in the dorsal horn of the lumbar spinal cord. Shifts in the incubation temperature are associated with the increased synthesis of c-Fos-like protein (a reliable marker of neuronal activity) in the sensory neurons innervating genetically female *L. olivacea* embryos. The intensity of the c-Fos expression in neurons was correlated with temperature changes (Jiménez-Trejo et al., 2011). In many reptiles, the perception in the brain (thermosensitive neurons in the preoptic area) of the environmental temperature leads to the expression of aromatase (Milnes et al., 2002; Willingham et al., 2000), which converts androgens into estrogen, determining the female sex when the embryo is genetically male.

It is noteworthy that in human embryos as well, bundles of autonomic nerve fibers arrive at the gonadal ridge 29–33 days after coitus, before sexual differentiation into the ovaries or testicles occurs (Møllgård et al., 2010).

New evidence shows that the sexual differentiation of the genital ridge is under brain control in birds as well; the sexually dimorphic expression of genes in the embryonic chick brain begins before the gonadal differentiation (Lee et al., 2009). Moreover, the effects of the temperature of incubation on the sex of chick embryos are transmitted to the next generation (Yılmaz et al., 2011). In the above-mentioned cases, the temperature seems to trigger a neural mechanism that overrides the chromosomal, or genetic, sex determination.

The role of the neural mechanisms in sex determination finds overwhelming support from numerous studies on the role of social factors in sex conversion in fish, which have led investigators to conclude the existence of a central nervous mechanism in the hypothalamus that is responsible for "the induction of the dramatic gonadal and behavioral transformations that are associated with sex change in hermaphroditic fish" (Elofsson et al., 1997b).

These same studies also lead to the conclusion on the primacy of changes in the brain over gonadal changes:

> *The initiation of the sex reversal is often controlled by social, behavioural factors, and since the only way behaviour can affect gonads is through the brain, here must be a central neuronal mechanism underlying the gonadal change [italics not in original].*
>
> *Elofssson et al. (1997a)*

Osteogenesis—Formation of Skeletal Bones

Central to the development of the skeleton in animals are three cell types: chondrocytes, the cell component of cartilage; osteoblasts, which are responsible for the deposition of the bone matrix; and osteoclasts, which specialize in bone resorption (Erlebacher et al., 1995). The first two come from a common progenitior cell, while osteoclasts derive from hematopoietic stem cells.

The evolution of bones from cartilaginous structures is an innovation associated with the appearance of vertebrates. It coincided with another crucial evolutionary innovation of vertebrates: appearance of the neural crest.

First, let us focus on the development of the craniofacial bones because its mechanism is better known. Cranial NCCs delaminate from the neural tube and migrate to the pharyngeal arches to transform themselves and the local cell populations into

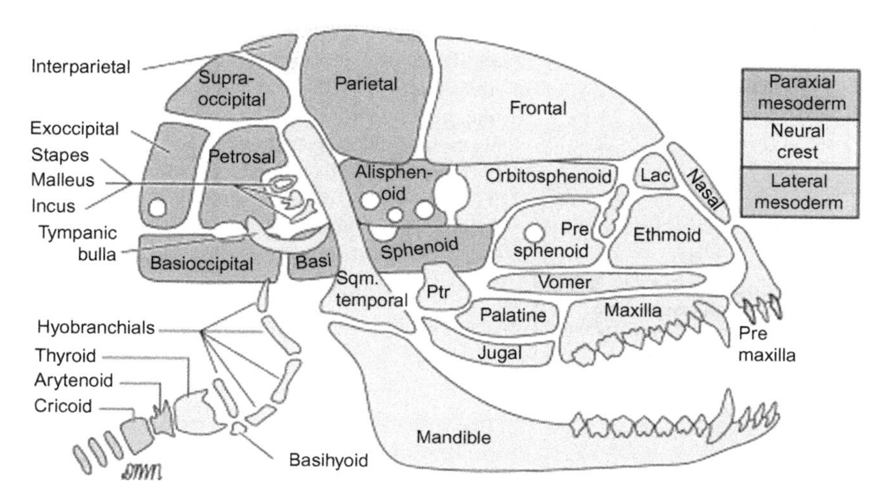

Figure 3.38 Schematic mouse skull showing the contributions of neural crest and paraxial and lateral mesoderms to the cranial skeleton. *Abbreviations*: Lac, lacrimal; Ptr, pterygoid. *Source*: From Noden and Trainor (2005).

cartilages and bones of the craniofacial region and entire the pharyngeal skeleton, including jaws, hyoid, and gill structures (Knight and Schilling, 2006) (Figure 3.38). The cranial crest takes over the functions of the mesoderm in this region, and its cells form skeletogenic cell types when they reach the presumptive bone regions.

The neural crest-derived mesenchyme is the source of bone-patterning information; it has an instructive function in shaping the skull, while the local epithelium "tends to play a more permissive role, heeding instructions from mesenchyme" (Jheon and Schneider, 2009). Before leaving the neural tube, NCCs are supplied with the migration program and the patterning information they enact in target sites:

> *... the proper program of events governing the migration of crest may need first to be established in the hindbrain, to allow migratory crest cells to interpret and respond to environmental signals set up through a series of tissue interactions [italics not in original].*
>
> *Trainor et al. (2002)*

In the above context, remember that the evolutionary precursor of the neural crest is the epidermal nerve plexus of protochordates (Gans and Northcutt, 1983) and in protochordates, the neural tube seems to have played a neural crest-like function (Meulemans and Bronner-Fraser, 2007).

During postnatal growth as well, bone development is under neural control via two pathways (Figure 3.39):

1. A direct pathway via the hypothalamic-pituitary axis, where release by specialized hypothalamic GHRH neurons induces pituitary secretion of the growth hormone, which stimulates bones to secrete IGF-1.

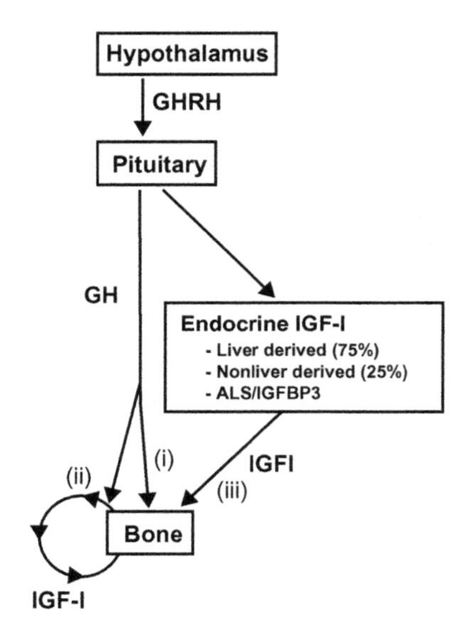

Figure 3.39 A model of GH-mediated regulation of postnatal longitudinal bone growth. *Abbreviations*: ALS (acid labile subunit), an insulin-like growth factor binding protein; GH, growth hormone; GHRH, growth hormone-releasing hormone; IGF-I, insulin-like growth factor; IGFBP3, IGF binding protein3.
Source: From Ohlsson et al. (2009).

2. An indirect pathway, where the axis stimulates the liver (and, to a lesser extent, other tissues) to secrete IGF-1, which stimulates bone growth.

Hormonal regulation of bone development is performed by the hypothalamus through three neuroendocrine axes: hypothalamic pituitary–gonadal (HPG), hypothalamic–pituitary–thyroid (HPT), and hypothalamic-pituitary (HP) (via action of the pituitary thyroid-stimulating hormone (TSH). Biologists marvel at the highly accurate symmetrical development of long bones. For instance, Wolpert et al. (1998) state:

> *In view of the complexity of the growth plate, it is remarkable that human bones in the limbs on either side of the body can grow for some 15 years independently of each other, and yet eventually match to an accuracy of about 0.2%.*

This precise symmetrical development implies that information on the development of both limbs is communicated to a higher organismic structure able to assess and compare limb sizes and send signals for their symmetrical development. It is the continual afferent input from bones that enables the CNS to assess their state, and, via the above-mentioned neuroendocrine axes HPG, HPT, and HP, send instructions about the size of their development. Besides, the peripheral sensory innervation of different limbs, which are connected via the dorsal root ganglia, represents another neural control system (Sample et al., 2010; Wu et al., 2009).

Neural Control of Bone Remodeling

Like all other somatic cells, osteoblasts and osteoclasts have a limited life span: about 3 months for osteoblasts and about 2 weeks for osteoclasts. The T account is

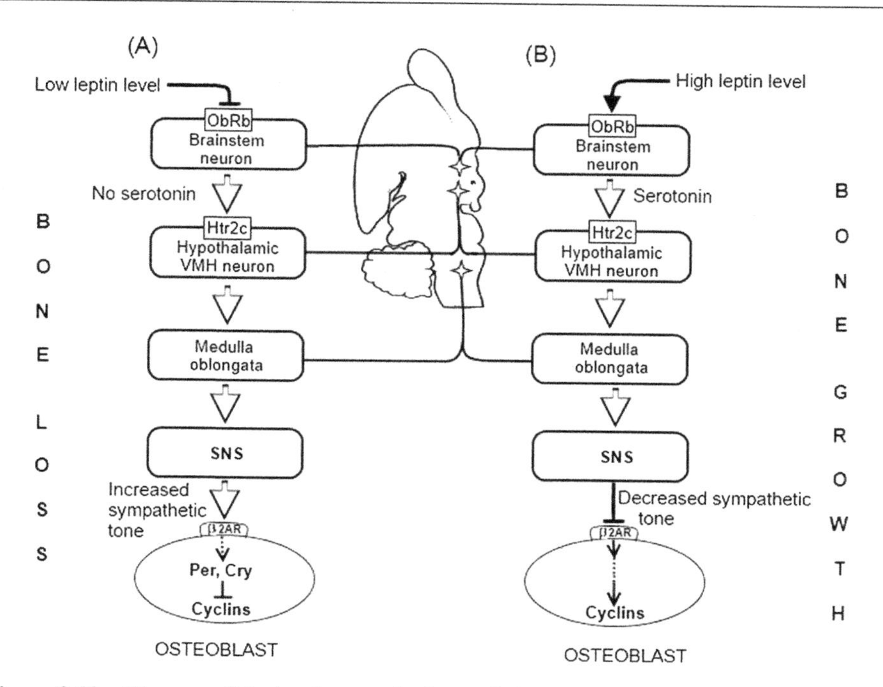

Figure 3.40 Diagram of the local sympathetic regulation of bone mass. (A) When leptin level is low (in nonobese individuals), it does not bind its receptor ObRb in brain stem neurons, which consequently secrete no serotonin, thus leading to strengthening of the sympathetic tone (increased noradrenalin secretion). The higher sympathetic tone of the SNS leads to the inhibition of osteoblast proliferation and increase of osteoclast proliferation, resulting in bone resorption and bone loss. (B) When leptin level is high (in obese individuals), it binds its receptors ObRb in brain stem serotonergic neurons, stimulating their electrical activity, secretion of serotonin, and its binding to Htr2c receptor in specific hypothalamic neurons. This leads to decreased activity of the sympathetic system (i.e., decreased secretion of noradrenaline), thus stimulating osteoblast proliferation, with bone growth as a result. *Abbreviations*: β2-AR, β2-adrenergic receptor; Htr2c, serotonin receptor; ObRb, leptin receptor; *Per* and *Cry*, period and cryptochrome genes; SNS, sympathetic nervous system; VMH, ventromedial hypothalamus.

again at work. The organism replaces the lost number of bone cells, so bones remain in a state of dynamic equilibrium, or homeostasis. Two neural mechanisms for the maintenance of bone homeostasis are operational in vertebrates: a central and a local mechanism.

The central neural mechanism of bone homeostasis is very complex and not yet known in full detail. It responds adaptively to the leptin blood level. When the leptin level is low, the leptin-sensitive neurons in the brain stem are not activated, and the hypothalamic VMH (ventromedial nucleus) starts a signal cascade that leads to the activation of osteoclasts and bone loss (Figure 3.40). When the leptin level is high, it binds to its receptor ObRb in neurons of the brain stem. These neurons secrete serotonin in response. The serotonin starts a signal cascade that leads to decreased sympathetic activity and consequent osteoblast proliferation and bone growth.

The local mechanism relies on the intense presence in bones of nerve endings, which release in bones a number of neurotransmitters and neuropeptides, such as serotonin, glutamate, catecholamine, vasoactive intestinal peptide (VIP), substance P (SP), neuropeptide Y (NPY), calcitonin-gene-related peptide (CGRP), etc., which are involved in bone formation and bone resorption (Lerner, 2002; Lundberg et al., 2000; Yeh et al., 2002).

How Animals Know When to Stop Growing—The Body Size Set Point

Thus far, we have dealt with the development of various organs and structures from a purely qualitative view, without considering the determination of the size or the limits of growth, an equally important aspect of development. Body size is a typical quantitative trait, and we will review briefly the mechanisms determining body size in animals, both in invertebrates and vertebrates.

Drosophila, the old workhorse of biological investigation, is one of the insects in which the mechanisms of neural control of body size are better known. During larval stages, *Drosophila's* growth is stimulated by the secretion of seven types of insulin-like neuropeptides, which come from seven median secretory neurons in the pars intercerebralis of its protocerebrum. Their receptors mediate the growth-stimulating effect of these neuropeptides. Ablation of these neurons reduces body size and the number of cells in the adult insect.

In some insects, the brain assesses the body size based on the information it receives from proprioceptors on the stretch caused by body growth. Increasing the stretch beyond a species-specific, neurally determined set point, as assessed by the insect's brain, signals for specific neurons to secrete allatostatins, which inhibit the secretion of the juvenile hormone (JH) by corpora allata. The same signal stimulates other neurons to secrete the neuropeptide PTTH, which induces secretion of the ecdysis hormone ecdysone by the prothoracic gland (PG). In higher dipterans, both the prothoracic gland and corpora allata are parts of a compound structure called the *ring gland* (Figure 3.41). Ecdysone and insulin-like neuropeptides (ILPs) have antagonist effects, and the silence of the ecdysone receptor (EcR), the mediator of the ecdysone growth-inhibiting action, leads to the production of larger *Drosophila* offspring (Colombani et al., 2005).

It is also suggested that larval growth stops when PG reaches a critical size and the higher level of ecdysteroids induces expression of the early gene for the transcription factor E74 and the beginning of the patterning of imaginal disks by ecdysone (Mirth et al., 2009).

Now we know that at a higher level of the neuroendocrine hierarchy of the neural regulation of body size, production of *Drosophila* ILPs (DILPs) is regulated by the secretion of the neurotransmitter serotonin (5-HT_{1A}), which, via its receptor, inhibits the secretion of DILP-2 by ILP neurons (Luo et al., 2012).

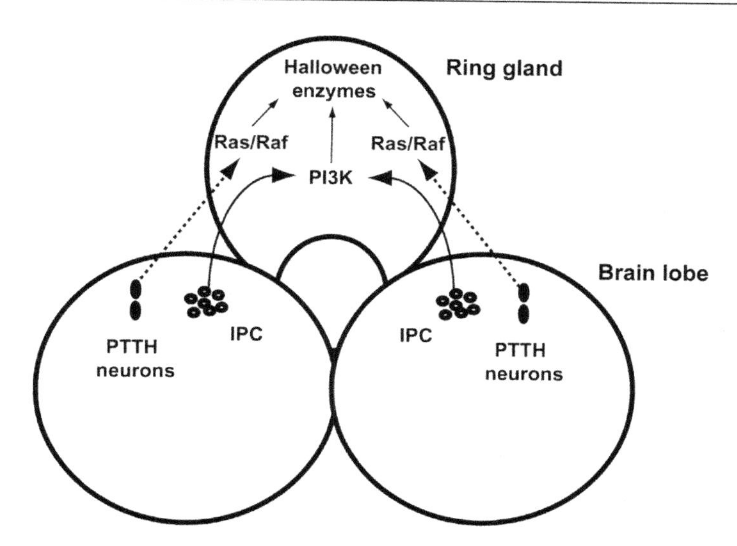

Figure 3.41 The regulation of ecdysone synthesis by insulin and PTTH. Brain lobes and the ring gland are indicated. Ecdysone synthesis is triggered when insulin released from the IPCs and PTTH released from the PTTH neurons activate the PI3K and Ras pathways, respectively, in the prothoracic gland. These two pathways, acting together, activate transcription of the PG-specific "Halloween genes," which encode the ecdysone biosynthetic enzymes, ultimately triggering ecdysone synthesis. Arrows indicate activation pathways for which there is experimental evidence in *Drosophila*; the hatched arrow indicates a speculative activation pathway.
Source: From Walkiewicz and Stern (2009).

The *Drosophila* nervous system responds to developmental disturbances, such as damages to imaginal disks, by postponing pupariation time to allow for regeneration to take place. Recently, it has been demonstrated that the CNS may use information on the increased secretion of the hormone peptide DILP8 by imaginal disks as a cue indicating growth status (Garelli et al., 2012) and the dose-dependent postponement of the growth and pupariation.

In the freshwater snail *Lymnaea stagnalis*, insulin-like neuropeptides control body growth. The snail's brain has about 200 giant neurons known as light green cells (LGCs), which produce four types of insulin-like neuropeptides that are transported to axon terminals and released to a neurohemal-like area. The cauterization of these giant neurons causes marked body growth retardation in juvenile, rapidly growing snails. Each of the lateral lobes of the brain contains an ectopic LGC connected to axons of giant neurons and has an inhibitory effect on body growth, as shown by the fact that their cauterization leads to the development of giant snails (Geraerts, 1976; Smit et al., 1998).

The neural control of body weight is also observed in lower animals such as planarians, where the adult animal size and tissue homeostasis are regulated by the neurally secreted, insulin-like neuropeptide Ilp-1. Disrupting the insulin signaling in the flatworm *Schmidtea mediterranea* leads to smaller degrown phenotypes similar to those produced by starvation (Miller and Newmark, 2012).

The evidence presented above points to the existence in the brain of insects and worms of set points that determine their characteristic body size. This evidence suggests these invertebrates somehow perceive the actual body size and assess it. When body size reaches the species-specific set point, the specialized neurons suppress or reduce the secretion of insulin-like neuropeptides, and animals cease growing.

How Long Animals Live—An Epigenetic Decision

The same insulin-like neuropeptides that determine animal growth negatively regulate life span and aging in animals. *C. elegans* has 37 genes that secrete insulin-like neuropeptides, some of which appeared approximately 600 Mya, and others evolved later through gene duplication (Nelson and Padget, 2003). Most evidence in this regard comes from the soil nematode worm *Caenorhabditis elegans* and *Drosophila*, but basic mechanisms seem to have been conserved across the animal world.

Under normal conditions, *C. elegans* development goes through four larval stages before adulthood. But, in a dramatic move, under challenging conditions (scarce food resources, crowding, high temperature, etc.), the worm halts at the third larval stage, switching to an alternative developmental program, or dauer (from German *dauer*, steady, permanent), to extend its life (Figure 3.42).

Crowding is the primary cue for switching to the dauer mode and delaying age-related physiological decline. These worms normally secrete a dauer pheromone (consisting of a number of components), and crowding leads to an increased concentration of the pheromone in the environment. Responding to this higher pheromone concentration, the worm shifts to dauer mode. It assesses the availability of food resources from the content of food odorants, as perceived by olfactory neurons. This is demonstrated by the fact that in the presence of these odorants, even in the absence of food, the worm switches to dauer mode.

During dauer, which may last for months, *C. elegans* stops food intake, reduces metabolism, and becomes resistant to external factors, including oxidative stress. The perception of unfavorable environmental factors (and especially detection by olfactory neurons of the dauer pheromone) indicating crowding are crucial for the switch to the dauer life cycle.

The worm has an extremely simple nervous system, consisting of only 302 neurons. The sensory input on environmental conditions is received by ciliated sensory neurons ASI, ADF, and ASG in the amphids (paired sensory organs in the head of the worm). Integrating the environmental stimuli into the amphid neurons induces the secretion of the neurotransmitter serotonin, which activates an insulin-like signaling by ASI and ASJ neurons and TGF-β by ASI neurons. Removal of these neurons causes growth to stop and the dauer program to be adopted, even in the presence of food (Fielenbach and Antebi, 2008), indicating that they are responsible and necessary for adopting the dauer program.

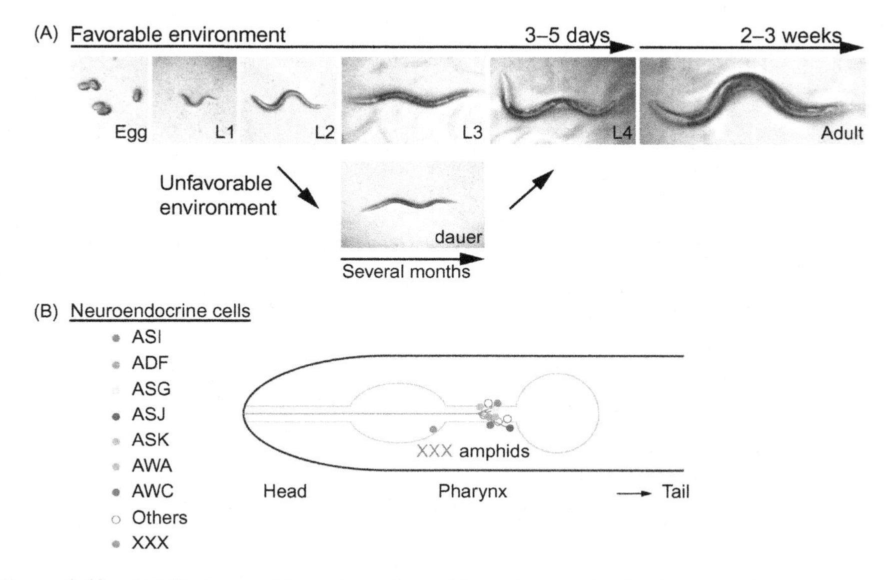

Figure 3.42 (A) *C. elegans* life cycle. In favorable environments, *C. elegans* undergoes reproductive development and progresses rapidly from embryo through four larval stages (L1–L4) to reach adulthood in 3–5 days (in temperatures of 15–20°C). Adults then live another 2–3 weeks. In unfavorable conditions, including overcrowding, limited food, and high temperature, *C. elegans* undergoes development to a specialized third larval stage called the *dauer diapause (L3d)*, which can live several months. Upon return to a favorable environment, dauer larvae recover and become reproductive adults with normal life spans. (B) Schematic of neuroendocrine cells. Integration of environmental cues (dauer pheromone, nutrients, and temperature) are transformed into endocrine signals by amphid neurons (ASI, ADF, ASG, ASJ, ASK, AWA, and AWC). Serotonergic signaling from ADF and cGMP signaling from ASJ and ASK influence the production of TGF-β from ASI and the ILPs from ASI, ASJ, and other cells. Insulin/IGF-1 signaling and TGF-β signaling converge on steroid hormone signaling in the XXX cells.
Source: From Fielenbach and Antebi (2008).

DAF-2 is the worm's only insulin/IGF-1-like receptor and is similar to vertebrate IGF receptors (Foxo). Mutations in the gene *daf*-2, or the loss of function of that gene, lead to doubling the life span of the mutant worm compared to the wild type (Kenyon et al., 1993). Mutations in *daf*-2 lead to the movement of DAF-16 into the cell nuclei, where it activates a signal cascade for a longer life span and stress resistance (Bartke, 2009). At a molecular level, the perception of unfavorable conditions or the simple perception of reduced food odorants activates the DAF-3/DAF-5 complex that specifies the dauer program (Fielenbach and Antebi (2008).

It is believed that amphid sensory neurons decode the environmental information on the availability of food resources, transforming it into instructions for adaptively switching between the normal and dauer programs of development (Pletcher, 2009).

In *Drosophila* as well, the insulin-like signaling pathway is fundamentally involved in determining the fly's life span. Ablation of the neurons that secrete insulin-like neuropeptides (Dilp2, Dilp3, and Dilp5) in the pars intermedialis of the

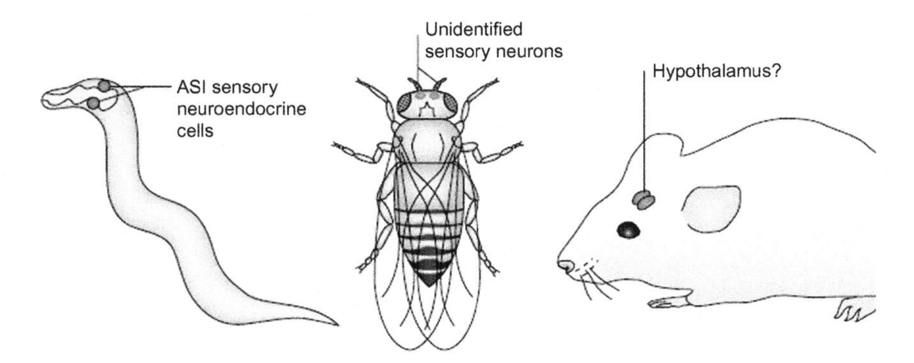

Figure 3.43 A neural basis for longevity induced by dietary restriction (DR). DR longevity is mediated by energy-sensitive central neurons in worms, flies, and possibly mammals. The ASI neurons mediate DR longevity in worms, and they signal during DR to cause increased respiration in peripheral tissues and long life spans. Unidentified olfactory neurons mediate DR longevity in flies. The hypothalamus may mediate DR longevity in mammals.
Source: From Bishop and Guarente (2007).

protocerebrum increases the life span of male flies by 10.5% and mated females by 33.5% (Broughton et al., 2005). Exposure of *Drosophila* flies to food-based odorants reduces their life span and a mutation in one of the olfactory neurons increases average life span by 60% or 30 days, compared to wild-type flies (Pletcher, 2012).

The role of the insulin/IGF-1 pathway in life spans under the conditions of dietary restriction seems to be evolutionarily conserved across the metazoan world. From *C. elegans* to humans, the activation of DAF-16/FOXO-like transcription factors promotes longer life (Kenyon, 2010). Longevity seems to be regulated by the same basic mechanisms in vertebrates, where the life-prolonging functions of the amphid neurons are taken over by the hypothalamus (Figure 3.43). The role of the pathway in mammals was demonstrated recently in the Ames and Snell dwarf mouse strain, which live about 70% longer than normal mice (Berryman et al., 2008), show reduced signs of aging, and are resistant to oxidative stress. The link between the pathway and the life span is mediated by the downstream effects on Forkhead box (FOX) transcription factors (Dantzer and Swanson, 2012).

There is evidence that aging is causally related to degrading changes in the nervous system. Similarly to humans, microscope electronic studies have shown that aging in *C. elegans* is associated with loss of synaptic integrity (Toth et al., 2012).

Metamorphosis: The Same Organism Produces Two Radically Different Body Plans

It was generally believed that differences in body plans are related to differences in genes. However, the gene sequencing of many animal species shows that almost the same developmentally essential and functionally unchanged genes are expressed across the animal taxa. It is impossible to reasonably relate changes in genes/regulatory sequences or the evolution of new genes to the evolution of animal Bauplans. To overcome this difficulty, the genetic toolkit hypothesis was developed. According to this theory, the patterns of gene expression, especially of *Hox* genes, rather than the number or the nature of genes, mold the Bauplan.

For the sake of argument, let us take for granted that *Hox* genes are the true determinants of the body plan. Rather than resolving anything, this assertion raises a formidable question. *Hox* genes, like all the nonhousekeeping genes, are induced by extracellular signals. The temporal and spatial patterns of *Hox* expression, and consequently body plans, are determined externally, not by *Hox* genes themselves. The question "What determines the body plan?" continues to haunt us. Whatever the real role of *Hox* genes in the development of the Bauplan might be, it is essential from a causal viewpoint to identify the ultimate origin of the information that determines the spatial and temporal patterns of the expression of the *Hox* gene (and of all the nonhousekeeping genes for that matter).

Metamorphosis is one of the most convenient biological phenomena for studying and identifying the source of information for determining Bauplan development. It gives us a glimpse into some aspects of development that are hidden in the egg or uterus. Metamorphosis is a widespread life history trait observed from the simplest metazoans (cnidarians) to vertebrates (amphibians), and a brief review of metamorphosis in some animal groups may be helpful in understanding the source of information that enables this amazing biological phenomenon.

The 1-mm-long, free-living larval planula of the cnidarian *Hydractinia echinata*, consisting of just 10,000 cells, has secretory neurons that specialize in detecting a specific bacterial cue in their environment. When the planula detects the cue, its neurons secrete a number of neuropeptides from a signal system (Schmich et al., 1998), represented by two opposing groups of neuropeptides, called GLWamide and RFamide neuropeptides, that stimulate and inhibit metamorphosis in *Hydra*, respectively (Katsukura et al., 2004).

The free-swimming molluscan larvae have a functioning nervous system that includes an apical sensory organ (ASO), whose neurons secrete various neurotransmitters (dopamine, norepinephrine, etc.). Larvae respond to specific chemical cues by entering metamorphosis and the receptors of these cues are ASO neurons. When the ASO neurons are destroyed by UV irradiation, the larvae do not respond to the cue (Hadfield et al., 2000). Experimental depletion of these neurotransmitters inhibits metamorphosis, and elevation of their levels induces it in competent *Phestilla* larvae (Pires et al., 1997). In the process of metamorphosis, ASO disappears (Ruiz-Jones and Hadfield, 2011; see Figure 3.44), suggesting that its function is to determine the time of metamorphosis and initiate it.

In insects, metamorphosis is determined primarily by the temporal and quantitative patterns of expression and interaction of juvenile hormone (JH) secreted by corpora allata and ecdysone secreted from the prothoracic gland. Both hormones are strictly cerebrally regulated by many neuropeptides, but primarily by PTTH for ecdysone and allatostatins and allatotropins for JH.

An essential behavioral component of insect metamorphosis is ecdysis, the shedding of the cuticula (the inflexible exoskeleton) to enable growth and morphological

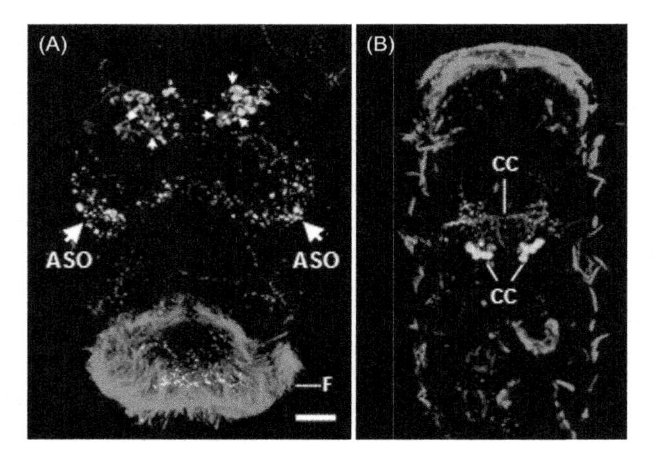

Figure 3.44 Serotonergic (yellow–green) and acetylated tubulin (red) immunoreactivity in metamorphosing and early-juvenile-stage *Phestilla sibogae*. (A) Four hours into metamorphosis: the five curled ampullary neurons (small arrowheads) appear to have retracted from the epidermal surface, and deterioration of serotonergic axons (large arrowheads) is evident. (B) Twenty-six hours into metamorphosis: the settled juvenile has elongated, and there is no sign of an ASO. F, foot; CG, cerebral ganglia; CC, cerebral commissure. (For interpretation of the references to color in this figure legend, the reader is referred to the web version of this book.)
Source: From Ruiz-Jones and Hadfield (2011).

modifications. Cuticle shedding is a motor behavior that, like any other behavior, is expected to be controlled and regulated by the nervous system. However, in the case of ecdysis behavior, Inka cells of an extraneural organ called the epitracheal gland are usually considered initiators of ecdysis behavior. This has been disproved: Inka cells in *Manduca sexta*, for example, are stimulated to produce pre-ecdysis-triggering hormone (PETH) and ecdysis-triggering hormone (ETH) by a neuropeptide called corazonin, which is secreted by lateral brain neurons. Corazonin, both *in vivo* and *in vitro*, stimulates Inka cells to secrete both PETH and ETH (Asuncion-Uchi et al., 2010; Kim et al. 2004). Responding to these hormones, the insect's brain expresses the ETH receptor and secretes the neuropeptide eclosion hormone (EH). The available evidence shows, as would be expected, that the first ecdysis signals the start in the insect's brain.

Certainly, programming of the innate ecdysis behavior is more complex and involves the activation of numerous neurons producing kinin, FMRFamides, EH, crustacean cardioactive peptide (CCAP), myoinhibitory peptides (MIPs), and bursicon (Truman, 2005).

Metamorphosis in ascidians is essentially different from all other organisms. While the transition of all other organisms from the larval stage to the adult stage is associated with the development of an adult structure that is evolutionarily more complex, adult ascidians are morphologically and behaviorally simpler organisms. The free-swimming ascidian larva, with only approximately 2600 cells, has a chordate Bauplan with a tail, fins, notochord, dorsal neural cord, and gill clefts. When the larva detects an environmental cue, it enters metamorphosis. In contrast with other metamorphosizing organisms, the transition from the larval stage to the juvenile stage in ascidians is associated with a developmental and behavioral degradation: larval muscles, the tail, fins, and the dorsal nerve cord are lost, and the brain becomes a neural ganglion. In response, they shift to a sedentary life by attaching to a solid substrate. For these reasons, it is said that ascidians go through a retrogressive metamorphosis. In the early 1980s, it was concluded that the metamorphosis in ascidians is controlled by the nervous system (Burke, 1983).

Free-swimming amphibian larvae have fishlike bodies, vertically flattened tails, internal gills, cartilaginous skeletons, etc. They are induced to enter metamorphosis by various internal cues, such as growth beyond a neurally determined threshold, and external cues, such as water temperature and availability, photoperiod, etc. (Denver, 1997). These external signals are received, integrated, and processed in the amphibian CNS. The information for starting metamorphosis in amphibians flows via the hypothalamus–pituitary–thyroid axis. In a simple form, the updated signal cascade for metamorphosis in amphibians looks as follows:

> Signals from the nonhypothalamic brain → hypothalamic thyroid-releasing hormone (TRH) → pituitary thyroid-stimulating hormone (TSH) → thyroid hormone (TH) → matrix metalloproteinases (MMPs) → integrins → signals for gene expression and apoptosis

Along the TRH, another hypothalamic neuropeptide, the corticotropin-releasing hormone (CRH) also stimulates secretion of TSH and the TH. Injections of this neuropeptide accelerate metamorphosis in several amphibian species (Denver, 1999).

The neuroendocrine mechanism of metamorphosis is conserved across amphibian taxa (Denver, 2002).

Metamorphosis induces radical changes in amphibian morphology, including the development of lungs, shortening of intestines, repositioning of eyes, restructuring of the nervous system to adapt for the new way of life, gradual loss of the tail, etc.

With the exception of ascidians, the development of the species-specific body plan during metamorphosis is preceded by an evolutionarily lower body plan. In contrast, the metamorphosis of ascidians leads to an adult morphology, physiology, and behavior that is simpler than that of the larva's. Metamorphosis is an amazing example of the dexterity of animals to switch to different developmental programs. This certainly contradicts the prevailing opinion that an egg or a zygote is provided with a program that determines development up to the adult stage. This gains more significance when one remembers the ease with which some metamorphosizing amphibians can switch to a direct mode of development, or even skip metamorphosis altogether. Is it possible that the same egg/zygote contains the programs for two different Bauplans, and sometimes even a program for skipping its species-specific Bauplan?

Metamorphosizing species, besides their own developmental program, have incorporated and executed ancestral developmental programs. Amazingly, like biological Houdinis, they shift the gears of development both forward (insects and amphibians) and backward (ascidians).

References

Alves, H.J., Alvares, L.E., Gabriel, J.E., Coutinho, L.L., 2003. Influence of the neural tube/notochord complex on *MyoD* expression and cellular proliferation in chicken embryos. Braz. J. Med. Biol. Res. 36, 191–197.

Aoki, F., Worrad, D.M., Schultz, R.M., 1997. Regulation of transcriptional activity during the first and second cell cycles in the preimplantation mouse embryo. Dev. Biol. 181, 296–307.

Ash, J.D., Overbeek, P.A., 2000. Lens-specific VEGF-A expression induces angioblast migration and proliferation and stimulates angiogenic remodeling. Dev. Biol. 223, 383–398.

Asuncion-Uchi, M., El Shawa, Martin, T., Fuse, M., 2010. Different actions of Ecdysis-Triggering Hormone on the brain and ventral nerve cord of the hornworm, *Manduca sexta*. Gen. Comp. Endocrinol. 166, 54–65.

Baker, V.H., Bronner-Fraser, M., 1997. The origins of the neural crest. Part II: an evolutionary perspective. Mech. Dev. 69, 13–29.

Barembaum, M., Bronner-Fraser, M., 2005. Early steps in neural crest specification. Semin. Cell Dev. Biol. 16, 642–646.

Bartke, A., 2009. Insulin and aging. Cell Cycle 7, 3338–3343.

Bayline, R.J., Duch, C., Levine, R.B., 2001. Nerve-muscle interactions reguiate motor terminal growth and myoblast distribution during muscle development. Dev. Biol. 231, 348–363.

Berryman, D.E., Christiansen, J.S., Johannsson, G., Thorner, M.O., Kopchick, J.J., 2008. Role of the GH/IGF-1 axis in lifespan and healthspan: lessons from animal models. Growth Horm. IGF Res. 18, 455–471.

Beyer, T., Danilchik, M., Thumberger, T., Vick, P., Tisler, M., Schneider, I., et al., 2012. Serotonin signaling is required for Wnt-dependent GRP specification and leftward flow in *Xenopus*. Curr. Biol. 22, 33–39.

Bininda-Emonds, O.R.P., Jeffery, J.E., Richardson, M.K., 2003. Inverting the hourglass: quantitative evidence against the phylotypic stage in vertebrate development. Proc. R. Soc. Lond. B 270, 341–346.

Birsoy, B., Kofron, M., Schaible, K., Wylie, C., Heasman, J., 2006. Vg1 is an essential signaling molecule in *Xenopus* development. Development 133, 15–20.

Bishop, N.A., Guarente, L., 2007. Genetic links between diet and lifespan: shared mechanisms from yeast to humans. Nat. Rev. Genet. 8, 835–844.

Blum, M., Andre, P., Muders, K., Schweickert, A., Fischer, A., Bitzer, E., et al., 2007. Ciliation and gene expression distinguish between node and posterior notochord in the mammalian embryo. Differentiation 75, 133–146.

Blum, M., Beyer, T., Weber, T., Vick, P., Andre, P., Bitzer, E., et al., 2009. *Xenopus*, an ideal model system to study vertebrate left-right asymmetry. Dev. Dyn. 38, 1215–1225.

Bossinger, O., Hoffmann, M., 2012. Development and cell polarity of the *C. elegans* intestine. In: Najman, S. (Ed.), Current Frontiers and Perspectives in Cell Biology. Rijek, Croatia, Intech, pp. 335–360.

Boveri, T., 1902. On the multipolar mitosis as a means of analysis of the cell nucleus (Über mehrpolige Mitosen als Mittel zur Analyse des Zellkerns). Verhandlungen der physikalisch-medizinischen Gesellschaft zur Würzburg. Neue Folge, 35, 67–90. Translated from the original German, 1964 in Foundations of Expmental Embryology, edited by Willier, B.H., Oppenheimer, J.M., Prentice Hall, Englewood Cliffs, NJ. According to Wieschaus, E., 1996. Embryonic transcription and the control of developmental pathways. Genetics 142, 5–10.

Broughton, S.J., Piper, M.D.W., Ikeya, T., Bass, T.M., Jacobson, J., Driege, Y., 2005. Longer lifespan, altered metabolism, and stress resistance in *Drosophila* from ablation of cells making insulin-like ligands. Proc. Natl. Acad. Sci. USA 102, 3105–3110.

Brueckner, M., 2001. Cilia propel the embryo in the right direction. Am. J. Med. Genet. 101, 339–344.

Bugnard, E., Zaal, K.J., Ralston, E., 2005. Reorganization of microtubule nucleation during muscle differentiation. Cell Motil. Cytoskeleton 60, 1–13.

Burke, R.D., 1983. Neural control of metamorphosis in *Dendraster excentricus*. Biol. Bull. 164, 176–188.

Burns, A.J., Thapar, N., Barlow, A.J., 2008. Development of the neural crest-derived intrinsic innervation of the human lung. Am. J. Respir. Cell Mol. Biol. 38, 269–275.

Butler, J., Cosmos, E., Brierley, J., 1982. Differentiation of muscle fiber types in aneurogenic brachial muscle of the chick embryo. J. Exp. Zool. 224, 65–80.

Cabej, N.R., 2012. Epigenetic Principles of Evolution. Elsevier Inc., London and Waltham MA, p. 182.

Carmeliet, P., Ferreira, V., Breier, G., Pollefeyt, S., Kieckens, L., Gertsenstein, M., et al., 1996. Abnormal vessel development and lethality in embryos lacking a single VEGF allele. Nature 380, 435–439.

Chan, A.P., Kloc, M., Larabell, C.A., LeGros, M., Etkin, L.D., 2007. The maternally localized RNA *Fatvg* is required for cortical rotation and germ cell formation. Mech. Dev. 124, 350–363.

Chin, E.R., Olson, E.N., Richardson, J.A., Yang, Q., Humphries, C., Shelton, J.M., et al., 1998. A calcineurin-dependent transcriptional pathway controls skeletal muscle fiber type. Genes Dev. 12, 2499–2509.

Colas, J.-F., Launay, J.-M., Maroteaux, L., 1999. Maternal and zygotic control of serotonin biosynthesis are both necessary for *Drosophila* germband extension. Mech. Dev. 87, 67–76.

Colombani, J., Bianchini, L., Layalle, S., Pondeville, E., Dauphin-Villemant, C., Antoniewski, C., et al., 2005. Antagonistic actions of ecdysone and insulins determine final size in *Drosophila*. Science 310, 667–670.

Consoulas, C., Levine, R.C., 1997. Accumulation and proliferation of adult leg muscle precursors in *Manduca* are dependent on innervation. J. Neurobiol. 32, 531–553.

Cossu, G., Borello, U., 1999. Wnt signaling and the activation of myogenesis in mammals. Embo J. 18, 6867–6872.

Currie, D.A., Bate, M., 1991. The development of adult abdominal muscles in *Drosophila*: myoblasts express twist and are associated with nerves. Development 113, 91–102.

Dammann, C.E., Nielsen, H.C., 1998. Regulation of the epidermal growth factor receptor in fetal rat lung fibroblasts during late gestation. Endocrinology 139, 1671–1677.

Dammann, C.E.L., Ramadurai, S.M., McCants, D.D., Pham, L.D., Nielsen, H.C., 2000. Androgen regulation of signaling pathways in late fetal mouse lung development. Endocrinology 141, 2923–2929.

Dantzer, B., Swanson, E.M., 2012. Mediation of vertebrate life histories via insulin-like growth factor-1. Biol. Rev. Camb. Philos. Soc. 87, 414–429.

Davies, W., Wilkinson, L.S., 2006. It is not all hormones: alternative explanations for sexual differentiation of the brain. Brain Res. 1126, 36–45.

Delaune, E., Lemaire, P., Kodjabachian, L., 2005. Neural induction in *Xenopus* requires early FGF signalling in addition to BMP inhibition. Development 132, 299–310.

Delbro, D.S., 2012. Do neuro-humoral signaling molecules participate in colorectal carcinogenesis/cancer progression? Neurogastroenterol. Motil. 24, 96–99.

Delfini, M.C., Duprez, D., 2004. Ectopic Myf5 or MyoD prevents the neuronal differentiation program in addition to inducing skeletal muscle differentiation, in the chick neural tube. Development 131, 713–723.

Delgadoa, G., Anderson, K.D., Cardoso, W.V., 2000. The molecular basis of lung morphogenesis. Mech. Dev. 92, 55–81.

Denver, R.J., 1997. Environmental stress as a developmental cue: corticotrophin-releasing hormone is a proximate mediator of adaptive phenotypic plasticity in amphibian metamorphosis. Horm. Behav. 31, 169–179.

Denver, R.J., 1999. Evolution of the corticotropin-releasing hormone signaling system and its role in stress-induced phenotypic plasticity. Ann. N.Y. Acad. Sci. 897, 46–53.

Denver, R.J., 2002. Acceleration of *Ambystoma tigrinum* metamorphosis by corticotropin-releasing hormone. J. Exp. Zool. 293, 94–98.

De Renzis, S., Elemento, O., Tavazoie, S., Wieschaus, E.F., 2007. Unmasking activation of the zygotic genome using chromosomal deletions in the *Drosophila* embryo. PLoS Biol. 5 (5), e117.

De Robertis, E.M., 2006. Spemann's organizer and self-regulation in amphibian embryos. Nat. Rev. Mol. Cell Biol. 7, 296–302.

De Robertis, E.M., Wessely, O., 2004. The molecular nature of Spemann's organizer. In: Grunz, H. (Ed.), The Vertebrate Organizer Springer-Verlag, Berlin-Heidelberg, pp. 55–71.

Dewing, P., Shi, T., Horvath, S., Vilain, E., 2003. Sexually dimorphic gene expression in mouse brain precedes gonadal differentiation. Mol. Brain. Res. 118, 89–90.

DiFiore, J.W., Wilson, J.M., 1994. Lung development. Semin. Pediatr. Surg. 3, 221–232.

Dowling, J.E., 2007. The Great Brain Debate—Nature or Nurture? Princeton University Press, Princeton, NJ, pp. 7–8.

Draizen, T.A., Ewer, J., Robinow, S., 1999. Genetic and hormonal regulation of the death of peptidergic neurons in the *Drosophila* central nervous system. J. Neurobiol. 38, 455–465.

Durbec, P.L., Larsson-Blomberg, L.B., Schuchardt, A., Costantini, F., Pachnis, V., 1996. Common origin and developmental dependence on c-ret of subsets of enteric and sympathetic neuroblasts. Development 122, 349–358.

Elinson, R.P., Kezmoh, L., 2010. Molecular haeckel. Dev. Dyn. 239, 1905–1918.

Elofssson, U., Winberg, S., Francis, R.C. Nilsson, G.E., 1997a. Neuronal correlates of socially induced sex change in teleost fish. In: Experimental Biology Online. The Society for Experimental Biology Annual Meeting, University of Kent at Canterbury, 7–11 April 1997, A8.20 (abstracts).

Elofsson, U., Winberg, S., Francis, R.C., 1997b. Sex differences in number of preoptic GnRH-immunoreactive neurons in a protandrously hermaphroditic fish, the anemone fish *Amphiprion melanopus*. J. Comp. Physiol. A 181, 484–492.

Emanuel, R.L., Torday, J.S., Mu, Q., Asokananthan, N., Sikorski, K.A., Sunday, M.E., 1999. Bombesin-like peptides and receptors in normal fetal baboon lung: roles in lung growth and maturation. Am. J. Physiol. 277, L1003–L1017.

Erlebacher, A., Filvaroff, E.H., Gitelman, S.E., Derynck, R., 1995. Toward a molecular understanding of skeletal development. Cell 80, 371–378.

Esser, S., Lampugnani, M.G., Corada, M., Dejana, E., Risau, W., 1998. Vascular endothelial growth factor induces VE-cadherin tyrosine phosphorylation in endothelial cells. J. Cell Sci. 11, 1853–1865.

Fernandes, J.J., Keshishian, H., 2005. Motoneurons regulate myoblast proliferation and patterning in Drosophila. Dev. Biol. 277, 493–505.

Fielenbach, N., Antebi, A., 2008. *C. elegans* dauer formation and the molecular basis of plasticity. Genes Dev. 22, 2149–2165.
Fishman, M.C., Chien, K.R., 1997. Fashioning the vertebrate heart: earliest embryonic decisions. Development 124, 2099–2117.
Freem, L.J., Delalande, J.M., Campbell, A.M., Thapar, N., Burns, A.J., 2012. Lack of organ specific commitment of vagal NCC derivatives as shown by back-transplantation of GFP chicken tissues. Int. J. Dev. Biol. 56, 245–254.
Fukuda, K., Kikuchi, Y., 2005. Endoderm development in vertebrates: fate mapping, induction and regional specification. Develop. Growth Differ. 47, 343–355.
Fukumoto, T., Kema, I.P., Levin, M., 2007. Serotonin signaling is a very early step in patterning of the left-right axis in chick and frog embryos. Curr. Biol. 15, 794–803.
Galis, F., Metz, J.A.J., 2001. Testing the vulnerability of the phylotypic stage: on modularity and evolutionary conservation. J. Exp. Zool. 291, 195–204.
Gans, C., Northcutt, R.G., 1983. Neural crest and the origin of vertebrates: a new head. Science 220, 268–274.
Garelli, A., Gontijo, A.M., Miguela, V., Caparros, E., Dominguez, M., 2012. Imaginal discs secrete insulin-like peptide 8 to mediate plasticity of growth and maturation. Science 336, 579–582.
Geraerts, W.P.M., 1976. Control of growth by the neurosecretory hormone of the light green cells in the freshwater snail *Lymnaea stagnalis*. Gen. Comp. Endocrinol. 29, 61–71.
Giet, R., Prigent, C., 2003. Contrôle de la détermination cellulaire par les centrosomes. Méd. Sci. 19, 656–658.
Gilbert, S.F., 2000. Developmental Biology, sixth ed. Sinauer Associates, Sunderland, MA, Gamete Fusion and the Prevention of Polyspermy. Available from: <http://www.ncbi.nlm.nih.gov/books/NBK10033/>.
Giraldez, A.J., 2010. microRNAs, the cell's nepenthe: clearing the past during the maternal-to-zygotic transition and cellular reprogramming. Curr. Opin. Genet. Dev. 20, 369–375.
Giraldez, A.J., Mishima, Y., Rihel, J., Grocock, R.J., Van Dongen, S., Inoue, K., et al., 2006. Zebrafish MiR-430 promotes deadenylation and clearance of maternal mRNAs. Science 312, 75–79.
Gore, A.V., Maegawa, S., Cheong, A., Gilligan, P.C., Weinberg, E.S., Sampath, K., 2005. The zebrafish dorsal axis is apparent at the four-cell stage. Nature 438, 1030–1035.
Grobstein, C., 1956. Trans-filter induction of tubules in mouse metanephrogenic mesenchyme. Exp. Cell Res. 10, 424–440.
Gros, J., Serralbo, O., Marcelle, C., 2009. WNT11 acts as a directional cue to organize the elongation of early muscle fibres. Nature 457, 589–593.
Guidobaldi, H.A., Teves, M.E., Uñates, D.R., Anastasía, A., Giojalas, L.C., 2008. Progesterone from the cumulus cells is the sperm chemoattractant secreted by the rabbit oocyte cumulus complex. PLoS One 3, e3040.
Hadfield, M.G., Meleshkevitch, E.A., Boudko, D.Y., 2000. The apical sensory organ of a gastropod veliger is a receptor for settlement cues. Biol. Bull. 198, 67–76.
Haeckel, E., 1866. Allgemeine Entwickelungsgeschichte der Organismen. Verlag von Georg Reimer, Berlin, p. 300: Die Ontogenesis ist die kurze und schnelle Recapitulation der Phylogenesis.
Hall, B.K., 2009. The Neural Crest and Neural Crest Cells in Vertebrate Development and Evolution. Springer, New York, p.118.
Hamatani, T., Carter, M.G., Sharov, A.A., Ko, M.S.H., 2004. Dynamics of global gene expression changes during mouse preimplantation development. Dev. Cell 6, 117–131.

Harland, R., Gerhart, J., 1997. Formation and function of Spemann's organizer. Annu. Rev. Cell Dev. Biol. 13, 611–667.

Hegstrom, C.D., Riddiford, L.M., Truman, J.W., 1998. Steroid and neuronal regulation of ecdysone receptor expression during metamorphosis of muscle in the moth, *Manduca sexta*. J. Neurosci. 18, 1786–1794.

Hernandes, L., da Silva Pereira, L.C.M., Alvares, E.P., 2003. Goblet cell number in the ileum of rats denervated during suckling and weaning. Biocell 27, 347–351.

Herzlinger, D., Qiao, J., Cohen, D., Ramakrishna, N., Brown, A.M., 1994. Induction of kidney epithelial morphogenesis by cells expressing Wnt-1. Dev. Biol. 166, 815–818.

Hogan, K.A., Ambler, C.A., Chapman, D.L., Bautch, V.L., 2004. The neural tube patterns vessels developmentally using the VEGF signaling pathway. Development 131, 1503–1513.

Holland, P.W.H., Graham, A., 1995. Evolution of regional identity in the vertabrate nervous system. Perspect. Dev. Neurobiol. 3, 17–27.

Hsieh, J.-C., 2004. Specificity of Wnt-receptor interactions. Front. Biosci. 9, 1333–1338.

Hutson, M.R., Kirby, M.L., 2007. Model systems for the study of heart development and disease. Cardiac neural crest and conotruncal malformations. Semin. Cell Dev. Biol. 18, 101–110.

Irie, N., Kuratani, S., 2011. Comparative transcriptome analysis reveals vertebrate phylotypic period during organogenesis. Nat. Commun. 2, 248.

Irie, N., Sehara-Fujisawa, A., 2007. The vertebrate phylotypic stage and an early bilaterian-related stage in mouse embryogenesis defined by genomic information. BMC Biol. 5 (1).

James, J.M., Gewolb, C., Bautch, V.L., 2009. Neurovascular development uses VEGF-A signaling to regulate blood vessel ingression into the neural tube. Development 136, 833–841.

James, J.M., Mukouyama, Y-s, 2011. Neuronal action on the developing blood vessel pattern. Semin. Cell Dev. Biol. 22, 1019–1027.

Jheon, A.H., Schneider, R.A., 2009. The cells that fill the bill: neural crest and the evolution of craniofacial development. J. Dent. Res. 88, 12–21.

Jiang, C., Lamblin, A.-F.J., Steller, H., Thummel, C.S., 2000. A steroid-triggered transcriptional hierarchy controls salivary gland cell death during *Drosophila* metamorphosis. Mol. Cell 5, 445–455.

Jiménez-Trejo, F., Olivos-Cisneros, L., Mendoza-Torreblanca, J., Díaz-Cintra, S., Meléndez-Herrera, E., Báez-Saldaña, A., et al., 2011. Sensory neurons in the spinal cord of nominal female embryos in the marine turtle *Lepidochelys olivacea* respond to shifts in incubation temperature: implications for temperature dependent sex determination. Adv. Biosci. Biotechnol. 2, 1–7.

Jordan, T., Jiang, H., Li, H., DiMario, J.X., 2004. Inhibition of ryanodine receptor 1 in fast skeletal muscle fibers induces a fast-to-slow muscle fiber type transition. J. Cell Sci. 117, 6175–6183.

Kaltschmidt, J.A., Brand, A.H., 2002. Asymmetric cell division: microtubule dynamics and spindle asymmetry. J. Cell Sci. 115, 2257–2264.

Kane, D.A., Kimmel, C.B., 1993. The zebrafish midblastula transition. Development 119, 447–456.

Katanaev, V.L., 2010. The Wnt/Frizzled GPCR signaling pathway. Biochemistry (Mosc.) 75, 1428–1434.

Katsukura, Y., Ando, H., David, C.N., Grimmelikhuijzen, C.J., Sugiyama, T., 2004. Control of planula migration by LWamide and RFamide neuropeptides in *Hydractinia echinata*. J. Exp. Biol. 207, 1803–1810.

Kenyon, C.J., 2010. The genetics of ageing. Nature 464, 504–512.

Kenyon, C., Chang, J., Gensch, E., Rudner, A., Tabtiang, R., 1993. A *C. elegans* mutant that lives twice as long as wild type. Nature 366, 461–464.

Kim, Y.-J., Spalovská-Valachová, I., Cho, K.-H., Zitnanova, I., Park, Y., Adams, M.E., et al., 2004. Corazonin receptor signaling in ecdysis initiation. Proc. Natl. Acad. Sci. USA 101, 6704–6709.

Kimelman, D., 2006. Mesoderm induction: from caps to chips. Nat. Rev. Genet. 7, 360–372.

Kimelman, D., Bjornson, C., 2004. Vertebrate mesoderm induction. In: Stern, C.D. (Ed.), Gastrulation: From Cells to Embryo Cold Spring Harbor Laboratory Press, Cold Spring Harbor, NY, pp. 363–372.

Kirschner, M.W., Gerhart, J.C., 2005. *Plausibility of Life—Resolving Darwin's Dilemma.* Yale University Press, New Haven, CT.

Knight, R.D., Schilling, T.F., 2006. Cranial neural crest and development of the head skeleton. Adv. Exp. Med. Biol. 589, 120–133.

Knox, S.M., Lombaert, I.M.A., Reed, X., Vitale-Cross, L., Gutkind, J.S., Hoffman, M.P., 2010. Parasympathetic innervation maintains epithelial progenitor cells during salivary organogenesis. Science 329, 1645–1647.

Kofron, M., Demel, T., Xanthos, J., Lohr, J., Sun, B., Sive, H., et al., 1999. Mesoderm induction in *Xenopus* is a zygotic event regulated by maternal VegT via TGFbeta growth factor. Development 126, 5759–5770.

Kormish, J.D., Gaudet, J., McGhee, J.D., 2010. Development of the *C. elegans* digestive tract. Curr. Opin. Genet. Dev. 20, 346–354.

Kotani, T., Kawakami, K., 2008. *Misty somites*, a maternal effect gene identified by transposon-mediated insertional mutagenesis in zebrafish that is essential for the somite boundary maintenance. Dev. Biol. 316, 383–396.

Ku, M., Melton, D.A., 1993. Xwnt-11: a maternally expressed *Xenopus Wnt* gene. Development 119, 1161–1173.

Kulesa, P.M., Bailey, C.M., Kasemeier-Kulesa, J.C., McLennan, R., 2010. Cranial neural crest migration: new rules for an old road. Dev. Biol. 344, 543–554.

Kuntz, S., Chesnel, A., Flament, S., Chardard, D., 2004. Cerebral and gonadal aromatase expressions are differently affected during sex differentiation of *Pleurodeles waltl*. J. Mol. Endocrinol. 33, 717–727.

Kuroda, H., Wessely, O., Robertis, E.M., 2004. Neural induction in *Xenopus*: requirement for ectodermal and endomesodermal signals via chordin, noggin, beta-catenin, and cerberus. PLoS Biol. 2, E92.

Lambert, J.D., Nagy, L.M., 2002. Asymmetric inheritance of centrosomally localized mRNAs during embryonic cleavages. Nature 420, 682–686.

Latham, K.E., Garrels, J.I., Chang, C., Solter, D., 1991. Quantitative analysis of protein synthesis in mouse embryos. I. Extensive reprogramming at the one- and two-cell stages. Development 112, 921–932.

Lee, S.I., Lee, W.K., Shin, J.H., Han, B.K., Moon, S., Cho, S., et al., 2009. Sexually dimorphic gene expression in the chick brain. before gonadal differentiation. Poult. Sci. 88, 1003–1015.

Lepikhov, K., Zakhartchenko, V., Hao, R., Yang, F., Wrenzycki, C., Niemann, H., et al., 2008. Evidence for conserved DNA and histone H3 methylation reprogramming in mouse, bovine and rabbit zygotes. Epigenet. Chromatin 1, 8.

Lee, Y.-H., Saint-Jeannet, J.-P., 2011. Cardiac neural crest is dispensable for outflow tract septation in *Xenopus*. Development 138, 2025–2034.

Lerner, U.H., 2002. Neuropeptidergic regulation of bone resorption and bone formation. J. Musculoskelet. Neuronal Interact. 2, 440–447.

Levin, M., Johnson, R.L., Stern, C.D., Kuehn, M., Tabin, C.A., 1995. Molecular pathway determining left-right asymmetry in chick embryogenesis. Cell 82, 803–814.

Liu, M., 2011. The biology and dynamics of mammalian cortical granules. Reprod. Biol. Endocrinol. 9, 149. doi: 10.1186/1477-7827-9-149.

Lund, E., Liu, M., Hartley, R.S., Sheets, M.D., Dahlberg, J.E., 2009. Deadenylation of maternal mRNAs mediated by miR-427 in *Xenopus laevis* embryos. RNA15, 2351–2363.

Lundberg, P., Lie, A., Bjurholm, A., Lehenkari, P.P., Horton, M.A., Lerner, U.H., et al., 2000. Vasoactive intestinal peptide regulates osteoclast activity via specific binding sites on both osteoclasts and osteoblasts. Bone 27, 803–810.

Lundgren, O., Jodal, M., Jansson, M., Ryberg, A.T., Svensson, L., 2011. Intestinal epithelial stem/progenitor cells are controlled by mucosal afferent nerves. PLoS One 6 (2), e16295.

Luo, J., Becnel, J., Nichols, C.D., Nässel, D.R., 2012. Insulin-producing cells in the brain of adult *Drosophila* are regulated by the serotonin 5-HT_{1A} receptor. Cell. Mol. Life Sci. 69, 471–484.

Magnusson, C., Svensson, A., Christerson, U., Tågerud, S., 2005. Denervation-induced alterations in gene expression in mouse skeletal muscle. Eur. J. Neurosci. 21, 577–580.

Mammoto, T., Ingber, D.E., 2010. Mechanical control of tissue and organ development. Development 137, 1407–1420.

Martin, A.C., 2010. Pulsation and stabilization: contractile forces that underlie morphogenesis. Dev. Biol. 341, 114–125.

McGhee, J.D., Fukushige, T., Krause, M.W., Minnema, S.E., Goszczynski, B., Gaudet, J., et al., 2009. ELT-2 is the predominant transcription factor controlling differentiation and function of the *C. elegans* intestine, from embryo to adult. Dev. Biol. 327, 551–565.

McGrath, J., Brueckner, M., 2003. Cilia are at the heart of vertebrate left–right asymmetry. Curr. Opin. Genet. Dev. 13, 385–392.

Meehan, R.R., Dunican, D.S., Ruzov, A., Pennings, S., 2005. Epigenetic silencing in embryogenesis. Exp. Cell Res. 309, 241–249.

Mellon Jr., D., 1999. Muscle restructuring in crustaceans: myofiber death, transfiguration and rebirth. Am. Zool. 39, 527–540.

Meulemans, D., Bronner-Fraser, M., 2007. Insights from amphioxus into the evolution of vertebrate cartilage. PLoS One 2 (8), e787. doi: 10.1371/journal.pone.0000787.

Mian, I., Pierre-Louis, W.S., Dole, N., Gilberti, R.M., Dodge-Kafka, K., Tirnauer, J.S., 2012. LKB1 destabilizes microtubules in myoblasts and contributes to myoblast differentiation. PLoS One 7 (2), e31583. doi: 10.1371/journal.pone.0031583.

Miller, C.M., Newmark, P.A., 2012. An insulin-like peptide regulates size and adult stem cells in planarians. Int. J. Dev. Biol. 56, 75–82.

Miller, D., Brinkworth, M., Iles, D., 2010. Paternal DNA packaging in spermatozoa: more than the sum of its parts? DNA, histones, protamines and epigenetics. Reproduction 139, 287–301.

Milnes, M.R., Roberts, R.N., Guillette, L.J., 2002. Effects of incubation and estrogen exposure on aromatase activity in the brain and gonads of embryonic alligators. Environ. Health Perspect 110 (Suppl. 3), 393–396.

Mirth, C.K., Truman, J.W., Riddiford, L.M., 2009. The ecdysone receptor controls the post-critical weight switch to nutrition-independent differentiation in *Drosophila* wing imaginal discs. Development 136, 2345–2353.

Miyagawa-Tomita, S., Waldo, K., Tomita, H., Kirby, M.L., 1991. Temporospatial study of the migration and distribution of cardiac neural crest in quail-chick chimeras. Am. J. Anat. 192, 79–88.

Miyanaga, Y., Torregroza, I., Evans, T., 2002. A maternal Smad protein regulates early embryonic apoptosis in Xenopus laevis. Mol. Cell Biol. 22, 1317–1328.
Møllgård, K., Jespersen, A., Lutterodt, M.C., Yding Andersen, C., Høyer, P.E., Byskov, A.G., 2010. Human primordial germ cells migrate along nerve fibers and Schwann cells from the dorsal hind gut mesentery to the gonadal ridge. Mol. Hum. Reprod. 16, 621–631.
Moon, R.T., Kimelman, D., 1998. From cortical rotation to organizer gene expression: toward a molecular explanation of axis specification in *Xenopus*. BioEssays 20, 536–545.
Morgan, H.D., Santos, F., Green, K., Dean, W., Reik, W., 2005. Epigenetic reprogramming in mammals. Hum. Mol. Genet. 14 (Suppl. 1), R47–R58.
Muelemans, D., Bronner-Fraser, M., 2004. Gene-regulatory interactions in neural crest evolution and development. Dev. Cell 7, 291–299.
Mukoyama, Y-s., Shin, D., Britsch, S., Taniguchi, M., Anderson, D.J., 2002. Sensory nerves determine the pattern of arterial differentiation and blood vessel branching in the skin. Cell 109, 693–705.
Müller, W.A., 1996. Developmental Biology. Springer, New York, NY, p. 90.
Muñoz-Sanjuán, I., Brivanlou, A.H., 2002. Neural induction, the default model and embryonic stem cells. Nat. Rev. Neurosci. 3, 271–280.
Nelson, D.W., Padget, R.W., 2003. Insulin worms its way into the spotlight. Genes Dev. 17, 813–818.
Nishida, H., 2002. Patterning the marginal zone of early ascidian embryos: localized maternal mRNA and inductive interactions. BioEssays 24, 613–624.
Noden, D.M., Trainor, P.A., 2005. Relations and interactions between cranial mesoderm and neural crest populations. J. Anat. 207, 575–601.
O'Farrell, P.H., Stumpff, J., Su, T.T., 2004. Embryonic cleavage cycles: how is a mouse like a fly? Curr. Biol. 14, R35–R45.
Ohlsson, C., Mohan, S., Sjögren, K., Tivesten, Å., Isgaard, J., Isaksson, O., et al., 2009. The role of liver-derived insulin-like growth factor-I. Endocr. Rev. 30, 494–535.
Osorio, L., Teillet, M.-A., Palmeirim, I., Catala, M., 2009. Neural crest ontogeny during secondary neurulation: a gene expression pattern study in the chick embryo. Int. J. Dev. Biol. 53, 641–648.
Oswald, J., Engemann, S., Lane, N., Mayer, W., Olek, A., Fundele, R., et al., 2000. Active demethylation of the paternal genome in the mouse zygote. Curr. Biol. 10, 475–478.
Oteíza, P., Köppen, M., Concha, M.L., Heisenberg, C.-P., 2008. Origin and shaping of the laterality organ in zebrafish. Development 135, 2807–2813.
Pan, J., Yeger, H., Cutz, E., 2004. Innervation of Pulmonary Neuroendocrine Cells and Neuroepithelial Bodies in Developing Rabbit Lung. J. Histochem. Cytochem. 52, 379–389.
Pan, J., Copland, I., Post, M., Yeger, H., Cutz, E., 2006. Mechanical stretch-induced serotonin release from pulmonary neuroendocrine cells: implications for lung development. Am. J. Physiol. 290, L185–L193.
Phillips, M.T., Kirby, M.L., Forbes, G., 1987. Analysis of cranial, neural crest distribution in the developing heart using quail-chick chimeras. Circ. Res. 60, 27–30.
Pieau, C., Dorizzi, M., 2004. Oestrogens and temperature-dependent sex determination in reptiles: all is in the gonads. J. Endocrinol. 181, 367–377.
Piran, R., Halperin, E., Guttmann-Raviv, N., Keinan, E., Reshef, R., 2009. Algorithm of myogenic differentiation in higher-order organisms. Development 136, 3831–3840.
Pires, A., Coon, S.L., Hadfield., M.G., 1997. Catecholamines and dihydroxyphenylalanine in metamorphosing larvae of the nudibranch *Phestilla sibogae* Bergh (Gastropod: Opisthobranchia). J. Comp. Physiol. A 181, 187–194.

Pletcher, S.D., 2009. The modulation of lifespan by perceptual systems. Ann. N.Y. Acad. Sci. 1170, 693–697.

Pletcher, S., 2012. Mechanisms of olfactory modulation of aging in *Drosophila*. 1 June 2007–31 May 2012 NIH RePORTER. Available from: <http://www.experts.umich.edu/grantDetail.asp?t=ep1&id=9240210&o_id=76&)>.

Pombero, A., Martinez, S., 2009. Telencephalic morphogenesis during the process of neurulation: an experimental study using quail-chick chimeras. J. Comp. Neurol. 512, 784–797.

Ralt, D., Goldenberg, M., Fetterolf, P., Thompson, D., Dor, J., Mashiach, S., et al., 1991. Sperm attraction of follicular factor(s) correlates with human egg fertilizability. Proc. Natl. Acad. Sci. USA 88, 2840–2844.

Read, A.T., Govind, C.K., 1997. Claw transformation and regeneration in adult snapping shrimp: test of the inhibition hypothesis for maintaining bilateral asymmetry. Biol. Bull. 193, 401–409.

Rebollo, E., Sampaio, P., Januschke, J., Llamazares, S., Varmark, H., González, C., 2007. Functionally unequal centrosomes drive spindle orientation in asymmetrically dividing *Drosophila* neural stem cells. Dev. Cell 12, 467–474.

Reinsch, S., Gönczy, P., 1998. Mechanisms of nuclear positioning. J. Cell Sci. 111, 2283–2295.

Reversade, B., Kuroda, H., Lee, H., Mays, A., De Robertis, E.M., 2005. Depletion of Bmp2, Bmp4, Bmp7 and Spemann organizer signals induces massive brain formation in *Xenopus* embryos. Development 132, 3381–3392.

Richardson, M.K., Keuck, G., 2002. Haeckel's ABC of evolution and development. Biol. Rev. 77, 495–528.

Rogers, C., Moody, S.A., Casey, E., 2009. Neural induction and factors that stabilize a neural fate. Birth Defects Res. C. Embryo Today 87, 249–262.

Rogers, C.D., Ferzli, G.S., Casey, E.S., 2011. The response of early neural genes to FGF signaling or inhibition of BMP indicate the absence of a conserved neural induction module. BMC Dev. Biol. 11, 74.

Roy, S., VijayRaghavan, K., 1998. Patterning muscles using organizers: larval muscle templates and adult myoblasts actively interact to pattern the dorsal longitudinal flight muscles of *Drosophila*. J. Cell Biol. 141, 1135–1145.

Ruiz-Jones, G.J., Hadfield, M.G., 2011. Loss of sensory elements in the apical sensory organ during metamorphosis in the nudibranch *Phestilla sibogae*. Biol. Bull. 220, 39–46.

Sabel, J.L., d'Alençon, C., O'Brien, E.K., Van Otterloo, E., Lutz, K., Cuykendall, T.N., et al., 2009. Maternal interferon regulatory factor 6 is required for the differentiation of primary superficial epithelia in *Danio* and *Xenopus* embryos. Dev. Biol. 325, 249–262.

Salame-Mendez, A., Herrera-Munoz, J., Moreno-Mendoza, N., Merchant-Larios, H., 1998. Response of diencephalon but not the gonad to female-promoting temperature with elevated estradiol levels in the sea turtle *Lepidochelys olivacea*. J. Exp. Zool. 280, 304–313.

Sample, S.J., Collins, R.J., Wilson, A.P., Racette, M.A., Behan, M., Markel, M.D., et al., 2010. Effects of ulna loading in male rats during functional adaptation. J. Bone Miner. Res. 25, 2016–2028.

Santos, F., Hendrich, B., Reik, W., Dean, W., 2002. Dynamic reprogramming of DNA methylation in the early mouse embryo. Dev. Biol. 241, 172–182.

Santos, G.C., Zucoloto, S., Garcia, S.B., 2000. Endocrine cells in the denervated intestine. Int. J. Exp. Pathol. 81, 265–270.

Sariola, H., Ekblom, P., Henke-Fahle, S., 1989. Embryonic neurons as in vitro inducers of differentiation of nephrogenic mesenchyme. Dev. Biol. 132, 271–281.

Sassoon, D., Kelley, D.B., 1986. The sexually dimorphic larynx of *Xenopus laevis*: development and androgen regulation. Am. J. Anat. 177, 457–472.
Sato, M., Yost, H.J., 2003. Cardiac neural crest contributes to cardiomyogenesis in zebrafish. Dev. Biol. 257, 127–139.
Sauer, H.W., Schwalm, F.E., Moritz, B.B., 1996. Morphogenesis, Seidel's legacy for developmental biology and challenge for molecular embryologists. Int. J. Dev. Biol. 40, 77–82.
Schatten, G., 1994. The centrosome and its mode of inheritance: the reduction of the centrosome during gametogenesis and its restoration during fertilization. Dev. Biol. 165, 299–335.
Schmich, J., Trepel, S., Leitz, T., 1998. The role of GLWamides in metamorphosis of *Hydractinia echinata*. Dev. Genes Evol. 208, 267–273.
Schmidt, M., Tanaka, M., Münsterberg, A., 2000. Expression of β-catenin in the developing chick myotome is regulated by myogenic signals. Development 128, 4105–4113.
Schneider, V.A., Mercola, M., 2001. Wnt antagonism initiates cardiogenesis in *Xenopus laevis*. Genes Dev. 15, 304–315.
Schneider, R.A., Helms, J.A., 2003. The cellular and molecular origins of beak morphology. Science 299, 565–568.
Scholz, B., Kultima, K., Mattsson, A., Axelsson, J., Brunström, B., Halldin, K., et al., 2006. Sex-dependent gene expression in early brain development of chicken embryos. BMC Neurosci. 7, 12.
Schweickert, A., Weber, T., Beyer, T., Vick, P., Bogusch, S., Feistel, K., et al., 2006. Cilia-driven leftward flow determines laterality in *Xenopus*. Curr. Biol. 17, 60–66.
Schwerson, T., Serbedzija, G., Bronner-Frazer, M., 1993. Regulative capacity of the cranial neural tube to form neural crest. Development 118, 1049–1061.
Seipp, S., Schmich, J., Leitz, T., 2001. Apoptosis: a death-inducing mechanism tightly linked with morphogenesis in *Hydractinia echinata* (Cnidaria, Hydrozoa). Development 128, 4891–4898.
Shook, D.R., Majer, C., Keller, R., 2004. Pattern and morphogenesis of presumptive superficial mesoderm in two closely related species, *Xenopus laevis* and *Xenopus tropicalis*. Dev. Biol. 270, 163–185.
Sinner, D., Kirilenko, P., Rankin, S., Wei, E., Howard, L., Kofron, M., et al., 2006. Global analysis of the transcriptional network controlling *Xenopus* endoderm formation. Development 133, 1955–1966.
Slusarski, D.C., Pelegri, F., 2007. Calcium signaling in vertebrate embryonic patterning and morphogenesis. Dev. Biol. 307, 1–13.
Smit, A.B., Van Kesteren, R.E., Li, K.W., Van Minnen, J., Spijker, S., Van Heerikhuizen, H., et al., 1998. Towards understanding the role of insulin in the brain: lessons from insulin-related signaling systems in the invertebrate brain. Progr. Neurobiol. 54, 35–54.
Snider, P., Olaopa, M., Firulli, A.B., Conway, S.J., 2007. Cardiovascular development and the colonizing cardiac neural crest lineage. Sci.World J. 7, 1090–1113.
Soriano, S., Kang, D.E., Fu, M., Pestell, R., Chevallier, N., Zheng, H., et al., 2001. Presenilin 1 negatively regulates β-catenin/T cell factor/lymphoid enhancer factor-1 signaling independently of β-amyloid precursor protein and notch processing. J. Cell Biol. 152, 785–794.
Spemann, H., Mangold, H., 1924. Über induktion von Embryonanlagen durch implantation artfremder organisatoren. Wilhelm Roux' Archiv für Entwicklung Mechanik der Organismen 100, 599–638.
Stern, C.D., 2004. Neural induction. In: Stern, C.D. (Ed.), Gastrulation: From Cells to Embryo Cold Spring Harbor Laboratory Press, Cold Spring Harbor, NY, p. 427.

Stern, C.D., 2005. Neural induction: old problem, new findings, yet more questions. Development 132, 2007–2021.

Streit, A., Berliner, A.J., Papanayotou, C., Sirulnik, A., Stern, C.D., 2000. Initiation of neural induction by FGF signalling before gastrulation. Nature 406, 74–78.

Stricker, S.A., 1999. Comparative biology of calcium signaling during fertilization and egg activation in animals. Dev. Biol. 211, 157–176.

Sugioka, K., Mizumoto, K., Sawa, H., 2011. Wnt regulates spindle asymmetry to generate asymmetric nuclear β-catenin in *C. elegans*. Cell 146, 942–954.

Tadros, W., Lipshitz, H.D., 2009. The maternal-to-zygotic transition: a play in two acts. Development 136, 3033–3042.

Tang, F., Kaneda, M., O'Carroll, D., Hajkova, P., Barton, S.C., Sun, Y.A., et al., 2007. Maternal microRNAs are essential for mouse zygotic development. Genes Dev. 21, 644–648.

Tao, Q., Yokota, C., Puck, H., Kofron, M., Birsoy, B., Yan, D., et al., 2005. Maternal Wnt11 activates the canonical Wnt signaling pathway required for axis formation in *Xenopus* embryos. Cell 120, 857–871.

Thatcher, E.J., Flynt, A.S., Li, N., Patton, J.R., Patton, J.G., 2007. MiRNA expression analysis during normal zebrafish development and following inhibition of the Hedgehog and Notch signaling pathways. Dev. Dyn. 236, 2172–2180.

Tierney, A.J., 1996. Evolutionary implications of neural circuit structure and function. Behav. Processes 35, 173–182.

Tobias, M.L., Marin, M.L., Kelley, D.B., 1993. The roles of sex innervation, and androgen in laryngeal muscle of *Xenopus laevis*. J. Neurosci. 13, 324–333.

Torday, J.S., Rehan, V.K., 2002. Stretch-stimulated surfactant synthesis is coordinated by the paracrine actions of PTHrP and leptin. Am. J. Physiol. Lung Cell. Mol. Physiol. 283, L130–L135.

Toth, M.L., Melentijevic, I., Shah, L., Bhatia, A., Lu, K., Talwar, A., et al., 2012. Neurite sprouting and synapse deterioration in the aging *Caenorhabditis elegans* nervous system. J. Neurosci. 32, 8778–8790.

Trainor, P.A., Sobieszczuk, D., Wilkinson, D., Krumlauf, R., 2002. Signalling between the hindbrain and paraxial tissues dictates neural crest migration pathways. Development 129, 433–442.

Truman, J.W., 2005. Hormonal control of insect ecdysis: endocrine cascades for coordinating behavior with physiology. Vitam. Horm. 73, 1–30.

Tzahor, E., Lassar, A.B., 2001. Wnt signals from the neural tube block ectopic cardiogenesis. Genes Dev. 15, 255–260.

Unguez, G.A., Zakon, H.H., 2002. Skeletal muscle transformation into electric organ in *S. macrurus* depends on innervation. J. Neurobiol. 53, 391–402.

Vainio, S.J., Itäranta, P.V., Peräsaari, J.P., Uusitalo, M.S., 1999. Wnts as kidney tubule inducing factors. Int. J. Dev. Biol. 43, 419–423.

van Mier, P., Joosten, H.W., van Rheden, R., ten Donkelaar, H.J., 1986. The development of serotonergic raphespinal projections in *Xenopus laevis*. Int. J. Dev. Neurosci. 4, 465–475.

Vespúcio, M.V.O., Turatti, A., Modiano, P., de Oliveira, E.C., Chicote, S.R.M., Pinto, A.M.P., et al., 2008. Intrinsic denervation of the colon is associated with a decrease of some colonic preneoplastic markers in rats treated with a chemical carcinogen. Braz. J. Med. Biol. Res. 41, 311–317.

Vieira, C., Pombero, A., García-Lopez, R., Gimeno, L., Echevarria, D., Martínez, S., 2010. Molecular mechanisms controlling brain development: an overview of neuroepithelial secondary organizers. Int. J. Dev. Biol. 54, 7–20.

Vogel, O., 1983. Pattern formation by interaction of three cytoplasmic factors in the egg of the leafhopper *Euscelis plebejus*. Dev. Biol. 99, 166–171.

von Baer, K.E., 1828.Über Entwickelungsgeschichte der Thiere. Beobachtung und Reflexion I. Bei den Gebrüdern Bornträgern, Königsberg, p. 199: daß der Embryo höherer Thiere die bleibenden Formen der niederen Thiere durchlaufe.

Vonica, A., Gumbiner, B.M., 2007. The *Xenopus* nieuwkoop center and Spemann-Mangold organizer share molecular components and a requirement for maternal Wnt activity. Dev. Biol. 312, 90–102.

Voytik, S.L., Przyborski, M., Badylak, S.F., Konieczny, S.F., 1993. Differential expression of muscle regulatory factor genes in normal and denervated adult rat hindlimb muscles. Dev. Dyn. 198, 214–224.

Walentek, P., Beyer, T., Thumberger, T., Schweickert, A., Blum, M., 2012. *ATP4a* is required for Wnt-dependent *Foxj*1 expression and leftward flow in *Xenopus* left-right development. Cell Rep. 1, 516–527.

Walkiewicz, M.A., Stern, M., 2009. Increased insulin/insulin growth factor signaling advances the onset of metamorphosis in drosophila. PLoS ONE 4 (4), e5072.

Walser, C.B., Lipshitz, H.D., 2011. Transcript clearance during the maternal-to-zygotic transition. Curr. Opin. Genet. Dev. 21, 431–443.

Wang, G., Scott, S.A., 2008. Retinoic signaling is involved in governing the waiting period for axons in chick hindlimb. Dev. Biol. 321, 216–226.

Warburton, D., Schwarz, M., Tefft, D., Flores-Delgadoa, G., Anderson, K.D., Cardoso, W.V., 2000. The molecular basis of lung morphogenesis. Mech. Dev. 92, 55–81.

Weaver, C., Kimelman, D., 2004. Move it or lose it: axis specification in *Xenopus*. Development 131, 3491–3499.

Weinstein, D.C., Hemmati-Brivanlou, A., 1999. Neural induction. Annu. Rev. Cell Dev. Biol. 15, 411–433.

Whitaker, M., 2008. Calcium signalling in early embryos. Philos. Trans. R. Soc. Lond. B Biol. Sci. 363, 1401–1418.

White, J.A., Heasman, J., 2008. Maternal control of pattern formation in *Xenopus laevis*. J. Exp. Zool. B Mol. Dev. Evol. 310, 73–84.

Wieschaus, E., 1996. Embryonic transcription and the control of developmental pathways. Genetics 142, 5–10.

Willert, K., Nusse, R., 1998. Beta-catenin: a key mediator of Wnt signaling. Curr. Opin. Genet. Dev. 8, 95–102.

Willingham, E., Baldwin, R., Skipper, J.K., Crews, D., 2000. Aromatase activity during embryogenesis in the brain and adrenal-kidney-gonad of the red-eared slider turtle, a species with temperature-dependent sex determination. Gen. Comp. Endocrinol. 119, 202–207.

Wilson, S.I., Edlund, T., 2001. Neural induction: toward a unifying mechanism. Nat. Neurosci. 4, 1161–1168.

Wolpert, L., Beddington, R., Lawrence, P., Jessell, T.M., 1998. Principles of Development. Current Biology Ltd. and Oxford University Press, Oxford-NewYork-Tokyo, p. 424.

Wright, S., 1941. The physiology of the gene. Physiol. Rev. 21, 487–527. According to Sapp, J., 1987. Beyond the Gene—Cytoplasmic Inheritance and the Struggle for Authority in Genetics. Oxford University Press, New York, p. 99.

Wu, Q., Sample, S.J., Baker, T.A., Thomas, C.F., Behan, M., Muir, P., 2009. Mechanical loading of a long bone induces plasticity in sensory input to the central nervous system. Neurosci. Lett. 463, 254–257.

Xu, S., Cheng, F., Liang, J., Wu, W., Zhang, J., 2012. Maternal xNorrin, a canonical Wnt signaling agonist and TGF-β antagonist, controls early neuroectoderm specification in *Xenopus*. PLoS Biol. 10 (3), e1001286.

Yamashita, Y.M., Fuller, M.T., 2008. Asymmetric centrosome behavior and the mechanisms of stem cell division. J. Cell Biol. 180, 261–266.

Yan, B., Neilson, K.M., Moody, S.A., 2009. *foxD5* plays a critical upstream role in regulating neural ectodermal fate and the onset of neural differentiation. Dev. Biol. 329, 80–95.

Yeh, L.C., Zavala, M.C., Lee, J.C., 2002. Osteogenic protein-1 and interleukin-6 with its soluble receptor synergistically stimulate rat osteoblastic cell differentiation. J. Cell Physiol. 190, 322–331.

Yılmaz, A., Tepeli, C., Garip, M., Çağlayan, T., 2011. The effects of incubation temperature on the sex of Japanese quail chicks. Poult. Sci. 90, 2402–2406.

Young, R.E., Pearce, J., Govind, C.K., 1994. Establishment and maintenance of claw bilateral asymmetry in snapping shrimps. J. Exp. Zool. 269, 319–326.

Zhang, J., Houston, D.W., King, M.L., Payne, C., Wylie, C., Heasman, J., 1998. The role of maternal VegT in establishing the primary germ layers in *Xenopus* embryos. Cell 94, 515–524.

Zhang, C., Basta, T., Fawcett, S.R., Klymkowsky, M.W., 2005. *SOX7* is an immediate-early target of VegT and regulates Nodal-related gene expression in *Xenopus*. Dev. Biol. 278, 526–541.

4 Living and Adapting to Its Own Habitat

Adaptation—Surviving in a Changing Environment

A living system, from birth to death, struggles against the thermodynamic forces of degradation to maintain its delicately balanced structure and functions. Even under ideal and constant conditions in the environment, living will ultimately depend on an organism's ability to avoid the "unavoidable" structural and functional decay by continually restoring degraded structures. Obviously, the task is much harder under natural conditions, which involve continual and often drastic changes in the environment. The Darwinian "struggle for life" or self-preservation is inherent to living systems but a related property of "self preservation" can be traced back to chemical systems in equilibrium, as described by Le Chatelier's principle that "in a system in equilibrium any change in the equilibrium displaces the equilibrium in the direction that counteracts the imposed change by reaching a new equilibrium." Both biological and chemical systems strive to counteract the imposed change, but an essential qualitative distinction is that biological systems *restore* the former equilibrium, rather than establish a new equilibrium as chemical systems do. The evolutionary progress of living systems, compared to the Le Chatelier's systems, relates to the evolution of a control system within the organism that involves a mechanism to restore the system's previous state.

The evolution of living organisms is characterized by an incessant adaptation and response to the changing environment. "Adapt or die" has been a basic law of the living world from the beginning. Successful adaptations and failure to adapt determined the evolutionary fate of the extant species populating the earth as well as the fate of long extinct species. Failure to adapt to changes in the environment has caused massive extinction events of species, higher taxa, and whole phyla. The fossil record shows us as much with certainty. If survival is the reason to exist, adaptation is the means to achieve and sustain survival in a continually changing world.

Adaptation increases the probability of survival and success of an organism in its environment. Despite the inherent tendency to adapt in response to the continually changing environment, living organisms are never perfectly adapted to it. Besides, inherent constraints exist in the capability of living organisms to adapt their phenotype to their environment.

The actual state of adaptation for an organism is rarely, if at all, the perfect adaptation to its particular environment. A cat would like to be able to dive for a fresh fish snack but, being a skilled four-legged hunter preying on mice and finless as she is, she has to be content with a dish of preserved fish. A wolf has sharper night sight

Building the Most Complex Structure on Earth. DOI: http://dx.doi.org/10.1016/B978-0-12-401667-5.00004-3

and a dog has a sharper sense of smell. Evolutionarily, it is not so unfortunate that that they did not share each other's best senses, because what they have are the best adaptations for their respective environments.

Because living organisms in nature do not always succeed in evolving adaptations that help their survival, biologists have concluded that there are constraints on the adaptive capability of living organisms.

From the early stages of individual development, cell differentiation is determined by the epigenetic information parentally provided to gametes in the form of cytoplasmic factors, centrioles, cytoskeletal elements, and so on. This epigenetic information is, to a certain degree, conservative and is the cause of the developmental constraints on early development. For reasons not yet clearly understood, there is a stage in metazoan development, the phylotypic stage (see p. 144, Chapter 3, section "The Phylotypic Stage"), that has been refractory to any change over the last 500 million years since the Cambrian period. The post-phylotypic stage that follows it is highly prone to adaptive morphological changes, as it may be reasonably concluded from the breathtaking diversity of extant and extinct of forms in the animal kingdom.

Constraints to adaptive morphological changes, however, also exist in the post-phylotypic stage. These constraints have determined the limits of morphological innovations and "imperfections" that are often observed in the metazoan adaptations.

In the course of evolution, living organisms came up with fascinating phenotypic adaptations in the form of discrete changes in the physiology, morphology, behavior, and life history that help them to sustain, maintain, or increase their fitness in the changing environment. Phenotypic adaptation in its broadest sense includes a number of biological phenomena such as the reaction norm and all the forms of the intragenerational and transgenerational plasticity that will be discussed in separate sections at the end of this chapter.

Most physiological adaptations take place at the molecular level and, conventionally, they are considered related to the occurrence of beneficial spontaneous mutations that lead to the evolution of new genes or improve the products of existing genes. But such favorable mutations are very rare and, as Dobzhansky famously remarked, the chances of finding a favorable mutation among spontaneous mutations are like finding a needle in a haystack. And it is well known why present products of genes, proteins, and RNAs are selected positively for improved biological functions during hundreds of millions to ~3 billion years; hence, any spontaneous change has a much higher probability of deteriorating rather than improving their functions. As rare as they are, beneficial mutations have only a slim chance of preservation in populations of dioecious species.

Living organisms have an inherent ability to adapt to various degrees of change in the environment, known as phenotypic plasticity, which implies phenotypic changes that arise in response to the changed conditions in the environment. These changes unfold as a series of continuous quantitative changes (norm of reaction) or as discrete qualitative changes. The latter implies the development of new traits, and hence will be termed as developmental plasticity, which will be dealt with later in this chapter. Both are epigenetically rather than genetically determined; what changes under the influence of changed conditions are not genes but the chemistry of the organism and the patterns of expression of genes.

Living organisms have evolved various morphological, physiological, behavioral, and life history adaptations.

Behaviors can be one or a number of sequential actions that the animal performs in response to an external or internal stimulus and in some cases even in the absence of a stimulus. Animal behaviors may be grouped into two main classes: innate and learned behaviors. Innate behaviors are inherited traits that organisms display without previous experience (see page 228, "Epigenetics of Behavior and Social Attachment in Animals").

Learned behaviors are the most plastic and rapidly changing of the phenotypic adaptations. Behavioral plasticity is especially the characteristic of metazoans, but plants and unicellulars also display surprising behavioral plasticity and flexibility in response to environmental conditions or stimuli. Behavior is often influenced by learning, especially in metazoans. The mechanics of animal behavior are similar in both innate and learned behaviors because both rely on neural circuits in the central nervous system (CNS), which generate and control the behavior (Baker et al., 2001; Gould, 1982). This is also why the innate behaviors are often influenced by learning and learned behaviors are influenced by innate behaviors. A clear example of learning influencing an innate behavior is the imprinting in goslings that, in the absence of the mother goose, learn to follow a moving object as they would follow their mother.

In an early evolutionary stage, behavioral adaptation relied heavily on the evolution of innate standard reactions or instincts rather than learning. This is the case with unicellulars and plants. However, there is evidence that unicellulars can learn from previous experience and conditions and can modify their behavior in response to later exposure to the same conditions. Learning implies remembering previous experiences.

The differentiation of the neuron and the evolution of the nervous net/CNS in metazoans during the Cambrian explosion provided metazoans with unprecedented behavioral plasticity, based on learning, which determined to a greater extent their evolutionary success. In mammals and especially in humans, the share of the learning in the overall behavioral inventory continued to increase in the course of vertebrate evolution. While it is known that this experience-induced memory in metazoans is stored in the nervous net/CNS, there is only speculation on the location or possible carriers of that memory in unicellulars and plants (see page 22, "The Control System in Unicellulars" and page 42, "The Plant Bauplan and Control System").

This section explores intragenerational adaptations and only tangentially consider the accompanying stress conditions, which arise in response to external/internal stimuli due to sudden changes in the environment. Intragenerational adaptations in metazoans are discrete changes in morphology, physiology, behavior, life history, or psychology arising in response to environmental stimuli, but they are not inherited in the offspring in the absence of the stimuli that first induced their emergence. The inherited adaptations will be discussed in the last chapter of this book.

First, to clarify the concept of stimulus as it will be used for the purpose of this discussion, it is an environmental action or condition that provokes a specific change in the phenotype. Living organisms are under the constant action of numerous environmental abiotic and biotic factors, but not all environmental factors serve as stimuli in metazoans. Moreover, a factor that serves as a stimulus for one species does not necessarily serve as a stimulus for another. Internal factors also may act as stimuli.

This last empirical observation is essential for understanding the nature of the stimulus and the organism–environment relationship. It implies that an environmental factor *per se* is not a stimulus. Whether an environmental factor or condition will be taken as a stimulus or not depends primarily on how it is perceived in the brain more than it does on the nature of the environmental factor itself. The decision to categorize it as a stimulus or not is made in the CNS. This is why the same agent is counted as a stimulus by organisms of one species but not by others. The organism responds adaptively when it perceives the agent as a stimulus.

How does the organism decide whether to take an agent/action as a stimulus or "ignore" it as "noise"? The answer requires an understanding of the pathway that links the external/internal stimulus to its perception in the CNS. In metazoans, certain cell types and organs (sense organs) are specialized to detect changes in the environment. The sensory organs and sensory cells are bombarded with streaming external factors and the organism must separate the environmental noise from the stimuli. Sensory cells have a background action potential that changes under the influence of external factors. They conduct this environmental data to respective centers in the CNS, where the degree of the environmental change is compared to species-specific set points. When the intensity of the change exceeds (or drops below) the limits of the set point, the organism responds with a phenotypic change. This is a quantitative criterion of the categorization of environmental factors and actions as stimuli. There are other factors in the environment, whose perceptions are taken as stimuli, regardless of their quantity (predators, light, darkness, etc.) and automatically elicit a phenotypic response.

Let us look at a few of the many examples. When the environmental temperature elevates to levels that challenge normal body temperature, a number of phenotypic changes occur in the organism to counteract and maintain normal temperature by decreasing thermogenesis. This implies that the organism has detected an increase in the body temperature above the species-specific set point, which in vertebrates, for example, resides in the hypothalamus. The increase in body temperature is assessed in the neurons of the preoptic area (POA). Besides the POA, brain stem centers and the spinal cord are also involved in the neuroendocrine mechanisms that slow down thermogenesis by inhibiting thyroid hormones and by increasing heat loss through vasodilatation, sweating, and panting. Finding places to cool off is often another beneficial adaptive behavioral change.

Morphological adaptations are characteristics of many lower metazoans (rotifers, snails, etc.). Upon detecting visually or olfactorily, the presence of predators in the environment may rapidly generate morphological changes that make it difficult to be eaten by predators. Curiously, the freshwater snail, *Helisoma trivolis*, specifies different changes in its morphology, morphometry, behavior, and life history, when it perceives each of its predators, the water bugs and crayfishes (Hoverman et al., 2005). When the pregnant viviparous lizard *Pseudemoia pagenstecheri* olfactorily perceives the presence of predator snakes, it gives birth to bigger bodied offspring with longer tails, thus improving their survival chance (Shine and Downes, 1999). Several polar predator and prey mammals (and even birds) in the winter months turn their color into white to adapt to the snow-covered landscape. For interested readers, there is ample evidence in the scientific literature on the intragenerational and TDP.

A behavioral adaptation is observed in some species of the crustacean *Daphnia* (Cladocera) that respond to the rising water temperature by swimming up (upward) and sinking deeper when the water temperature drops (Gerritsen, 1982).

Finally, certain tadpoles display an adaptive change in their life history in response to environmental stimuli. Tadpoles of the western toad (*Bufo boreas*) respond adaptively upon detecting chemical cues of their predators, aquatic insect backswimmers of *Notonecta* species in their environment; they accelerate their growth and development to reach the metamorphosis stage earlier, thus reducing their exposure to predators (Chivers et al., 1999).

Most stimuli act as stressors. Although adaptive responses to particular stimuli are generally specific, they are commonly associated with a general and nonspecific stress response, which evolved as an adaptive mechanism. The stress response is a systemic general response regardless of the stressors that trigger it. It involves physiological, psychological, life history, and rarely even morphological changes. The CNS is the central venue of the stress response. Acute stress activates the locus coeruleus and the brain stem, which stimulate the sympathetic-adrenomedullary system, with resulting "fight or flight" behavior intended for the survival of the individual. It is associated with a number of physiological and behavioral changes.

Homeostatic mechanisms are strained and homeostasis may be disturbed, under the action of stressors. The disturbed homeostasis is assessed in the brain and the adaptive stress response starts in the higher brain centers. In vertebrates, the stress condition may also be triggered before stressors affect homeostasis. It may be triggered as a direct response to the perception of threatening situations, while in higher vertebrates, especially in humans, stress conditions may be caused not only by real, but even by imagined stressors.

The stress response may be immediate or delayed. The immediate stress response results from the perception of a threat or a threatening condition.

Neural Control of Gene Expression

Since 1930s, the genecentric view has dominated theoretical biology. The enthusiasm generated by the landmark discoveries in molecular genetics silenced dissenting biologists. However, the expected crowning of the genecentric age, the era of genome sequencing failed to impress; essentially the same genes and the same "genetic toolkit" were used across metazoan taxa, from primitive worms to insects and vertebrates.

Now, almost by consensus, biologists believe that it is the spatiotemporal patterns of gene expression, rather than the number or nature of genes, that determine differences among species and higher taxa. Differential expression of genes is also responsible for differentiation of distinct types of cells that have the same genes. Biologists need to know why differentiated cells of various types express different genes and have different temporal and spatial patterns of gene expression.

Expression of nonhousekeeping genes, which the cell expresses for the organism's benefit, rather than its own, is induced by signals of extracellular origin. So, where is the ultimate source of signals or instructions for their expression?

The most plausible answer from the genecentric view, that the instructions come from the cell's genome, is rejected simply because these instructions come from outside rather than from within the cell. The remaining alternative answer from that viewpoint is that some cells have information or send instructions to other cells on what genes they have to express for the benefit of the organism. But there are a number of problems that render this hypothesis implausible and/or nonverifiable.

First, the hypothetical cell(s) that would know what other cells must do are not known as of yet.

Second, the hypothetical cells need to know the cells that must express which genes.

Third, the hypothetical cells must know what chemical to secrete to induce expression of relevant genes in other cells.

Fourth, given the inducer is a nonhousekeeping gene, the instructing cell's decision to secrete the inducer has to be implemented by secretion of another specific inducer by a third type of cell. Obviously, cells of this third type will need another type of cell to secrete the relevant inducer. Thus, the second alternative genecentric hypothesis leads us to a logical dead end of endless tautological rounds of reasoning. Each of the above arguments rejects the alternative gene-centered explanation of the mechanism of the expression of nonhousekeeping genes if the three seemingly insurmountable difficulties did not exist.

There are two ways to approach the problem of the origin of instructions for the expression of nonhousekeeping genes in metazoans. First, starting from the reliable premise that nonhousekeeping genes are induced by extracellular signals, it is important to trace the signal cascade back upstream to the ultimate source of information for expression of nonhousekeeping genes and visualize the flow of information for their expression.

The majority of the known inducers consist of hormones, growth factors, secreted proteins, neurotransmitters, neuropeptides/neurohormones, and so on. It has been known for at least more than half a century that secretion of hormones of the target endocrine glands (i.e., thyroid, adrenals, pineal, pancreas, and gonads) is stimulated by pituitary hormones and that for this reason they are known as "stimulating hormones." Each of these pituitary hormones is induced by a specific neurohormone secreted by the hypothalamus, one of the evolutionarily earliest structures of the vertebrate brain. In turn, hypothalamic nuclei seem to secrete their neurohormones by processing information from other brain areas. The secretion of growth factors and proteins is mainly induced by hormones of the target endocrine glands and pituitary as well as neurohormones. This reliably leads to the idea that the ultimate source of information for expression of nonhousekeeping genes in vertebrates may be the CNS (Cabej, 2005, 2008).

Signals from the CNS are at the origin of processes of development, growth, and metamorphosis in invertebrates and the neural net in lower vertebrates, such as cnidarians (Cabej, 2008, pp. 139–201, 2010, 2012, pp. 149–204).

It is well known that differences in expression of nonhousekeeping genes are responsible for the differentiation of all types of cells starting from a single cell, zygote, or egg, during individual development. Bearing in mind that the ultimate source of information for expression of these genes is the CNS (cytoplasmic factors during the early development up to the phylotypic stage), a hierarchical scheme of the control of gene expression with the CNS as the controller is easy to envisage. The advent of the neural regulation of gene expression during the Cambrian explosion was crucial for the evolution of metazoans.

Now, in order to corroborate the above conclusion on the neural origin of information for gene expression, we will approach the problem in the reverse direction. Metazoans respond to environmental stimuli by changing their morphology, physiology, behavior, or life history. These adaptive changes involve changes in the normal patterns of gene expression in relevant tissues and organs. Now we will try to follow the pathway from the environmental stimuli to gene expression.

The next section shows how the organism sends instructions to genes in a language that genes could understand.

Making Environmental Signals Intelligible to Genes

In everyday parlance, environmental stimuli is said to induce or even regulate the expression of specific genes. This notion is so engraved in the biological conceptual system that it comes as a revelation when, upon closer scrutiny, it turns out that no external stimuli that could directly induce the expression of any gene are known. No biotic or abiotic agent *per se* (the viruses' case is irrelevant) is capable of inducing expression of any gene. Yet a clear correlation and causal relationship between particular environmental agents and expression of particular genes is empirically demonstrated to exist in more than an adequate number of cases. How is it possible for environmental agents to "exert" this long-range action on gene expression, while unable to induce expression of genes in direct contact with them? What is that which translates the language of external agents into instructions that are intelligible to genes?

A closer examination of this biological "*alibi*" will reveal to us where the environmental agents acquire "meaning" and how this meaning is translated into a language that genes could understand. This will shed light into the black box of how genetically incomprehensible environmental messages are translated into intelligible instructions for gene expression.

For simplicity, this mechanism is illustrated using the case of higher metazoans, whose integrated control system is known best. In a typical case, in homeothermic animals, such as birds and mammals, the environmental temperature and other environmental agents such as the photic signals, food, presence of predators, preys, and other social stimuli are detected by the animal's senses of vision, hearing, smell, and so on. All of the information related to these external stimuli is received by sensory neurons, which convert them into specific electrical signals and transmit them to respective centers of the brain for further processing. The processing of the electrical input in these brain centers interprets the stimulus (i.e., its perception and categorization). The electrical spikes train from the receptive sensory neurons is processed, perceived, categorized, and interpreted in neural circuits to provide it a meaning (i.e., to find out what the stimulus means for the organism). The final stage of neural processing is determining the appropriate adaptive response to the anticipated effects of the stimulus itself or of the environmental effects it might presage. The final product of the neural processing is an output that, in the form of a chemical signal, triggers an "adaptive" signal cascade, ultimately leading to the expression of one or a number of specific genes related to the specific adaptation. The signal cascade usually induces expression of genes not in all cells of the organism but in particular cells, tissues, or organs. This implies the presence of another mechanism for preventing the expression of genes in other cells (see p. 200, Restricting Gene Expression to Relevant Cells Alone: Binary Neural Control of Gene Expression).

The processing is a neural codification or a translation of environmental stimuli into messages that are intelligible to genes. The output of the processing comes in the form of a chemical that triggers a specific signal cascade.

The above explanation raises questions with possible teleological implications. Indeed, even the idea that the unconscious brain can find these adaptive responses seems to unavoidably imply a purpose in neural processing and, syllogistically, the action of a teleological factor. A discussion of this subject is outside the scope of this book.

It is of paramount importance to animal evolution that the nervous system by using specific signal cascades or gene regulatory networks (GRNs) can adaptively and "at will" relate naturally unrelated external agents to virtually any gene. This manipulative expression of genes, as opposed to the classical modes of gene expression, represents a landmark in the history of life that contributed essentially to the unprecedented explosive evolution in the animal world during and after the Cambrian explosion.

This mechanism enables the brain to choose from the available off-the-shelf signal cascades and GRNs leading to phenotypic results that adapt the organism to the environmental stimulus. The neural processing is very costly, but its evolution proves that the advantages it offered outweighed its price tag.

The neural manipulation of gene expression in animals is illustrated here with a few examples.

Astatotilapia burtoni is a cichlid fish in Lake Tanganyika, East Africa. Its males appear in dominant or subordinate forms, according to their social status. Dominant males are brightly colored, whereas subordinates' colors are dull and they flee when dominant males attack. However, upon perceiving a social opportunity, the subordinate individual switches to the dominant males' reproductive and behavioral physiology. This is associated with rapid changes in expression of genes in the three levels of the hypothalamic–pituitary–gonadal (HPG) axis, which is typical for the dominant males. Because of this social ascent, it expresses the immediate-early gene (IEG) *Egr-1*, also known as *zenk*, in neurons of the POA and in GnRH1 neurons of the hypothalamus, in 20–30 min. At the same time, the fish increases secretion of the neurohormones kisspeptin and GnRH1. The changes in gene expression as a result of the change in social status extend farther down the axis to the pituitary and gonads. In the pituitary, the social ascent increases the expression of gonadotropin-releasing hormone (GnRH) receptors to the level of dominant males and induces the expression of follicle-stimulating hormone (FSH) and luteinizing hormone (LH) genes. In ovaries and testes, social information leads to the expression of FSH and androgen receptors, stimulating reproductive activity. Changes in gene expression and behavior are followed by morphological changes (Maruska and Fernald, 2011).

In another example, the adipose tissue in mammals is composed of white adipose tissue (WAT) and brown adipose tissue (BAT). The latter consists of brown adipose cells characterized by numerous lipid droplets and a greater number of mitochondria. BAT is the main source of nonshivering heat in hibernating animals and in human newborns. In the latter, BAT represents about 25% of a newborn's weight. Activation of BAT prevents hypothermia in neonates. Production of the uncoupling protein-1 (UCP-1), also known as thermogenin, triggers mitochondrial thermogenesis (Figure 4.1) in adipocytes. BAT thermogenesis is activated by a long signal cascade starting in the CNS, after information on the cold thermosensory nerve endings reaches the dorsal root ganglion (DRG) (Morrison, 2004; Bartness and Song, 2005).

The simplified neural cascade for nonshivering thermogenesis looks as follows:

> *Cold temperature → cold thermoreceptors → DRG → DH (dorsal horn) → GABAergic neurons in a still unidentified region of the brain → neurons of the POA (hypothalamic preoptic area) → neurons of the DMH (hypothalamic dorsomedial nucleus) → sympathetic motor neurons in rostral raphe pallidus nucleus of the brain stem → autonomic motor neurons of the IML (intermediolateral cell column) → epinephrin released by postganglionic sympathetic nerves → a signal transduction pathway in brown adipocytes → expression of UCP-1 gene in brown adipocytes → adipocyte mitochondrial thermogenesis.*

Both approaches, from the expression of nonhousekeeping genes up and from the environmental stimuli down, show that the epigenetic information used to express these genes is generated by processing stimuli in neural circuits. The odyssey of the encoded stimulus from the sensory receptors through various centers of the spinal cord and brain is necessary for computing the adaptive response. The output of the stimulus processing in the neural circuit is a specific chemical, which activates a specific signal cascade and ultimately expresses one or a number of genes, rather than roaming aimlessly. Evolution is parsimonious. Natural selection does not allow the wasteful expenditure of energy that is not outweighed by a benefit.

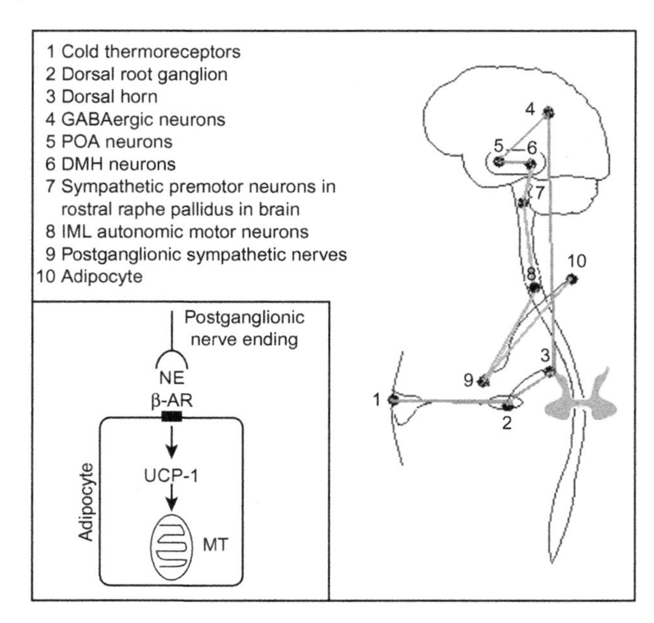

Figure 4.1 The simplified neural circuit for processing the cold temperature signal transmitted by cold thermosensory nerve endings. Numbers in the figure show the sequential steps of processing the stimulus in the nervous system. The output of the processing is NE released by the postganglionic sympathetic nerves on individual adipocytes. By binding β-AR on the adipocyte membrane, NE triggers a signal transduction pathway that leads to induction of UCP-1, which in turn induces mitochondrial thermogenesis. *Abbreviations*: POA, hypothalamic preoptic area; DMH, hypothalamic dorsomedial nucleus; GABA, gamma-aminobutyric acid; IML, intermediolateral nucleus; NE, norepinephrine; β-AR, beta-adrenergic receptor; UCP-1, uncoupling protein-1 gene(s); MT, mitochondrion.
Source: From Cabej (2012).

Restricting Gene Expression to Relevant Cells Alone: Binary Neural Control of Gene Expression

As mentioned earlier, different cell types express different genes and have different temporal patterns of gene expression. Since most extracellular signals that induce expression of nonhousekeeping genes circulate with body fluids, a mechanism must exist that prevents the expression of these genes in cells throughout the animal body and restricts their expression to certain cell types or tissues.

But how to tame the inducers that circulate freely throughout the body to selectively act on relevant cells only?

Two decades ago, it was discovered that the nervous system induces gene expression through a novel mechanism, by inducing/suppressing expression of specific receptors, the mediators of the inductive action of extracellular signals. The mechanism involves local nerves that induce expression of specific receptors in the region they innervate. This remarkable mechanism of adjacent neural control of gene expression complements the global neural control of gene expression performed by the CNS, primarily via the hypothalamic–pituitary target endocrine gland axes.

This is a binary neural control of gene expression and is the only mechanism of spatial restriction of gene expression known as of now (Cabej, 2008, 2012, pp. 243–247), which is of paramount importance for the life and evolution of metazoans.

Experimental evidence on the adjacent neural control of gene expression is still modest for the simple reason that it was not on the researchers' agendas (rarely can you expect to find something you are not looking for). But the evidence is certainly adequate to demonstrate that the mechanism exists and evolution could not afford to waste a mechanism of such paramount importance for development.

The binary neural control of gene expression is observed in both invertebrates, suggesting that it may have evolved more than 500 million years ago, during the Cambrian explosion.

Let us look at the mechanism as it unfolds in a few selected examples.

In the moth *Manduca sexta* (tobacco hornworm), the dorsal external oblique 1 (DEO1) muscle consists of five muscle fibers. During metamorphosis, not all but one of these muscle fibers are lost. The surviving muscle fiber develops into the adult DEO1. In 1996, Hegstrom et al. (1998) observed that the only difference related to the fate of the surviving muscle fiber from the four others was that the axonal arbor of the motoneuron that innervates the larval DEO1 retracts from all the larval fibers except from the one that eventually turns into the adult DEO1. Investigators explained why the action of the circulating hormone ecdysone (ECD) is restricted to only one of the five fibers of the DEO1 muscle: it is the only fiber that expresses the ECD isoform receptor EcR-B1:

> *Innervation regulates the choice of EcR isoforms expressed in growing muscle. This choice may then determine the nature of the response of the muscle to changing steroid titers.*
>
> *Hegstrom et al. (1998)*

Binding of circulating ECD to the receptor isoform EcR-B1 expressed by myoblasts of the muscle fiber stimulates their proliferation and survival to become the adult DEO1. Myoblasts of the four other muscle fibers, which lost contact with the axonal arbor of the motoneuron, express another isoform of the ECD receptor, EcRA, which stimulates their programmed death and elimination. Thus, local innervation determines that the circulating ECD performs its proliferating action on one muscle fiber/myoblast but not on others. It is demonstrated, in both *in vivo* and *in vitro* experiments, that local nerves perform their muscle-forming action by releasing diffusible chemicals or by transmitting electrical signals (Curie and Bate, 1991).

In 1999, Nijhout marveled on the subject:

> *It is as if tissues somehow "know" when the hormonal signal will come and become receptive to it only at that time.*
>
> *Nijhout (1999)*

Experiments by Hegstrom et al. have shown that the innervation "knows" the time when the muscle fiber will express the ECD receptor to become receptive to the hormone.

FSH is the pituitary hormone induced by the hypothalamic neurohormone GnRH. Although FSH via blood and other body fluids circulates throughout the body, its action is restricted to a circumscribed group of granulosa cells in the ovarian follicle only. The FSH action is restricted in only granulosa cells, because it is there that the local sympathetic innervation releases the neurotransmitters norepinephrine (NE) and vasoactive intestinal peptide (VIP), which induce expression of the FSH receptor (Mayerhof et al., 1997). Then, granulosa cells transform androgens into estradiol that stimulates the development of the ovarian follicle.

Besides the above mechanisms of inducing the expression of receptor molecules, local innervation is involved in other ways of gene expression. This is experimentally demonstrated not only directly but also indirectly in experiments of denervation where it is observed that denervated tissues cannot express genes that they normally do. A typical example is the development of the laryngeal muscle in the African clawed frog, *Xenopus laevis*. In juvenile individuals of both sexes, the muscle is similar, but after metamorphosis the muscle in males takes its male-characteristic form, as a result of both the global action of androgens and the male-specific local innervation:

> *[I]nnervation and androgen effects on fiber size are independent and additive. Fiber number is regulated by the interaction of nerve and hormone.*
>
> *Tobias et al. (1993)*

The Source of Information for Selecting Sites of Histone Modification

Chromosomes are transient structures that form during cell division apparently to facilitate equal distribution of genes to daughter cells. Chromosomes are transcriptionally inert because of their highly compact state. Chromatin organization of the genome is required for gene expression. Switching from chromosomal to chromatin structures after cell division, and the reverse, results primarily from the epigenetic changes in the nucleosomal histone molecules (Figure 4.2).

Chromatin structure is less compact and resembles loose threads of DNA and histone proteins. Elementary structural units of chromatin are nucleosomes (Figure 4.3). Each nucleosome contains a DNA segment wound around an octamer consisting of one pair of four core histone molecules, H2A, H2B, H3, and H4, creating the image of a bead on a string. N-terminal tails of these histones may be acetylated, methylated, phosphorylated, and so on. On the outside, the nucleosome contains another histone molecule, H1 (LaVoie, 2005).

Chromatin appears in two main forms, heterochromatin and euchromatin. In the first form, the DNA/genes are tightly packed as a result of methylation and

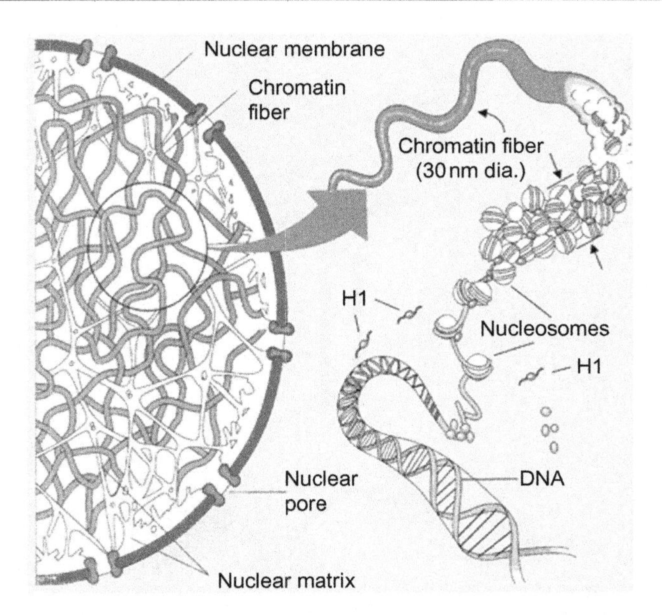

Figure 4.2 Modification of the histone components of nucleosomes helps regulate DNA accessibility by promoting folding or unfolding of chromatin fibers and by recruiting chromatin remodeling complexes and other factors to specific genomic loci.
Source: From Mizzen (2012).

deacetylation of histones by histone methyltransferases (HMTs) and deacetylases (HDACs), respectively. In euchromatin, genes are looser or "open," hence accessible to transcription factors, as a result of the action of histone acetylases (HATs), coactivator complexes, and RNA polymerase. Reciprocal transition from heterochromatin to euchromatin is presented in Figure 4.3.

Epigenetic changes in histones occur in strictly determined sites of the DNA molecule, "where the organism needs them," rather than in random places, clearly suggesting that some form of information is used to produce these epigenetic marks. The question is: are histones authentic generators of the information, or are they the conveyors of epigenetic information for gene expression coming from any upstream sources?

Now, there is adequate empirical evidence to answer the above question.

The source of information for histone acetylation, for example, is determined conveniently in the case of nuclear receptors (TRs) that mediate the transcriptional action of steroid, retinoid, thyroid, and other lipophilic endocrine hormones as well as vitamins, dietary lipids, and so on (Sonoda et al., 2008).

Usually the hormone forms a complex with its receptor and recruits one or more coregulators (coactivators or corepressors). In the case of the hormone estradiol, E2, the estradiol receptor complex (E2-ER) migrates to the nucleus where it enables acetylation and phosphorylation of itself and of histones, recruits its coactivators,

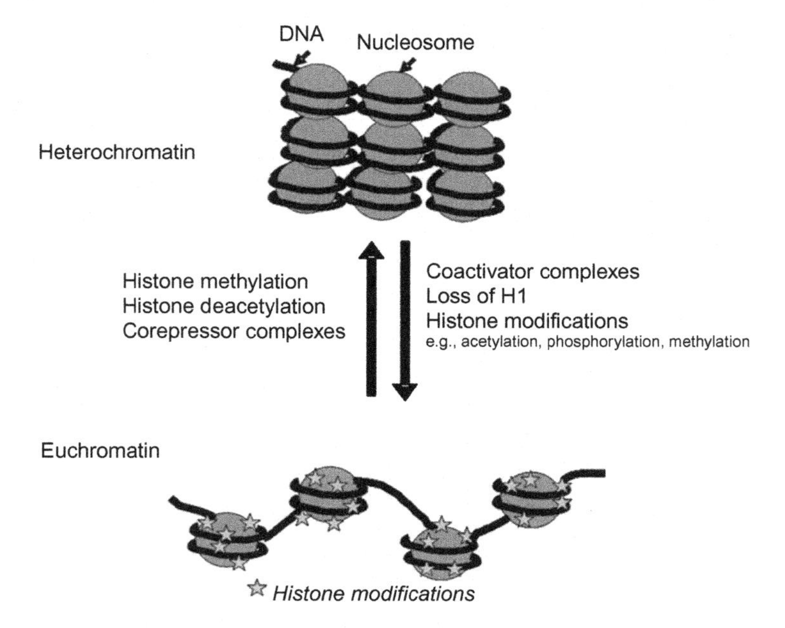

Figure 4.3 Schematic representation of transition euchromatin–heterochromatin.
Source: From Adcock et al. (2006).

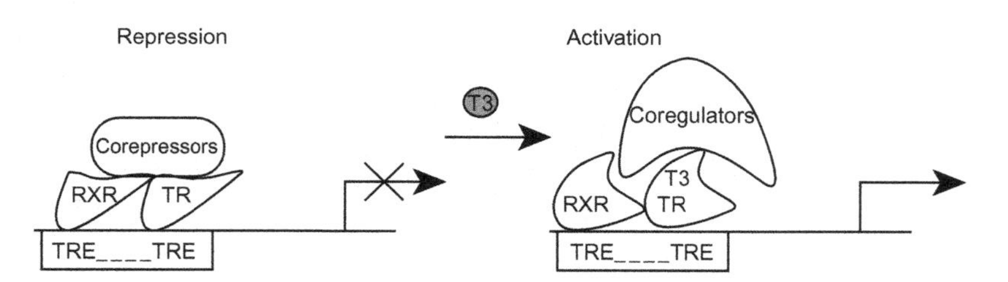

Figure 4.4 Mechanism of TR transcriptional control. *Abbreviations*: TR, thyroid hormone; T3, triiodothyronine; RXR, retinoid X receptor; TRE, thyroid response element.
Source: From Moore and Guy (2005).

dimerizes, and binds to the ER response element, thus inducing transcription of estrogen target genes (Kim et al., 2001; Ruh et al., 1999).

Upon entering the cell nucleus, the thyroid hormone (TH), especially in its triiodothyronine (T3) form, binds a TR isoform, changes its conformation, detaches corepressors, and recruits coactivators (Figure 4.4). The TR-RXR dimer binds to the T3 response elements (TREs) in the promoters of TR-regulated genes and recruits coactivators (enzymes of acetylation, deacetylation, phosphorylation, etc.), making possible the expression of TH-regulated genes. In the absence of the TH, its nuclear receptor (TR) forms a heterodimer with the retinoic X receptor (RXR) and in this

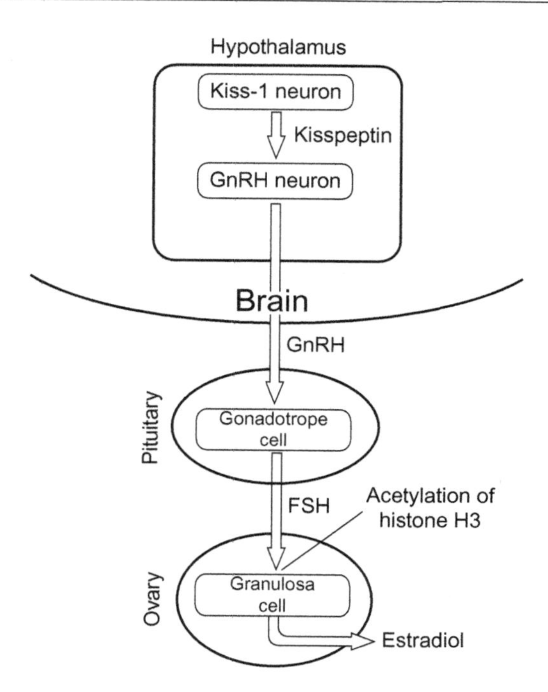

Figure 4.5 Neural control of the signal cascade that leads to acetylation of histone H3 and, consequently, induction of FSH-responsive genes and the synthesis of estradiol by granulosa cells. Note that the epigenetic information necessary for H3 acetylation flows from hypothalamic neurons, via the FSH-secreting pituitary cells, to granulosa cells of the ovarian follicle, where it induces FSH-responsive genes and their differentiation and proliferation. *Abbreviations*: FSH, follicle-stimulating hormone; GnRH, gonadotropin-releasing hormone.

form binds to TREs thus preventing the transcription of the TH-regulated genes (Moore and Guy, 2005).

Folliculogenesis is regulated by the pituitary FSH. The hormone binds its specific receptor in the cell membrane of the granulosa cells, which surround the oocyte. There it activates the protein kinase A (PKA) transduction pathway that phosphorylates and acetylates histone H3. The resulting chromatin remodeling enables transcription of FSH-responsive genes in granulosa cells, inducing their proliferation and differentiation (LaVoie, 2005; Salvador et al., 2001). Tracing back the signal cascade that induces granulosa cell proliferation, differentiation, and estradiol secretion leads us to a neural source of the information (Figure 4.5).

Injection of agonists of receptors for the neurotransmitters dopamine (DA), muscarinic acetylcholine (mACh), and glutamate (GLU) caused chromatin remodeling by inducing a specific modification (phosphorylation) of histone H3 in specific hippocampal neurons. It was concluded that "phosphorylation of histone H3 is coupled to increased neuronal activity and is directly linked to IEG transcription". (Crosio et al., 2003).

From Where Does the Epigenetic Information for DNA Methylation Come?

DNA methylation, the addition of a methyl (CH_3) group to a cytosine or adenine ring in the DNA molecule, is responsible for the gene imprinting and inactivation of the X chromosome. It is also involved in cell differentiation during individual development and has a pivotal role in establishing gene expression patterns for each cell type. Methylation is conserved during the cell division so that the epigenetic marks serve daughter cells to "remember," maintain, and transmit in future generations the parental cell type. DNA methylation is reversible under the influence of chromatin structure, but it also may induce the methylation of histones (Meaney and Szyf, 2005a).

DNA must be methylated at strictly determined sites, "where it is needed," to perform its function in gene imprinting, otherwise, reproduction and development would be impossible. This suggests that information is used for methylating the right base out of billions of bases of the DNA molecule. Obviously, this information is different from the information for protein biosynthesis; hence, it comes from sources that are external to DNA itself.

Let us begin with an example of a phenomenon where the flow of information for DNA methylation may be followed upstream to what seems to be the generator of that information in response to a social stimulus. The example shows how an early life epigenetic reprogramming, without any change in genetic information, may switch the offspring between two alternative behaviors.

According to the level of licking and grooming (LG) of the puppies during the first postnatal week, rat mothers display high- or low-LG behavior and they transmit this trait to the offspring. In general, low-LG rat mothers and their offspring are more fearful in response to stress compared to high-LG mothers. Rat puppies of high-LG mothers that experience the affectionate maternal care during the first week of neonatal life, as adults, tend to show the same high-LG behavior to their own offspring. When during the first week of life, the puppies of high-LG mothers are fostered to low-LG/maltreating mothers; as adults, they display to their own offspring the low-LG/maltreating behavior of the foster mother rather than the high-LG behavior of the biological mother (Meaney and Szyf, 2005a,b). The epigenetically determined behavior overrides the genetic behavior of the biological mother:

> *[I]ndividual differences in patterns of gene expression and behavior can be directly linked to maternal care over the first week of life.*
>
> *Meaney and Szyf (2005b)*

At the cellular level, relevant changes in the cross-fostered offspring consist of modification of the structure and function of the CA1 neuronal network of the hippocampus, indicating that the maternal care modulates the development and function of the hippocampus in rats (Champagne et al., 2008).

At the molecular level, the offspring of both high- and low-LG mothers show no differences at birth in the stress level and have the same methylation status in the exon 1_7 of the glucocorticoid receptor (GR) gene. The differences in puppies of high-LG mothers from those of low-LG mothers appear between the first and sixth day and include increased synthesis of the neurotransmitter serotonin and GR expression, particularly in the hippocampus area. This coincides with a period of time in which differences in the maternal LG behavior are apparent, clearly suggesting that "the sensory input associated with maternal LG selectively alters 5-HT activity in specific brain regions" (Meaney and Szyf, 2005b).

Investigators relate this to the presence of an enzymatic machinery of methylation/demethylation in hippocampal neurons (Meaney and Szyf, 2005a). Figure 4.6 schematically depicts the structure of the epigenetic machinery responsible for maternal care and the flow of the epigenetic information along the machinery. It shows that the afferent information on maternal tactile stimuli induces the secretion of serotonin in the puppy's hippocampal neurons, which is followed by the demethylation of the nerve-growth-factor-inducible factor-A (NGFI-A), also known as zenk or egr-1, and of the exon 1_7 promoter of the *GR* gene. This enables binding of NGFI-A to the exon promoter, thus inducing GR expression (Figure 4.6). The increased synthesis of GR, in a feedback, acts as an antistress factor. It decreases the secretion of the stress hormone/neurotransmitter corticotropin-releasing hormone (CRH) and the activity of the hypothalamic–pituitary–adrenal (HPA) axis. The sensitivity of the brain to stressful stimuli is diminished.

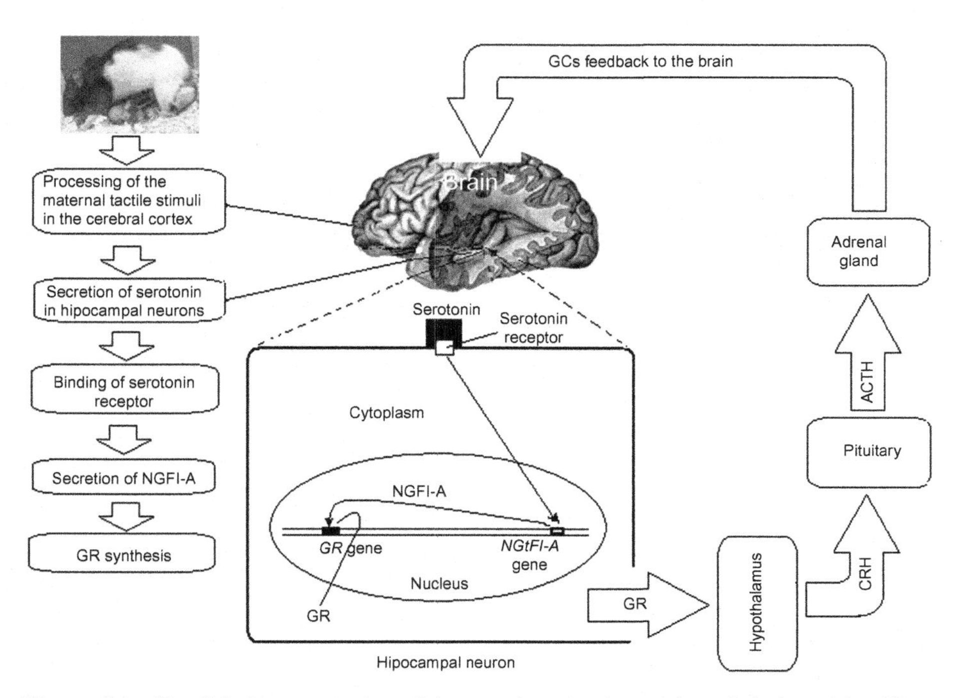

Figure 4.6 Simplified representation of the neural mechanism of demethylation of the *GR* gene and GR synthesis in high-LG mother rats. *Abbreviations*: ACTH, adrenocorticotropic hormone; CRH, corticotropin -releasing hormone; GC, glucocorticoids; GR, glucocorticoid receptor; NGFI-A, nerve growth factor-induced clone A.

In response to social stimuli, the puppy's brain translates afferent information of the maternal LG behavior into efferent information for demethylation of specific genes in hippocampal neurons and for modifying the function of the HPA axis and the stress response. Figure 4.6 reveals the main features of the methylation/demethylation in hippocampal machinery that translates social stimuli into instructions for inducing changes in gene expression and producing phenotypic (behavioral) changes without changes in genes.

Early life separation of baby mice from their mothers leads to lifelong stress, characterized by increased activity of the HPA axis. The hyperactivity of the HPA axis is related to increased production of the neuropeptides CRH and arginin-vasopressin (AVP), with the latter strengthening the action of the former. Experimental evidence shows that stress induces hypomethylation of a specific region of the *AVP* gene enhancer in hypothalamic paraventricular nucleus (PVN) neurons. The hypomethylation reduces the binding of the phosphorylated methyl-CpG-binding protein 2 (MeCP2) to the hypomethylated region of the enhancer of the *AVP* gene. The release of the enhancer of *AVP* gene from MeCP2 occupancy allows its transcription (Dulac, 2010), lifelong increased HPA activity, and stress condition.

> *[T]his vicious circle does not only uncouple MeCP2 occupancy from the initial stimulus but ultimately leads to the hard-coding of the early life experience at the level of DNA methylation. The mice are permanently caught by their adverse early life experience and their latent scars become easily unmasked by further stress exposure.*
>
> *Murgatroyd et al. (2010)*

Recently, the adaptive regulation of the methylation of genes in response to social stimuli is also observed in honey bees (*Apis mellifera*). A correlation is observed to exist between the task (nursing or foraging) and the level of methylation of genes in the brain. So, e.g., the protein kinase C binding protein-1 (*PKCbp1*) gene is consistently higher in the brains of foragers than nurses. Correlations between the methylation level and the task are also identified for other genes in the honey bee brains. It is believed that higher methylation in foraging honey bees is related to the higher cognitive demands of foraging and to production of task-specific protein isoforms through manipulative splicing (Lockett et al., 2012).

In the 1960s, Altman (1963) published experimental evidence to "support the possibility that new neurons may be formed in forebrain structures, both in rodents and carnivores." It took a number of other investigations during the rest of the twentieth century before the dogma that no new neurons produced in adult animals were invalidated. Today, it is known that adult neurogenesis may be induced by electrically stimulating particular brain regions. In adult rodents, for example, deep brain electrical stimulation induces a group of cells in the subventricular and the dentate gyrus of the hippocampal formation to differentiate into neurons (Toda et al., 2008). Social stimuli in the form of an enriched environment and learning also stimulate adult neurogenesis in mammals (Abrous et al., 2005; Zhong et al., 2007).

Clocks and Calendars for Timing Phenotypic Changes

Not all changes in the environment are unpredictable. There are two major periodical changes in the environment that result from geophysical factors, the rotation of the Earth around its own axis and around the Sun and, to a lesser extent, the rotation of the moon around the Earth. The first two determine diurnal (from Latin *dies* denoting day and *urnus* denoting time) and annual seasonal cycles, respectively.

Diurnal and annual cycles involve climatic changes that in moderate climates are drastic enough to pose serious physiological problems. Living organisms have had to cope with these problems from the dawn of life on Earth. In order to adapt their physiology to these drastic changes in the temperature, irradiation, humidity, and predation related to the day–night and annual cycles, living organisms evolved circadian (from Latin *circa* denoting around, roughly) clocks and circannual "calendars," which synchronize their physiology and behavior to these cycles.

The evolution of these timing devices allows living organisms to avoid troublesome surprises, predict the approach of unfavorable conditions in the environment and, in anticipation, develop morphological, physiological, or life history adaptations.

At the most basic level, adaptation of the phenotype to daily and seasonally changing conditions requires the daily and seasonal adjustment of temporal patterns of gene expression. Biological systems needed to perceive time, so they evolved a "sense" of time by "inventing" a biological clock, a timing device that would allow them to forecast and prepare for approaching changes in the environment. This enabled them to program and harmonize their physiology and behavior with cyclically changing conditions in the environment. The pressure for biological clocks has been so strong that they evolved with the photosynthetic prokaryotes, cyanobacteria, one of the most ancient forms of which evolved probably 3.5 billion years ago (Ditty et al., 2003).

How important is the advent of the biological clock to the evolution of living systems is shown by its presence in cyanobacteria, the oldest of extant unicellulars. These prokaryote cells perform two basic, but incompatible processes: photosynthesis and nitrogen fixation. The enzyme nitrogenase, which fixes atmospheric nitrogen, is sensitive to the oxygen generated during photosynthesis. Given that in unicellular organisms, these processes cannot be spatially separated, the bacteria temporally separate photosynthesis and nitrogen fixation, by undertaking the first in the day and the second at night (Kondo and Ishiura, 2000). These simple prokaryotes also use the circadian rhythms for cell division (Sweeney and Borgese, 1989). Molecular mechanisms of circadian clocks in the living world are, to a great extent, conserved (Paranjpe and Sharma, 2005).

Where is the timing device in unicellulars located? Unlike metazoans, where the central clock is located in hypothalamic neurons, it is impossible to answer this question for unicellulars. There is abundant information on changes in gene expression related to the function of the biological clock in unicellulars. It is suggested that the cytoskeleton may play a role in the circadian rhythm since both are closely related with the cell cycle (Shweiki, 1999). Indeed, migration of organelles

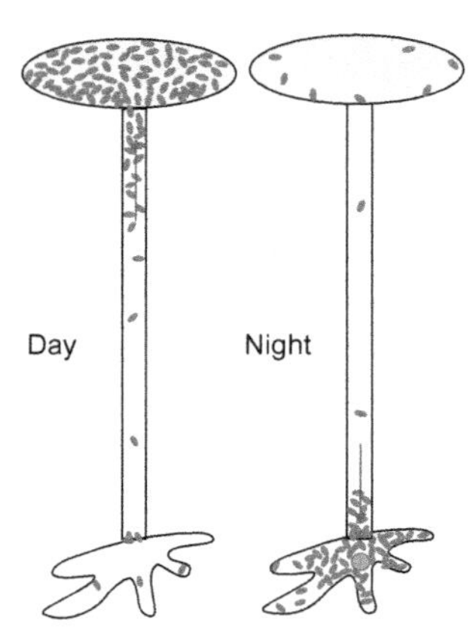

Figure 4.7 Circadian chloroplast migration in an *Acetabularia mediterranea* cell. Accumulation during the day in the hat and upper stalk (left), and during the night in the rhizoid and lower part of the stalk.
Source: From Engelmann (2009).

or changes in the shape of cells in response to day–night cycles is carried out by the cytoskeleton. At night, in the unicellular alga, *Acetabularia*, for example, chloroplasts are concentrated in the rhizoid, but when the sun rises, these same chloroplasts migrate to the hat to use its energy for photosynthesis (Figure 4.7). It is the cytoskeleton that is responsible for their movement.

The above hypothesis finds some indirect empirical support from the discovery that resetting the biological clock during the light-induced phase-shift in mammals is accompanied by the expression of IEGs, including a new IEG, that produces activity-regulated cytoskeleton-associated (Arc) protein (Lyford et al., 1995). This protein is enriched in dendrites along the cytoskeletal F-actin that is believed to be involved in resetting the circadian clock, especially during the phase delay. Under light exposure, the *Arc* gene expresses in the hypothalamic suprachiasmatic nucleus (SCN), in response to electrical signals coming from the retinal ganglion, primarily via the retinohypothalamic tract (RHT) (Nishimura et al., 2003). The *Arc* gene, along the protein kinase D (PKD), is activated as a result of neuronal activity and the discovery that PKD is involved in cytoskeletal remodeling by interacting directly with the F-actin of the cell cortex (Ziegler et al., 2007) corroborated by the previous evidence on the Arc's possible role in cytoskeletal remodeling.

Based on this role of the Arc protein in the circadian clock and on its role in determining the position of organelles within the cell, in regulating gene expression, DNA replication, chromosome segregation, and cell cycle in general (see page 22, "The Control System in Unicellulars"), it is plausible that the cytoskeleton may be an essential component of the circadian clock and circadian rhythms in unicellulars.

While unicellulars evolved a diurnal timetable for their daily physiological activities, multicellulars, both animals and plants, had to evolve an additional annual calendar for their physiology that includes patterns of gene expression.

Circadian rhythms are crucial in plant life. Matching biological rhythms to environmental periods is positively related to photosynthetic output (McClung, 2006). In the small flowering plant, *Arabidopsis thaliana*, for example, plants with circadian clocks better matching the environment contain more chlorophyll, fix more carbon, grow faster, and survive more often than other plants (Dodd et al., 2005). Expression of about 10–15% of genes in this plant is regulated by the plant's central circadian mechanism (McClung, 2008).

Differential exposure to light (many cells are not exposed at all) and temperature of different cells in multicellulars makes the diurnal harmonization of their function impossible. An evolutionary pressure for harmonizing functions of all cells led to the evolution of the central neural clock, adjusting and synchronizing the clocks of all cells, throughout the animal body, from insects to birds and mammals and vertebrates in general.

The *Drosophila* brain contains ~100,000 neurons. About 150 express the canonical clock machinery. These neurons control and regulate the circadian rhythms of the fly's physiology and behavior. Based on their anatomical location, these clock neurons are divided into seven groups (Nitabach and Taghert, 2008). Four are in the lateral brain and include large ventrolateral neurons (l-LNvs), small ventrolateral neurons (s-LNvs), dorsolateral neurons (LNds), and lateral posterior neurons (LPNs). Almost all of the l-LNvs and s-LNvs express the neuropeptide pigment-dispersing factor (PDF), and they are the only PDF-expressing neurons in the fly brain.

Light information is conveyed to *Drosophila* clock neurons via three pathways: retinal photoreceptors of compound eyes, brain photoreceptive neurons Hofbauer–Buchner, and the photopigment cryptochrome (CRY) protein. The first two pathways transmit light information to the clock neurons through synaptic connections (Nitabach and Taghert, 2008).

Circadian rhythm is driven by transcription factors clock (CLK) and cycle (CYC) that induce expression of period (PER) and timeless (TIM) and other genes. The products of these genes, mRNAs and proteins, fluctuate in functional levels, owing to post-translational modifications (Figure 4.8). The proteins PER and TIM suppress the expression of genes *clk* and *cyc*, leading to the depletion of PER and TIM and, consequently, to the resumption of *clk* and *cyc* expression in 24 h cycles (Nitabach and Taghert, 2008).

Although the *Drosophila* has circadian clocks in many tissues, the central neuronal clock is required for some peripheral clocks. The central and peripheral clocks communicate with each other at least at the physiological and behavioral levels. The circadian mechanism in *Drosophila* is involved in "hundreds of behavioral and metabolic events that are adjusted on a daily basis" (Bradshaw and Holzapfel, 2010). Flies with normal circadian rhythms live longer than flies with disturbed rhythms (Kumar et al., 2005). Mating clock-deficient flies result in 40% less progeny compared to normal flies (Paranjpe and Sharma, 2005).

Looking at the figure, there is no explanation for the initial (and for the latter, for that matter) expression of the *clk* and *cyc* genes. These genes are present in all *Drosophila* cells but expressed only by the clock neurons in response to electrical signals coming to the clock neurons from photoreceptors. Why do not other cells

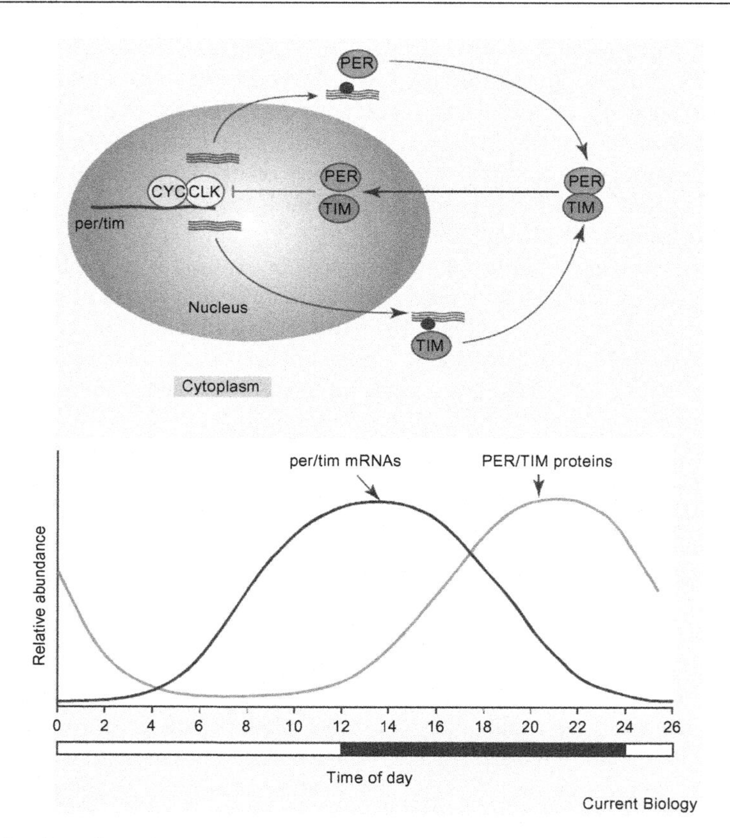

Figure 4.8 Overview of the *Drosophila* molecular circadian oscillator. Expression of the *per* and *tim* genes is promoted by the heterodimeric CLK–CYC transcription factors and reaches a peak late in the day. Translation of *per* and *tim* RNAs leads to the gradual accumulation and dimerization of PER and TIM proteins within the cytoplasm. The protein levels peak at night, during which time they separately enter the nucleus to inhibit further CLK–CYC transcriptional activity as shown. *Abbreviations*: CLK, clock; CYC, cycle; per, period; tim, timeless. *Source*: From Nitabach and Taghert (2008).

in *Drosophila*, including photoreceptor neurons, which are exposed to daylight, express *clk* and *cyc*? As a first impression, it seems paradoxal that *clk* and *cyc* genes are expressed by clock neurons not directly exposed to the light, yet not expressed in cells and neurons that are exposed. As it was explained earlier in this chapter, genes do not understand nature's language; they have no hunch on what dawn and nightfall presage for the fly's physiology, survival, and reproduction. To make these environmental messages intelligible to genes, the messages must be translated into a language the genes understand and the polyglot translator is in the brain.

The hierarchical character of the circadian mechanisms is clearer in mammals, where the central neural clock controls and resets the peripheral clocks. The timing

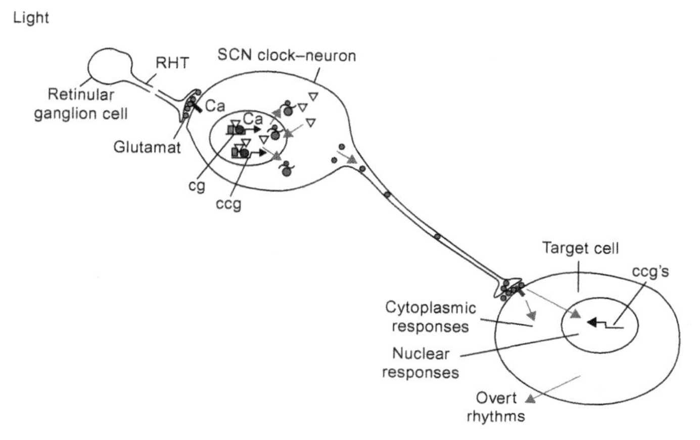

Figure 4.9 Events between light perception, the SCN clock neuron, and the target cell: Light is perceived in ganglion cells of the retina. Neurotransmitter GLU (violet) is secreted and reacts with receptors (black bar). The expression of mPer and mCry of the clock-gene (cg) is turned on via a negative feedback loop with amplifying factors CLOCK (green) and BMAL (blue). Ca2+ is involved in it. Clock-protein-mRNA (red circles with ~) are produced, leaving the nucleus and synthesized clock-protein (triangles) in the cytoplasm. It enters the nucleus, interacts with mPER, and facilitates its translation by inhibiting the CLOCK (green) and BMAL (blue) dependent transcription: The mRNA concentration decreases. After a time delay, negative acting complexes are inactivated and the gene expression begins again. The next round of negative and positive acting factors boost the rhythmic expression of clock-controlled genes (ccg's). The product, clock-controlled proteins contain information concerning time of day and pass it to the SCN neurons, via synaptic or paracrine signals, to target cells. Target-specific circadian outputs via cytoplasmic reactions or reactions in the nucleus affect secondary ccg's. An example is the *N*-acetyltransferase. It controls the melatonin synthesis. *Abbreviation*: BMAL, a protein that dimerizes with CLOCK; RHT retinohypothalamic tract; GLU, glutamate; Cry, cryptochrome; SCN, suprachiasmatic nucleus. *Source*: From Engelmann (2009).

system comprises the input pathways, the SCN, which serves as the circadian pacemaker, and the output pathways with peripheral clocks in cells and organs.

In mammals as well, the most important zeitgebers (time givers, synchronizers) used by the circadian mechanism are the light and, to a lesser degree, restricts feeding (Challet et al., 2003), temperature, and humidity, while rhythmic activities of prey, predators, parasites, intra- and inter-species competition, and availability of food are used to fine-tune them (Sharma and Chandrashekaran, 2005). The SCN central clock responds directly to photic signals coming from the retinal neurons via the RHT as well as to the nonphotic signals coming from the geniculo-hypothalamic tract and the raphe nuclei (Dibner et al., 2010) (Figure 4.9). Photic signaling from RHT includes release of the neurotransmitter GLU and neuropeptide pituitary adenylate cyclase-activating protein (PACAP). Phase shifts in the SCN are associated with changes in clock-gene expression, especially *PER1* and *PER2*.

Recent evidence suggests that a number of other neural and nonneural tissues exhibit circadian rhythms in gene clock expression, hormone secretion, and electrical activity in mammals (Guilding and Piggins, 2007). These brain and peripheral circadian clocks are self-sustained and may be reset by restricted feeding (Challet et al., 2003). However, changes in the peripheral clocks result from signals generated in response to light in the SCN, rather than any direct effect of light.

That the SCN serves as pacemaker and synchronizer implies the transmission of relevant information to other brain and peripheral clocks. Indeed, SCN neurons project to other brain clock regions and via circulating hormones, and the autonomic nervous system controls the circadian rhythms of physiology in all the organs including the pineal gland and the function of the adrenal cortex (Dibner et al., 2010). This brain center serves as "a master pacemaker for the organism to drive rhythms in activity and rest, feeding, body temperature, and hormones" (Mohawk et al., 2012).

In birds, the timing system appears more complicated than that of mammals. This is important especially for migratory birds, such as arctic birds that need to adjust their timing system quickly to accommodate changing conditions during migration (Gwinner and Brandstatter, 2001).

Most multicellular organisms live more than a year to experience another disturbing geophysical phenomenon, the circannual cycle, which is accompanied by changes in temperature, photoperiod, sources of food, humidity, and predation that are all more drastic than those associated with the circadian cycle.

So living longer came not for free. To survive and thrive, plants and animals evolved a "long-term" circannual temporal structure, or calendar, alongside the circadian timetable. This new temporal structuring device would allow them to anticipate and be prepared for extreme conditions in the environment, by developing adaptive seasonal changes in physiology, behavior, morphology, and life cycle.

All plants and animals that outlive the year have evolved a circannual photoperiodic calendar, which lets them program the phenotypic changes necessary to adapt to seasonal environmental changes.

Despite the wealth of information on the functions of this circannual clock, little is known of its mechanics compared to the circadian clock. Day length is the environmental cue that multicellular organisms use to set up their photoperiodic calendar. There is sufficient reason to believe that the structures responsible for establishing the circannual calendar are localized in the brain (Figure 4.10).

What is certain is that the input and output pathways of the photoperiodic calendar end in the brain and begin from the brain. This makes it obvious that the circannual calendar structure is located in the brain. The input pathway starts with the perception of the photoperiodic cue (i.e., the length of the day and night in retinal neurons) and transmission of this photic information to the SCN, or with direct brain photoreception in some insects. The output pathway for photoperiodic changes in animal physiology, behavior, and life history starts from brain neurosecretory cells that send instructions for adaptive phenotypic changes to effectors.

The photoperiodic calendar's crucial role is revealed in the course of the individual development of diapausing insects and some other arthropods. Depending on the photoperiod and related temperature, at particular stages of their development, these

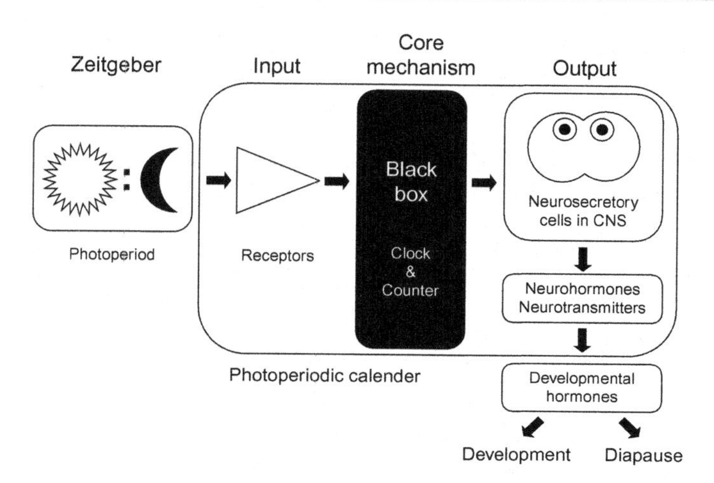

Figure 4.10 Schematic structure of the photoperiodic calendar. Environmental signals such as photoperiod are perceived by poorly characterized receptors and processed in a core mechanism composed of a photoperiodic clock and counter. Physiological principles underlying functionality of a core mechanism remain unresolved in insects. Most of the current knowledge comes from experiments using a black box approach. Output pathways include neurosecretory cells that produce neurohormones, neurotransmitters, and mitogens, which in turn influence biosynthesis and the release of developmental hormones, the presence/absence or titer of which determine developmental destiny (direct development vs. diapause). *Source*: From Košťál (2011).

insects decide on whether to continue developing or stop growing and fall into a safer dormant state until a more conducive season arrives. Their decisions are based on the photoperiodic cues, shortening of the day length, and decline of temperature, which presage the approach of winter. On identifying these cues, the animal turns off its development, then turns it back on when it perceives a sufficient change in the day length and temperature, which presage the approach of spring.

A simplified schema of the mechanics of the photoperiodic calendar in determining the diapause in insects would look as follows:

Photic cues on the length of the day are received by photoreceptive neurons in the retina (directly from the brain in some insects). Shortening of days are perceived and counted in the brain. When the number of the shortening days reaches a set point, particular neurons in the pars intercerebralis and pars lateralis of the insect brain start a neuroendocrine cascade. Juvenile hormone (JH) and ecdysteroids act as downstream signals, inducing diapause with interruption of growth and development while maintaining a minimal metabolism. The structure of the insect's counter to the shortening days in its brain, which may be at the heart of the photoperiodic calendar, represents a black box (Košťál, 2011).

The present state of knowledge on the photoperiodic calendar and the circadian clock does not allow for conclusions about whether there is a single system that performs both functions of the circadian clock and photoperiodic calendar or whether there are two separate, but cooperating systems (Košťál, 2011).

Discrete Phenotypic Changes Are Designed by the Brain

Diapause—the Predictive Form of Insect Larval Dormancy

An interesting form of adaptation in plants and animals is dormancy, the suspension of embryonic growth and development accompanied by a dramatic decline in metabolic activity during or just before the winter, until better environmental conditions arrive. In the animal world, embryonic dormancy is a characteristic aspect of the arthropod life cycle, especially insects. It is also observed in various vertebrate species of fish, reptiles, and mammals. The main form of dormancy in insects is the diapause that can occur in all the stages of insect development. Larval diapause in insects is controlled by the brain's circannual calendar, which is based primarily on the perception of the shortening of the day length and the decline in the environmental temperature. The insect diapause is a predictive form of dormancy, unlike *aestivation*, the consequential dormancy of insects, that is arrested by the development in response to the unpredictable adverse conditions of hot weather, especially desiccation—by using physiological and morphological adaptations to prevent dehydration.

The photoperiod is the main diapause-inducing cue in diapausing insects. The decline in environmental temperature is another diapause-inducing cue in insects, not only in concert with the photoperiod in temperate latitudes, but also in itself in equatorial regions (Denlinger et al., 2011).

The effect of photoperiod as a diapause-inducing cue in insects implies that they measure the day length and count the number of shortening days (Denlinger et al., 2011). It is believed that the neurons in which photoperiodic information is processed and the neurons that secrete prothoracicotropic hormone (PTTH) are in the same brain region (Denlinger et al., 2011).

While diapause-inducing cues are generally common in insects, the molecular mechanisms of diapause are widely diverse. A common feature of all mechanisms is that they are neurohormonal mechanisms involving ecdysteroids and JH, as well as neurohormones pheromone biosynthesis activation neuropeptide (PBAN) and diapause hormone (DH)/DH-like hormones. With ecdysteroids and JH being products of the lower limbs of the brain/prothoracic gland and brain/corpora cardiaca, respectively, it is reasonable to suggest that insect diapause is cerebrally regulated.

It is noteworthy that these same hormones—ecdysteroids and JH—in different species can play different and even opposing roles in diapause. If at first this seems paradoxical, it is worth remembering that the nervous system can relate any external or internal stimulus to any gene as shown earlier in this chapter (see page 197, "Making Environmental Signals Intelligible to Genes").

Needless to say, growth inhibiting ecdysteroids are not produced in larval stages. During diapause, the prothoracic gland, in response to brain PTTH, secretes ecdysteroids to maintain a specific level of the hormone in the hemolymph.

In general, it is to be expected that ecdysteroids would act as inducers of diapause and this seems to be the case in many insect species, including the gypsy moth, *Lymnatria dispar*, an obligatory diapausing insect species. The insect's prothoracic gland synthesizes ecdysteroids throughout diapause, but its synthesis drops by the end of the diapause period to allow it to continue with its development. In *Chilo* stemborers, and other insects, diapause is controlled by JH; stimulated by the secretion of brain allatotropins, corpora allata (CA) synthesizes JH during diapause and JH levels drop by the end of diapause. In other insects, application of JH leads to termination of diapause (Denlinger et al., 2011).

In *Bombyx mori* larvae, the diapause is regulated by brain hormones DH and PBAN. Although DH does not induce diapause in other insects, a DH-like neuropeptide is secreted with the neuropeptide PBAN from a single mRNA in the brains and especially the subesophageal ganglion (SOG) of insects, such as *Heliothis virescens* (Fabricius) (Xu and Denlinger, 2003) and *Helicoverpa armigera* (Hübner) (Zhang et al., 2004). But, contrary to its well-known diapausing effect in *B. mori*, in *H. virescens*, secretion of DH and PBAN stimulates development and their suppression induces diapause.

Parents Transmit Epigenetic Instructions for Switching Their Offspring to Alternative Life Histories

Epigenetic information determines gene expression, cell differentiation, and intragenerational phenotypic changes in physiology, morphology, and behavior. All the examples of epigenetically induced phenotypic changes presented so far are not heritable (i.e., they are not inherited) in the offspring in the absence of the stimuli that induced their original expression. The following two examples demonstrate that parents can transmit to their offspring new traits that the parents did not possess and how they generate and pass these instructions down to the next generation.

The silkworm is a facultative diapausing insect, but normally it does not determine itself whether it will enter diapause or not. Whether it will enter diapause or not depends on the secretion of DH, an oocyte-targeting neuropeptide. Under conditions of the summer-long photoperiod and high temperature, the photoperiodic clock in the insect's brain induces an increased firing activity of the labial secretory (Lb) neurons in the SOG (Ichikawa, 2003). These neurons are exclusive producers of DH (Denlinger et al., 2011). Epigenetically determined differences in the firing activity of the Lb neurons in diapausing and nondiapausing pupae are correlated with differences in their DH content. A dopaminergic circuit (Noguchi and Hayakawa, 2001) and the product of the gene *Pitx* (Shiomi et al., 2007) also stimulate these neurons to synthesize DH from the transcript of the diapause hormone–pheromone biosynthesis activating neuropeptide (*DH-PBAN*) gene, which, via hemolymph, targets ovaries (Kamei et al., 2011; Kitagawa et al., 2005; Shiomi et al., 2007). The diapause-inducing action of the DH is mediated by its receptor that is expressed in the oocyte (Homma et al., 2006) (Figure 4.11).

In the ovary of the silkmoths that produce diapause-producing eggs, DH increases trehalase activity and the amount of glucose in the oocyte as a result of the uptake from the hemolymph (Kamei et al., 2011). At the beginning of diapause, glycogen is converted into sorbitol and glycerol, which prevent the freezing of eggs in cold weather (Kihara et al., 2009).

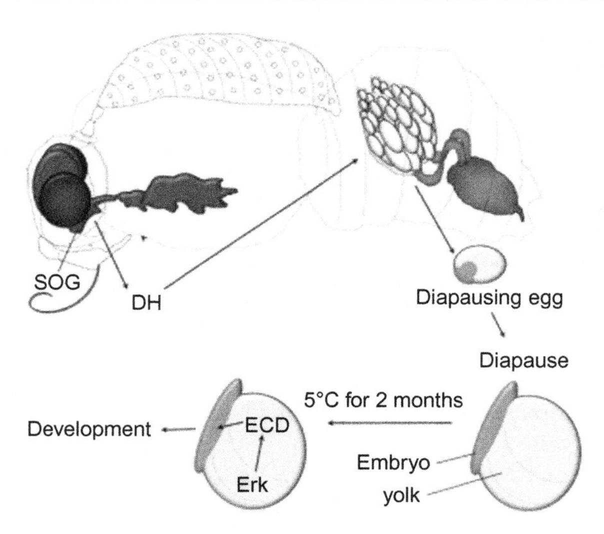

Figure 4.11 Maternal control of embryonic diapauses in the silkworm *B. mori*. DH is produced by the SOG of the mother that developed under the long photoperiod and high temperature of summer. DH signals in the ovaries where it induces the development of diapause-fated eggs. In these eggs, embryonic diapause occurs after the development of the cephalic lobe and the following mesoderm segmentation. The embryos require 2–3 months at low temperatures (about 5°C) before they can arrest diapause and resume development at high temperature. In this phase, the ERK pathway is activated in the yolk cells and it promotes the production of 20-OH ECD. ECD signals to the embryo inducing the removal of the block imposed to development. *Abbreviations*: DH, diapause hormone; ECD, ecdysone; ERK, extra-cellular signal-regulated kinase; SOG, suboesophageal ganglion.
Source: From Schiesari et al. (2011).

The flesh fly, *Sarcophaga bullata*, offers another interesting example of the maternal neural control of diapause in insects. Diapausing mothers of *S. bullata*, which as pupae are reared under short photoperiod (SP), prevent the "normal" appearance of diapause in the offspring, even when the offspring are exposed to diapause-inducing SPs. The inhibition of diapause in the offspring is determined by the mother's photoperiodic experience (Henrich and Denlinger, 1982). Since the primary environmental cue that induces diapause is the length of the day and because this is perceived in the insect's brain, one can expect that either the mother's brain must be responsible for inducing/suppressing the diapause or it is the site where the causal chain of events leading to induction or inhibition of diapause begins. This theoretical expectation is verified empirically.

Transplantation of ovaries between long-day history and short-day history larvae does not influence the diapause fate of the progeny, implying that the information for diapause is still retained within the brain at the end of larval stage (Rockey et al., 1991). Diapause can be suppressed by brain extracts of mothers reared during an SP and by administration of the neurotransmitter gamma aminobutyric acid (GABA). It is noteworthy that the GABAergic circuit is responsible for inhibiting DH synthesis in the silkworm, *B. mori* (Shimizu et al., 1989). The neurotransmitter also drastically reduces the probability of diapause in flesh flies kept in strong diapause-inducing environments (long-day 12:12h at 20°C) and thus expected to diapause (Rockey et al., 1991). Diapause induction depends on the activation of a dopaminergic circuit and inhibition of diapause starts with the activation of the GABAergic system in the brains of SP flesh fly mothers (Webb and Denlinger, 1998).

Switching to the new life history character without affecting the genetic information (genes/regulatory sequences) implies that some form of epigenetic information, transmitted to the offspring via the egg, is responsible for the switch. It is believed that the mother provides this information to the egg in the form of a maternal cytoplasmic factor. The nature of the factor as of yet is not known, but the GABA signal may transmit the diapause-suppressing information from the brain by inducing the secretion of a relevant factor in the egg. Investigators believe that "the information transfer from mother's brain (the site of the photoperiodic reception) to her ovaries occurs sometime after pupariation but before the second day of the adult life" (Henrich and Denlinger, 1982).

Most recently, studies on the endoparasitoids of insect eggs *Trichogramma* spp. have shown that short-day-induced diapause is passed on beyond the first generation, to the second and third generations and, weaker, to the fourth and fifth generations (Reznik et al., 2012).

A Quick Change of the Phenotype—Phase Transition in Locusts

Locusts of the species *Locusta migratoria* (Linnaeus, 1758) and *Schistocerca gregaria* (Forskål) have a unique and fascinating ability to switch from the solitarious to the gregarious phases involving discrete changes in multiple morphological, physiological, behavioral, and life history traits. The changes are so extensive that biologists initially believed they were different species (DeLoof et al., 2006). It goes without saying that these are discrete changes, not related to the norm of reaction.

Phase transition in locusts is induced by two distinct sensory neural pathways:

1. The visual-olfactory, or cerebral pathway, which is activated by perceiving other conspecific locusts in the vicinity.
2. The tactile or thoracic pathway, which is activated by the stimulation of hind leg mechanoreceptors, as occurs in nature, by jostling with other locusts.

The earliest phenotypic change is the gregarious and migrating behavior, which locusts display within 2h of perceiving the phase transition-inducing stimuli. The change in body coloration takes ~24h, whereas morphological changes in wings, hind legs, and so on may take several generations to be fully expressed (Burrows et al., 2011).

Morphological changes in transiting from the solitarious to gregarious phase include alterations in head and brain size, in the number of sensilla, morphometry, and cuticle color (from green to dark), shorter wings and hind legs, smaller eyes and antennae. Although long-term gregarious locusts are smaller in size, their brains are 30% larger than the brain of the long-term solitarious locusts. Differences are also observed in the proportions of various regions of the brain during the two phases. The increase in the brain size and changed proportions of brain parts in gregarious

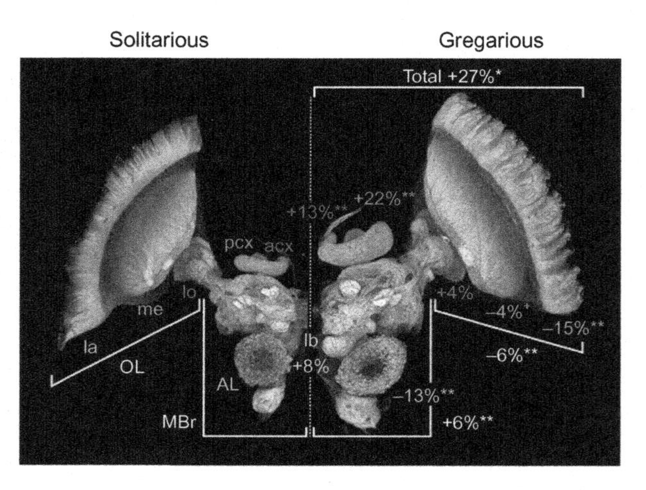

Figure 4.12 Body and brain changes between solitarious and gregarious locusts. Differences in brain size and proportions of particular regions in solitarious (left) and gregarious (right) brains of adult locusts are shown. Regions in the midbrain (MBr) include the olfactory antennal lobe (AL) and three neuropils in the mushroom body: the olfactory primary calyx (pcx), the gustatory accessory calyx (acx), and the multimodal lobes (lb). The optic lobe (OL) comprises three successive visual neuropils: the lamina (la), the medulla (me), and the lobula (lo). Absolute total brain size is 27% larger in gregarious locusts. The remaining numbers refer to the differences in proportions of different brain regions relative to the total brain size. Positive numbers indicate that a region is disproportionally larger in gregarious locusts than in solitarious locusts (** $P < 0.01$; * $P < 0.05$; + $P < 0.1$).
Source: From Burrows et al. (2011).

locusts relate to the increased need for higher neural processing (Burrows et al., 2011) (Figure 4.12).

The primary behavioral change is the aggregation behavior, the tendency to join groups of other conspecifics and migrate by flying up to hundreds of kilometers a day. Just 2 h of crowding is sufficient to induce a gregarious behavior similar to that of the long-term gregarious locusts. Physiological changes involve numerous changes in gene expression, but very characteristic and relevant to phase transition, which seem to be the changes in 11 of 13 neurotransmitters and neuromodulators analyzed in the nervous system of solitary and gregarious locusts (Rogers et al., 2004). Eight of these neurotransmitters and neuromodulators were upregulated in gregarious locusts (Burrows et al., 2011).

It is firmly established that transitioning from the solitarious to the gregarious form is determined, initiated, and regulated by neural mechanisms, which within hours are activated in response to various sensory (e.g., visual, olfactory, auditory, and tactile) stimuli under natural and laboratory conditions, when the intensity of the stimuli reaches the necessary threshold.

Given that all the perceptions (i.e., visual, olfactory, tactile, and auditory) inducing phase transition take place in the locust brain, it would make sense to believe that the brain is where the relevant phenotypic changes start or where the relevant information

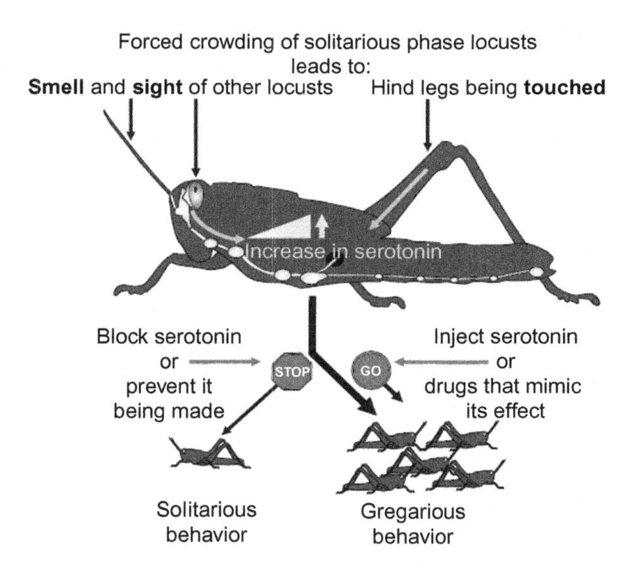

Figure 4.13 Changes from solitarious to gregarious behavior occur rapidly and are mediated by serotonin. Diagrammatic summary of the behavioral gregarization pathway and the role of serotonin is shown.
Source: From Burrows et al. (2011)

comes from. The gregarious behavior is experimentally demonstrated to be related to changes in the activity of the flight-related neurons. So, for example, spontaneous electrical activity of the flight-related tritocerebral commissure dwarf (TCD) interneuron is stronger and gregarious than solitarious locusts (Fuchs et al., 2003).

The olfactory (and visual) stimuli in the form of electrical spike trains are transmitted from the sensory organs to be processed in labyrinthine neural circuits already described in detail (Bargmann, 2006; Laurent et al., 1998). As mentioned earlier, the most characteristic changes related to the transition from the solitarious to the gregarious phase are observed in the function of neurons and neural circuits. Among the mentioned dramatic changes in the synthesis of the neurotransmitters and neuromodulators in the nervous system, the most conspicuous and the most relevant seem to be the rapid increase of serotonin in the thoracic ganglia in the first 1–4 h following stimulation. This increase in the serotonin level induces gregarious behavior and the intensity of this gregarious behavior is proportional to the increase in serotonin level. This neurotransmitter effect is verified by the experimental administration of serotonin into the thoracic ganglia that induces the transition to the gregarious phase even in locusts that have never encountered other locusts (Figure 4.13). To the contrary, the mediator of the serotonin action prevents transition to gregarious phase in locusts by blocking the serotonin synthesis or serotonin receptor (Burrows et al., 2011).

Besides the serotonergic pathway, the dopaminergic pathway is also important, most likely synergistic, in establishing gregarious behavior in locusts and the injection of DA also induces gregarious behavior (Burrows et al., 2011; Ma et al., 2011).

At a global genomic level, changes between the two phases are observed in 1444 (15.8%) of 9154 genes analyzed (Burrows et al., 2011). Gregarious locusts also show a higher DNA methylation level in both repetitive and protein-encoding regions (Robinson et al., 2011).

Phase transition is also transmitted to the offspring, but the transgenerational phenotypic changes will be dealt with in the following sections.

Transgenerational Developmental Plasticity—Insights into the Nature of Evolutionary Morphological Change

A basic tenet of the new synthesis is that any evolutionary changes in phenotype result from the action of natural selection over mutations in genes/regulatory sequences or over existing genetic variability. At present the biology has no tools for telescoping the mechanisms of evolutionary changes that occurred in the past, but many gene mutations can be related to the evolution of physiological traits at the molecular level.

The situation is different when it comes to the evolution of morphological, behavioral, and life history traits. Extensive experimental attempts over about 100 years to produce evolutionary morphological changes through gene mutations have failed in the sense that they invariably resulted in harmful effects on the animal's fitness and survival. Being harmful for the organism, such changes would normally be eliminated rather than spread in natural populations.

The century-long empirical failure to relate the evolution of genes to morphological, behavioral, and life history traits is not a surprise from the perspective of modern biology knowledge. As Nijhout (1990) points out, the only thing that genes code for is the primary structure of proteins. Morphologies do not arise from proteins, but from specific patterns of spatial arrangement of cells of different types. Almost by consensus, it is now admitted that cell differentiation, the formation of numerous different types of cells of the same genotype, is an epigenetic process.

The empirical and theoretical failure to explain the evolution of phenotype at a supramolecular level by changes in genes, along with the new concept of the common metazoan genetic toolkit of several hundred genes from several dozen gene families (Rokas, 2008), contributed to the development of a new field of biological study. This area, known as evolutionary developmental biology or evolution of development (evo-devo), intended to look for the sources of evolutionary change in individual development.

Chapter 3 showed that the development of the metazoan embryo from the unicellular stage (egg or zygote) to the phylotypic stage is determined by the epigenetic information in the form of cytoplasmic factors parent(s) provide to the zygote. After the phylotypic stage, the development of different species depends on the specific patterns of expression of basically the same genetic toolkit, leading to cell differentiation and organogenesis.

An interesting observation pointed out in Chapter 3 is that at the phylotypic stage, the embryo has an operative CNS, which is the first organ system to develop in all metazoan embryos. The development of embryonic structures proceeds from signal cascades or elements of the CNS, suggesting that the CNS develops before any other organ system and that this development is not accidental. All of the above clearly support the evo-devo concept of the developmental rather than the genetic origin of evolutionary change.

Changes in developmental pathways do occur, but they are neither frequent nor predictable, so that the appearance of the evolutionary change in embryos cannot be examined. However, direct examination of the mechanism of the evolutionary change may not be as secretive as it appears. Another approach seems to offer biologists relevant clues, if not the real developmental mechanism of evolutionary change. The evolutionary change is a new/modified trait transmitted to the offspring and maintained for more than one generation in the absence of the agent that initially induced the change. If this is true, then the described cases of the TDP fall in the category of evolutionary changes. Their study may provide *in vivo* models of the mechanisms of evolutionary change or, at least, provide important clues of its mechanism.

Here are a few examples of TDP.

The water flea, *Daphnia pulex*, in response to detecting kairomones (chemical cues) released by the predatory phantom midge *Chaoborus flavicans* larvae in the environment, develops an outgrowth called neckspine with a varying number of teeth, delays its reproductive maturity to increase its body size, and may change its life history. The formation of the neckteeth diminishes the danger of juvenile *Daphniae* being ingested by the predator larvae. The water flea is more sensitive to the kairomone during embryonic development. At this stage, the embryonic crest epithelium thickens and continues developing neckteeth until the third instar stage. The embryo can form neckteeth via maternal signals deposited in the eggs of mothers that have perceived the presence of the kairomone in the environment or embryos can themselves respond to the kairomone by developing neckteeth (Imai et al., 2009).

Experimental studies have outlined the mechanism of the development of neckteeth in *D. pulex* (Figure 4.14).

Antennae of the *D. pulex* carry asthetascs (olfactory receptors), which contain, among others, chemoreceptors for kairomones. The information they transmit is integrated in the deuterocerebrum (midbrain) (Hanazato and Dodson, 1992), which innervates the antennulae. This is the region of the brain where kairomone is perceived (Hallberg et al., 1992). Neural signals secreted in the Daphnia's brain stimulate expression of the *DD1* gene and secretion of endocrine hormones, including insulin and JH. The latter, downstream, induces the expression of morphogenetic genes (*Hox3*, *DD2*, *DD3*, etc.) determining formation of neckteeth structure (Miyakawa et al., 2010).

Below the site of the neckteeth development, a high concentration of polyploid cells is observed, but the role of these cells in neckteeth formation is not established (Weiss et al., 2012).

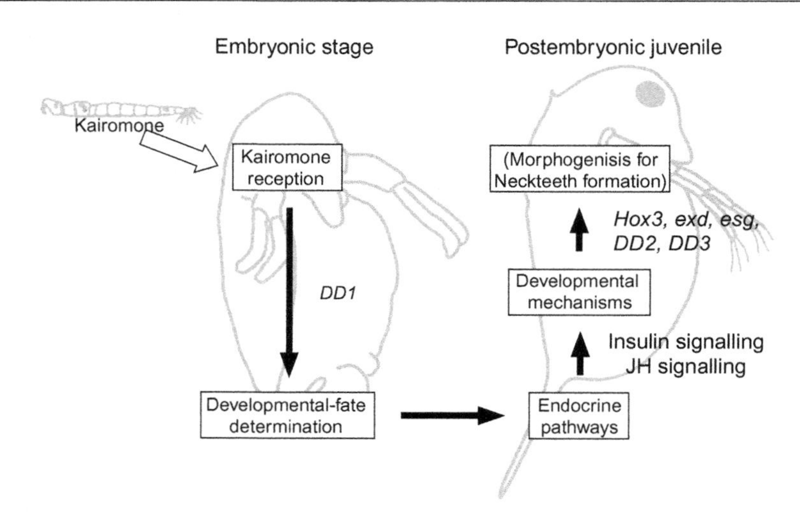

Figure 4.14 Schematic diagram showing the process of defense morph formation with the putative involved genes and biological pathways suggested by the present study. *DD1* is thought to be involved in kairomone reception and/or fate determination during the embryonic stage. The other genes are considered to be involved in the morphogenesis of postembryonic juveniles.
Source: From Miyakawa et al. (2010).

It is worthwhile noting that the full expression of transgenerational predator-induced defenses in some Daphnia species requires more than a generation of exposure to predator cues. For instance, *Daphnia mendotae* forms the biggest size of its round helmet in the third generation of exposure to chemical cues of the largest planktonic cladoceran, *Leptodora kindtii* (Tanner and Branstrator, 2006).

Populations of *Daphnia magna*, a native Daphnia species of north-west America switch to one of the reproductive modes to better adapt themselves according to conditions in the environment. Under favorable conditions, they expand rapidly as diploid all-female populations that reproduce parthenogenetically, but when conditions deteriorate (crowding, depletion of food resources, deterioration of the food quality, etc.) or presage deterioration (i.e., shortening photoperiod), they enter the sexual reproduction phase by giving birth to male and female individuals (Figure 4.15). This is very adaptive: by mating with males, females produce fertile diapausing eggs resistant to freezing and desiccation, that can survive for decades before hatching into male individuals. Producing male individuals during this phase is correlated with an increased level of methyl farnesoate (MF), a crustacean JH (Olmstead and LeBlanc, 2007).

All stressful stimuli are perceived in the brain where they activate the cholinergic system, which inhibits the synthesis and secretion by neurons of the X-organ-sinus gland complex of mandibular organ-inhibiting hormone 1 (MO-IH-1) and mandibular organ-inhibiting hormone 2 (MO-IH-2). These neurohormones derepress the gene for MF in the mandibular organ. The synthesis and secretion of MF, which binds its

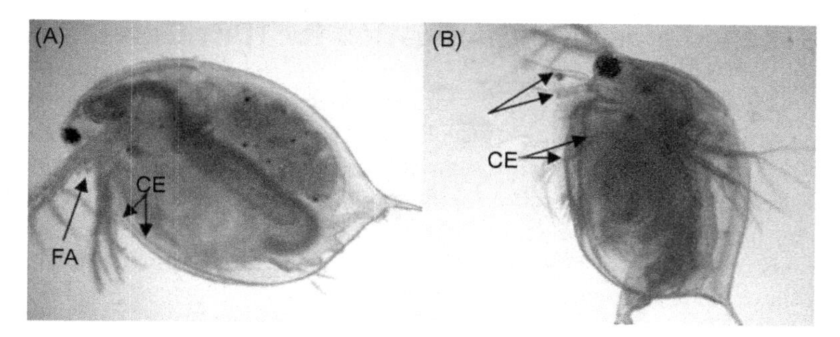

Figure 4.15 Female (A), male (B) *D. magna.* Differentiating sex characteristics include the pair of minute first antennae (FA) of the females, the elongated FA of the males. The bivalved-like carapace of the female has two uniform, symmetrical edges (CE). Both CEs of the male are asymmetrical and are edged by setae.
Source: From Olmstead and LeBlanc (2007).

receptor in the oocyte, activates a transduction pathway that induces the expression of sex determining genes (*drmt?, sex?, fem?*) (LeBlanc, 2007).

When attacked by its predators, the pea aphid, *Acyrthosiphon pisum*, releases an alarm pheromone, (E)-β-farnesene (EBF), which triggers escape behavior in the colony by dropping off the plant. The increased contact with other aphids as a result of jostling and escape response and the direct action of the EBF induce increased proportion of winged individuals in the offspring. Both inducing stimuli, the tactile stimulus and the reception of the pheromone via olfactory neurons, are processed in the brain, but the pathway from the brain to the aphid ovary, ovarioles, and oocyte is not known. Experimental exposure of aphids to EBF alone increases the proportion of winged individuals in the offspring (Kunert et al., 2005).

TDP is also known in the plant kingdom where, like metazoans, it is an adaptive response to stressful conditions in the environment. Seedlings of the third generation of the annual plant, *Polygonum persicaria*, grown in dry soil produce longer roots that extend deeper and faster into dry soil than the seedings of adequately watered plants. They also produce a greater biomass and have greater survivorship (Herman et al., 2012).

The American bellflower, *Campanulastrum americanum*, is a 5–15 cm tall, biennial or annual plant in North America. The type of life history that the offspring adopts is determined maternally. When the mother plant is grown understory, it produces seeds that adopt the annual life history, which is adaptive under the maternal shady-light environment; plants grown in a light gap produce seed that adopts the biennial life history, which is also adaptive under this light environment. The fitness of plants that are appropriately cued to their light environment is 3.4 times greater than plants not grown in the maternal light environment (Galloway and Etterson, 2007).

Maternal environmental effects are greater in dormancy and germination of seeds in plants (Donohue, 2009).

Where Is the Epigenetic Information for TDPs Generated?

This is a logical question. A new trait requires information for a new spatiotemporal pattern of arrangement of millions of cells of several types (morphological and/or life history change) or the reorganization or addition of new neurons to form a special neural circuit (behavioral and/or life history change). As explained in Chapter 2, the genetic information for protein biosynthesis does not play a factor. Theoretical speculations aside, it is firmly and unequivocally determined that no changes in genes occur or are related with TDPs.

Reported cases of TDP arise in response to stressful stimuli or conditions and contribute to an organism's fitness. They are almost always adaptive, rather than random changes.

The pathway from receiving the environmental stimulus, to its processing in the animal's brain, to the maternal/paternal factors in the egg cell that determine the phenotypic change in the offspring, is complex and, in many essential details, unknown. All that can safely be done here is to review the empirical evidence and then look for possible leads that may allow for the outlining of the mechanics of a stimulus' translation, via individual development, into a new phenotypic character in the offspring. The importance of researching TDPs, from a theoretical point of view, can hardly be overstated because it offers a mechanism of induction in the offspring for one to several generations of a trait that the parents, themselves, lack. The transmission of the new trait involves no change in genes, defying the basic neo-Darwinian tenet that genes are the only determinants of inherited traits.

The stimulus has no direct relevant effect on any cell or gene. It is in the brain where it is decoded and the response is determined. So, for example, decapitated females of the cowpea aphid, *Aphis craccivora* Koch (Johnson and Birks, 1960) and of *Megoura viciae* Buckton of the Aphidinae subfamily (Lees, 1967) that have been cued (by crowding-related tactile stimulation) to produce winged offspring, produce only wingless insects, clearly indicating that signals for wing development originate in the brain. This was the investigators' assessment:

> *Wing dimorphism is apparently controlled by a diffusible "alata-determiner" liberated from the head of the parent.*
>
> *Lees (1967)*

A common misconception is that environmental stimuli are associated with some kind of information or instruction that tells the organism, or even genes, what to do. This is not the case. The stimulus poses a problem that the organism must cope with. As already pointed out, no environmental stimulus, *per se*, can randomly, let alone adaptively, induce the expression of any gene or trigger any signal cascade. The epigenetic information to trigger a signal cascade or induce a particular gene is generated after processing the stimulus in the brain. That information is the output of the neural processing released as a chemical/electrical signal. The probability of inducing the TDP here increases from 0 (the probability that the environmental stimulus can induce it) to 1 and satisfies Shannon's definition of information.

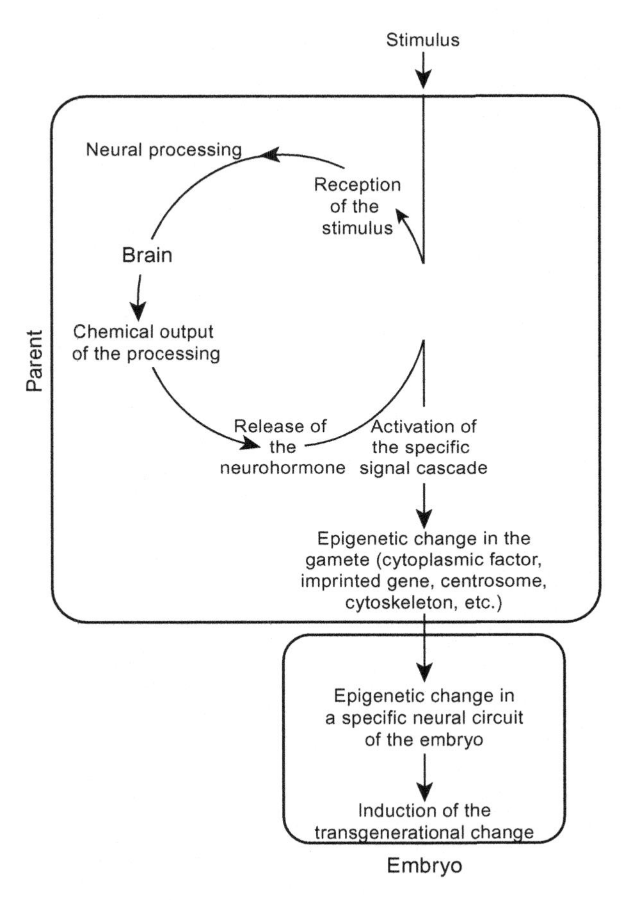

Figure 4.16 Diagrammatic representation of the "stimulus detour" mechanism of the induction of transgenerational developmental change. The neural reception of the stimulus and its processing in the neural circuit results in a chemical that induces the secretion of a neurohormone, which activates a specific signal cascade leading to an adaptive epigenetic change in gamete(s). The neural manipulation of the stimulus establishes a previously nonexisting causal relationship between the stimulus and the signal cascade that causes the transgenerational change.
Source: From Cabej (2012, p. 396).

Receiving the stimulus (i.e., visual, olfactory, tactile, and auditory), the sensory receptors/organs transform it into an electrical message. It is in this form that they transmit it to a specific center in the CNS. The brain then leads the encoded stimulus through a labyrinthine chain of connections across various centers for carrying out the neural processing (Figure 4.16). In the meantime, the involved neural circuits reconfigure their synaptic morphology (Choi et al., 2005), resulting in the modification of computational properties of neural circuits and specific changes in the chemical output (Getting, 1989).

The energetic cost of processing the stimulus in the brain is not negligible but neural processing is conserved in the course of evolution because of some benefit offsetting the cost. The benefit is that the processing figures out ways to adaptively (in the case of the TDPs, also predictively) respond to the challenges the stressful stimuli pose to the organism.

Thus, the information for the adaptive responses that lead to phenotypic (i.e., morphological, physiological, behavioral, and life history) changes in cases of TDP is generated through processing of stressful environmental stimuli in the brain.

Epigenetics of Behavior and Social Attachment in Animals

Here, behavior is defined as any motor action or sequence of motor actions the organism performs in response to an external/internal stimulus. Animal behaviors fall into two groups: innate behaviors, or instincts, and learned behaviors.

Innate behaviors are performed in perfect form from the beginning. An ameba engulfing debris, a newborn calf suckling its mother's teats, or a plant turning its flowers or leafs in the direction of the sun, are all examples of innate behaviors that these organisms perform for the first time without previous experience. Innate behaviors rely on the presence of inborn fixed action patterns (FAPs) that are activated in response to specific stimuli. FAPs result from the activation of specific neural circuits. Biologists have identified neural circuits for a number of innate behaviors (Marin-Burgin et al., 2006) and the location of some others in the brain (Balaban, 1997; Long et al., 2001; Teillet et al., 2005). How these neural circuits are hardwired prenatally (i.e., independent of experience) is one of modern biology's enigmas.

Darwin believed innate behaviors evolved in the past from learned behaviors:

> *Some intelligent actions, after being performed during several generations, become converted into instincts and are inherited.*
>
> *Darwin (1874)*

Darwin's idea still thrives in modern biology. Some empirical evidence seems to prove he was right. It is reported that the Australian native black snake *Pseudechis porphyriachus* started preying on the cane toad, *Bufo marinus*, when the toad was first introduced in the fifth continent. The prey was always lethal to the snake, but now, 23 generations later, the snake has developed an innate avoidance behavior toward the toad (Phillips and Shine, 2006). Solitarious desert locusts *S. gregaria* (Forskål) prefer to live alone and rest in their niches. They do not fly, except when disturbed. When they switch to the gregarious phase, they prefer to live in groups and fly with other locusts. This learned behavior is transmitted to the next generation as an innate behavior.

The fact that animals perform many behaviors such as chewing, breathing, swallowing, swimming, flying, crawling, and burrowing in the absence of stimuli and most of them cannot be modified by the sensory input led to the concept of the hardwired central pattern generators (CPGs), which are responsible for the above and other rhythmic behaviors.

Among the most complex innate behaviors regulated by CPGs are the electrogenic (production of electric discharges for communication with conspecifics, navigation, or prey localization) and electroreceptive phenomena in fish. Many of these fish have even evolved complex mechanisms of avoiding the jamming electric signals emitted by conspecifics by changing their electric signals' frequency (Heiligenberg et al., 1996; Metzner, 1999). In the vertebrate brain, the hypothalamus plays a central role in regulating innate behaviors and its lesion abolishes innate defensive behavior and avolitional eating (Manoli et al., 2008). For instance, activation of the hypothalamic ventromedial region (VMH) sends permissive signals to the midbrain centers for lordosis and regulates the reproductive behavior, including its synchronization with ovulation (Pfaff, 2005).

Learned behaviors in metazoans are acquired by experience during their lifetime. Like innate behaviors, learned behaviors are patterned by specific neural circuits, but, unlike them, they may be perfected over time. All learned behaviors rely on the presence of innate motor programs or program preadaptations in the CNS. This may explain why some learned behaviors, through repeated practicing, may be performed automatically, unconsciously, like innate behaviors.

> *The emerging picture, then, whether it be the wholly innate performance of spiders, the goal-directed learning of infants, or the plastic learning we see in piano playing, is that all are routines stored, either from the outset or ultimately, as discrete neural programs.*
>
> *Gould (1982)*

Learning new behaviors may be facilitated by the flexibility of neural circuits, which may modify connections between neurons in order to generate new behaviors. Neural circuits sometimes are conserved after the loss of a behavior. When ancestral conditions in the environment recur, learning new behavior may result from activating the inactive ancestral circuit.

Although learned behaviors are acquired through experience, they also have a heritable component. Since any inherited trait is, even unconsciously, considered a function of genes, let us briefly look at the role genes play in behavior. As mentioned earlier, almost all behavior implies a motor action. It is not easy to see how one or a number of genes can determine an adaptive action in response to an external stimulus. Even a simple behavior, such as walking, involves activation and coordination of numerous leg muscles and joints and even most muscles and joints of the animal body. The commands for activation and coordination of muscles come from the CNS, not from genes. This is not to say that genes are not involved in walking (gene defects may prevent animals and humans from walking), but simply that they do not have the information necessary for walking or any other behavior. It may be argued that genes may be necessary for the establishment or functioning of neural circuits determining behaviors, but so are many metabolites:

> *Any behavior requires the functioning of a multicellular circuit beginning with input to the nervous system, propagation and interpretation of that input in the CNS, and output via neurons that direct a response via neuromuscular, or neuroendocrine systems, or both.*
>
> *Baker et al. (2001)*

Figure 4.17 Prairie voles parenting their puppies. http://jeancmiranda.blogspot.com/2011/06/as-sustancias-quimicas-e-as-estrategias.html

There are numerous examples demonstrating that a species performs an innate behavior that another closely related species does not and cannot although the latter has all the genes needed to perform the behavior.

We will illustrate this with the case of the social bonding that only some voles display. The prairie vole, *Microtus ochrogaster* (family Muridae, subfamily Arvicolinae), is a native small mammal of North America. This is a monogamous species that forms enduring pair bonds between mates (Young et al., 1998, 2001). They share a nest during the breeding season and remain together for life (Wang and Aragona, 2004). The male partner displays parental care and mate guarding. A closely related vole species, the montane vole, *Microtus montanus*, is polygamous and does not display parental care (Figure 4.17).

The social attachment in the prairie vole is related to the brain reward system with the nucleus accumbens (NAcc) as a central player. The most relevant changes at the molecular level occur in NAcc and related brain regions and involve neuropeptides oxytocin and AVP and DA. Within 3 days of cohabitating with a female, an increase of AVP mRNA in the bed nucleus of the stria terminalis (BST) and a decrease in the lateral septum (LS) occurs in males. The expression of receptors for AVP and for oxytocin is different in the relevant regions in monogamous and polygamous voles (Wang and Aragona, 2004). Prairie voles have more receptors for vasopressin in the NAcc (Figure 4.18). DA activates D2-type receptors and is crucial to the role of NAcc in vole monogamous behavior (Wang and Aragona, 2004). DAergic neurons are present in other adjacent brain regions (Young et al., 2008).

Experimental evidence showed that DA antagonists block mating-induced partner preferences, whereas DA agonists induce it in the absence of mating (Aragona et al., 2003). Expression of the DA receptor D2R in NAcc facilitates bond formation (Young et al., 2011).

No functional differences exist between neuropeptides AVP and oxytocin between monogamous and polygamous voles. The only differences between them are differences in the *expression* of receptors for AVP, oxytocin and DA (i.e., the epigenetic differences in their brains).

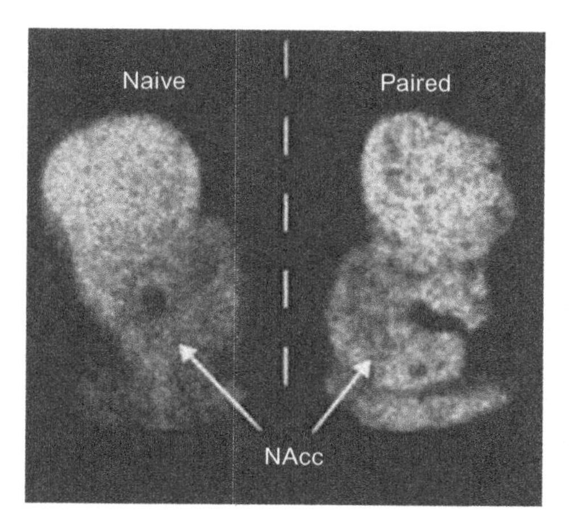

Figure 4.18 Photomicrographs showing increased D1-like receptor binding in the NAcc of male prairie voles that were pair bonded for 2 weeks (Paired) in comparison to sexually naive males (Naive).
Source: From Young et al. (2008).

References

Abrous, D.N., Koehl, M., Le Moal, M., 2005. Adult neurogenesis: from precursors to network and physiology. Physiol. Rev. 85, 523–569.

Adcock, I.M., Ford, P., Ito, K., Barnes, P.J., 2006. Epigenetics and airways disease. Respir. Res. 7, 21.

Altman, J., 1963. Autoradiographic investigation of cell proliferation in the brains of rats and cats. Anat. Rec. 145, 573–591.

Aragona, B.J., Liu, Y., Curtis, J.T., Stephan, F.K., Wang, Z., 2003. A critical role for nucleus accumbens dopamine in partner-preference formation in male prairie voles. J. Neurosci. 23, 3483–3490.

Baker, B.S., Taylor, B.J., Hall, J.C., 2001. Are complex behaviors specified by dedicated regulatory genes? Reasoning from *Drosophila*. Cell 105, 13–24.

Balaban, E., 1997. Changes in multiple brain regions underlie species differences in a complex, congenital behavior. Proc. Natl. Acad. Sci. U.S.A. 94, 2001–2006.

Bargmann, C.I., 2006. Comparative chemosensation from receptors to ecology. Nature 444, 295–301.

Bartness, T.J., Song, C.K., 2005. Innervation of brown adipose tissue and its role in thermogenesis. Can. J. Diabetes 29, 420–428.

Bradshaw, W.E., Holzapfel, C.M., 2010. What season is it anyway? Circadian tracking vs. photoperiodic anticipation in insects. J. Biol. Rhythms 25, 155–165.

Burrows, M., Rogers, S.M., Ott, S.R., 2011. Epigenetic remodelling of brain, body and behaviour during phase change in locusts. Neural Syst. Circuits 1, 11.

Cabej, N.R., 2005. Neural Control of Development. Albanet, Dumont, NJ, pp. 37–47.

Cabej, N.R., 2008. Epigenetic Principles of Evolution. Albanet, Dumont, NJ, pp. 39–52.

Cabej, N.R., 2010. Epigenetic Principles of Evolution. Albanet, Dumont, NJ, pp. 211–212.

Cabej, N.R., 2012. Epigenetic Principles of Evolution. Elsevier Inc., London/Waltham, MA, p. 47.

Challet, E., Caldelas, I., Graff, C., Pévet, P., 2003. Synchronization of the molecular clockwork by light- and food-related cues in mammals. Biol. Chem. 384, 711–719.

Champagne, D.L., Bagot, R.C., van Hasselt, F., Ramakers, G., Meaney, M.J., de Kloet, E.R., et al., 2008. Maternal care and hippocampal plasticity: evidence for experience-dependent structural plasticity, altered synaptic functioning, and differential responsiveness to glucocorticoids and stress. J. Neurosci. 28, 6037–6045.

Chivers, D.P., Kiesecker, J.M., Marco, A., Wildy, E.L., Blaustein, A.R., 1999. Shifts in life history as a response to predation in western toads (*Bufo boreas*). J. Chem. Ecol. 25, 2455–2463.

Choi, S.-Y., Chang, J., Jiang, B., Seol, G.-H., Min, S.-S., Han, J.-S., et al., 2005. PLC and LTD in visual cortex. J. Neurosci. 25, 11433–11443.

Crosio, C., Heitz, E., Allis, C.D., Borrelli, E., Sassone-Corsi, P., 2003. Chromatin remodeling and neuronal response: multiple signaling pathways induce specific histone H3 modifications and early gene expression in hippocampal neurons. J. Cell Sci. 116, 4905–4914.

Curie, D., Bate, M., 1991. The development of adult abdominal muscles in *Drosophila*: myoblasts express *twist* and are associated with nerves. Development 113, 91–102.

Darwin, C., 1874. The Descent of Man and Selection in Relation to Sex, second ed. John Murray, London, Revised and augmented, p. 67.

DeLoof, A., Claeys, I., Simonet, G., Verleyen, P., Vandersmissen, T., Sas, F., et al., 2006. Molecular markers of phase transition in locusts. Insect Sci. 13, 3–12.

Denlinger, D.L., Yocum, G.D., Rinehart, J.P., 2011. Hormonal control of diapause. In: Gilbert, L.I. (Ed.), Insect Endocrinology Elsevier, Amsterdam, pp. 430–463.

Dibner, C., Schibler, U., Albrecht, U., 2010. The mammalian circadian timing system: organization and coordination of central and peripheral clocks. Annu. Rev. Physiol. 72, 517–549.

Ditty, J.L., Williams, S.B., Golden, S.S., 2003. A cyanobacterial circadian timing mechanism. Annu. Rev. Genet. 37, 513–543.

Dodd, A.N., Salathia, N., Hall, A., Kévei, E., Tóth, R., Nagy, F., et al., 2005. Plant circadian clocks increase photosynthesis, growth, survival, and competitive advantage. Science 309, 630–633.

Donohue, K., 2009. Completing the cycle: maternal effects as the missing link in plant life histories. Phil. Trans. R. Soc. B. 364, 1059–1074.

Dulac, C., 2010. Brain function and chromatin plasticity. Nature 465, 728–735.

Engelmann, W., 2009. Cell clocks. Available from: <http://tobias-lib.uni-tuebingen.de/volltexte/2009/3993/pdf/cellclocks20090517UBT.pdf>

Fuchs, E., Kutsch, W., Ayali, A., 2003. Neural correlates to flight-related density-dependent phase characteristics in locusts. J. Neurobiol. 57, 152–162.

Galloway, L.F., Etterson, J.R., 2007. Transgenerational plasticity is adaptive in the wild. Science 318, 1134–1136.

Gerritsen, J., 1982. Behavioral response of *Daphnia* to rate of temperature change: possible enhancement of vertical migration. Limnol. Oceanogr. 27, 254–261.

Getting, P.A., 1989. Emerging principles governing the operation of neural networks. Ann. Rev. Neurosci. 12, 185–204.

Gould, J.L., 1982. Ethology—The Mechanisms and Evolution of Behavior. W.W. Norton & Co., New York, p. 177.

Guilding, C., Piggins, H.D., 2007. Challenging the omnipotence of the suprachiasmatic timekeeper: are circadian oscillators present throughout the mammalian brain? Eur. J. Neurosci. 25, 3195–3216.

Gwinner, E., Brandstatter, R., 2001. Complex bird clocks. Phil. Trans. R. Soc. Lond. B 356, 1801–1810.

Hanazato, T., Dodson, S., 1992. Complex effects of a kairomone of *Chaoborus* and an insecticide on *Daphnia pulex*. J. Plankton Res. 14, 1743–1755.
Hallberg, E., Johansson, K.U., Elofsson, R., 1992. The Aesthetasc concept: Structural variations of putative olfactory receptor cell complexes in Crustacea. Microsc. Res. Tech. 22, 325–335.
Hegstrom, C.D., Riddiford, L.M., Trurnan, J.W., 1998. Steroid and neuronal regulation of ecdysone receptor expression during metamorphosis of muscle in the moth, *Manduca sexta*. J. Neurosci. 18, 1786–1794.
Heiligenberg, W., Metzner, W., Wong, C.J.H., Keller, C.H., 1996. Motor control of the jamming avoidance response of *Apteronotus leptorhynchus*: evolutionary changes of a behavior and its neuronal substrates. J. Comp. Physiol. A 179, 653–674.
Henrich, V.C., Denlinger, D.L., 1982. A maternal effect that eliminates pupal diapause in progeny of the flesh fly, *Sarcophaga bullata*. J. Insect Physiol. 28, 881–884.
Herman, J.J., Sultan, S.E., Horgan-Kobelski, T., Riggs, C., 2012. Adaptive transgenerational plasticity in an annual plant: grandparental and parental drought stress enhance performance of seedlings in dry soil. Integr. Comp. Biol. 52, 77–88.
Homma, T., Watanabe, K., Tsurumaru, S., Kataoka, H., Imai, K., Kamba, M., et al., 2006. G protein-coupled receptor for diapause hormone, an inducer of *Bombyx* embryonic diapause. Biochem. Biophys. Res. Commun. 344, 386–393.
Hoverman, J.T., Auld, J.R., Relyea, R.A., 2005. Putting prey back together again: integrating predator-induced behavior, morphology, and life history. Oecologia 144, 481–491.
Ichikawa, T., 2003. Firing activities of neurosecretory cells producing diapause hormone and its related peptides in the female silkworm, *Bombyx mori*. I. Labial cells. Zool. Sci. 20, 971–978.
Imai, M., Naraki, Y., Tochinai, S., Miura, T., 2009. Elaborate regulations of the predator-induced polyphenism in the water flea *Daphnia pulex*: kairomone-sensitive periods and life-history tradeoffs. J. Exp. Zool. A Ecol. Genet. Physiol. 311A, 788–795.
Johnson, B., Birks, P.R., 1960. Studies on wing polymorphism in aphids. I. The developmental process involved in the production of the different forms. Entomol. Exp. Appl. 3, 327–339.
Kamei, Y., Hasegawa, Y., Niimi, T., Yamashita, O., Yaginuma, T., 2011. Trehalase-2 protein contributes to trehalase activity enhanced by diapause hormone in developing ovaries of the silkworm, *Bombyx mori*. J. Insect Physiol. 57, 608–613.
Kihara, F., Itoh, K., Iwasaka, M., Niimi, T., Yamashita, O., Yaginuma, T., 2009. Glycerol kinase activity and *glycerol kinase-3* gene are up-regulated by acclimation to 5°C in diapause eggs of the silkworm, *Bombyx mori*. Insect Biochem. Mol. Biol. 39, 763–769.
Kim, M.Y., Hsiao, S.J., Kraus, W.L., 2001. A role for coactivators and histone acetylation in estrogen receptor a-mediated transcription initiation. EMBO J. 20, 6084–6094.
Kitagawa, N., Shiomi, K., Imai, K., Niimi, T., Yaginuma, T., Yamashita, O., 2005. Establishment of a sandwich ELISA system to detect diapause hormone, and developmental profile of hormone levels in egg and subesophageal ganglion in the silkworm, *Bombyx mori*. Zool. Sci. 22, 213–221.
Kondo, T., Ishiura, M., 2000. The circadian clock of cyanobacteria. BioEssays 22, 10–15.
Koštál, V., 2011. Insect photoperiodic calendar and circadian clock: independence, cooperation, or unity? J. Insect Physiol. 57, 538–556.
Kumar, S., Mohan, A., Sharma, V.K., 2005. Circadian dysfunction reduces lifespan in *Drosophila melanogaster*. Chronobiol. Int. 22, 641–653.
Kunert, G., Otto, S., Roese, U.S.R., Gershenzon, J., Weisser, W.W., 2005. Alarm pheromone mediates production of winged dispersal morphs in aphids. Ecol. Lett. 8, 596–603.
Laurent, G., MacLeod, K., Stopfer, M., Wehr, M., 1998. Spatiotemporal structure of olfactory inputs to the mushroom bodies. Learn. Memory 5, 124–132.

LaVoie, H.A., 2005. Epigenetic control of ovarian function: the emerging role of histone modifications. Mol. Cell. Endocrinol. 243, 12–18.

LeBlanc, G.A., 2007. Crustacean endocrine toxicology: a review. Ecotoxicology 16, 61–81.

Lees, A.D., 1967. The production of the apterous and alate forms in the aphid *Megoura viciae*, with special reference to the role of crowding. J. Insect Physiol. 13, 289–318.

Lockett, G.A., Kucharski, R., Maleszka, R., 2012. DNA methylation changes elicited by social stimuli in the brains of worker honey bees. Genes Brain Behav. 11, 235–242.

Long, K.D., Kennedy, G., Balaban, E., 2001. Transferring and inborn auditory, perceptual predisposition with interspecies brain transplants. Proc. Natl. Acad. Sci. U.S.A. 98, 5862–5867.

Lyford, G.L., Yamagata, K., Kaufmann, W.E., Barnes, C.A., Sanders, L.K., Copeland, N.G., et al., 1995. *Arc*, a growth factor and activity-regulated gene, encodes a novel cytoskeleton-associated protein that is enriched in neuronal dendrites. Neuron 14, 433–445.

Ma, Z., Guo, W., Guo, X., Wang, X., Kang, L., 2011. Modulation of behavioral phase changes of the migratory locust by the catecholamine metabolic pathway. Proc. Natl. Acad. Sci. U.S.A. 108, 3882–3887.

Manoli, D.S., Meissner, G.W., Baker, B.S., 2008. Blueprints for behavior: genetic specification of neural circuitry for innate behaviors. Trends Neurosci. 29, 444–451.

Marin-Burgin, A., Eisenhart, F.J., Kristan Jr., W.B., French, K.A., 2006. Embryonic electrical connections appear to prefigure a behavioral circuit in the leech CNS. J. Comp. Physiol. A Neuroethol. Sens. Neural Behav. Physiol. 192, 123–133.

Maruska, K.P., Fernald, R.D., 2011. Social regulation of gene expression in the hypothalamic–pituitary–gonadal axis. Physiology 26, 412–423.

Mayerhof, A., Dissen, G.A., Costa, M.F., Ojeda, S.R., 1997. A role for neurotransmitters in early follicular development: induction of functional follicle-stimulating hormone receptors in newly formed follicles of the rat ovary. Endocrinology 138, 3320–3329.

McClung, C.R., 2006. Plant circadian rhythms. Plant Cell 18, 792–803.

McClung, C.R., 2008. Comes a time. Curr. Opin. Plant Biol. 11, 514–520.

Meaney, M.J., Szyf, M., 2005a. Maternal care as a model for experience-dependent chromatin plasticity? Trends Neurosci. 28, 456–463.

Meaney, M.J., Szyf, M., 2005b. Environmental programming of stress responses through DNA methylation: life at the interface between a dynamic environment and a fixed genome. Dialogues Clin. Neurosci. 7, 103–123.

Metzner, W., 1999. Neural circuitry for communication and jamming avoidance in gymnotiform electric fish. J. Exp. Biol. 202, 1365–1375.

Miyakawa, H., Imai, M., Sugimoto, N., Ishikawa, Y., Ishikawa, A., Ishigaki, H., et al., 2010. Gene up-regulation in response to predator kairomones in the water flea, *Daphnia pulex*. BMC Dev. Biol. 10, 45.

Mizzen, C., 2012. Chromatin Structure, Function and Metabolism. Available from: <http://mcb.illinois.edu/faculty/profile/cmizzen>

Mohawk, J.A., Green, C.B., Takahashi, J.S., 2012. Central and peripheral circadian clocks in mammals. Annu. Rev. Neurosci. 35, 445–462.

Moore, J.M.R., Guy, R.K., 2005. Coregulator interactions with the thyroid hormone receptor. Mol. Cell. Proteomics. 4, 475–482.

Morrison, S.F., 2004. Activation of 5-HT1A receptors in raphe pallidus inhibits leptin-evoked increases in brown adipose tissue thermogenesis. Am. J. Physiol. Regul. Integr. Comp. Physiol. 286, R832–R837.

Murgatroyd, C., Wu, Y., Bockmühl, Y., Spengler, D., 2010. Genes learn from stress—how infantile trauma programs us for depression. Epigenetics 5 (3), 194–199.

Nijhout, H.F., 1990. Metaphors and the role of genes in development. BioEssays 12, 441–446.
Nijhout, H.F., 1999. Polyphenic development in insects. Bioscience 49, 181–192.
Nishimura, M., Yamagata, K., Sugiura, H., Okamura, H., 2003. The activity-regulated cytoskeleton-associated (*Arc*) gene is a new light-inducible early gene in the mouse suprachiasmatic nucleus. Neuroscience 116, 1141–1147.
Nitabach, M.N., Taghert, P.H., 2008. Organization of the *Drosophila* circadian control circuit. Curr. Biol. 18, R84–R93.
Noguchi, H., Hayakawa, Y., 2001. Dopamine is a key factor for the induction of egg diapause of the silkworm, *Bombyx mori*. Eur. J. Biochem. 268, 774–780.
Olmstead, A.W., LeBlanc, G.A., 2007. The environmental-endocrine basis of gynandromorphism (intersex) in a crustacean. Int. J. Biol. Sci. 3, 77–84.
Paranjpe, D.A., Sharma, V.K., 2005. Evolution of temporal order in living organisms. J. Circadian Rhythms 3, 7.
Pfaff, D., 2005. Hormone-driven mechanisms in the central nervous system facilitate the analysis of mammalian behaviours. J. Endocrinol. 184, 447–453.
Phillips, B.L., Shine, R., 2006. An invasive species induces rapid adaptive change in a native predator: cane toads and black snakes in Australia. Proc. Biol. Sci 273, 1545–1550.
Reznik, S.Y., Vaghina, N.P., Voinovich, N.D., 2012. Multigenerational maternal effect on diapause induction in *Trichogramma* species (Hymenoptera: Trichogrammatidae). Biocontrol Sci. Technol. 22, 429–445.
Robinson, K.L., Tohidi-Esfahani, D., Lo, N., Simpson, S.J., Sword, G.A., 2011. Evidence for widespread genomic methylation in the migratory locust, *Locusta migratoria* (Orthoptera: Acrididae). PLoS ONE 6 (12), e28167.
Rockey, S.J., Yoder, J.A., Denlinger, D.L., 1991. Reproductive and developmental consequences of a diapause maternal effect in the flesh fly, *Sarcophaga bullata*. Physiol. Entomol. 16, 477–483.
Rogers, S.M., Matheson, T., Sasaki, K., Kendrick, K., Simpson, S.J., Burrows, M., 2004. Substantial changes in central nervous system neurotransmitters and neuromodulators accompany phase change in the locust. J. Exp. Biol. 207, 3603–3617.
Rokas, A., 2008. The origins of multicellularity and the early history of the genetic toolkit for animal development. Annu. Rev. Genet. 42, 235–251.
Ruh, M.F., Tian, S., Cox, L.K., Ruh, T.S., 1999. The effects of histone acetylation on estrogen responsiveness in MCF-7 cells. Endocrine 11, 157–164.
Salvador, L.M., Cottom, J., Park, Y., Maizels, E.T., Jones, J.C., Schillace, R.V., et al., 2001. Follicle-stimulating hormone stimulates protein kinase A-mediated histone H_3 phosphorylation and acetylation leading to select gene activation in ovarian granulosa cells. J. Biol. Chem. 276, 40146–40155.
Schiesari, L., Kyriacou, C.P., Costa, R., 2011. The hormonal and circadian basis for insect photoperiodic timing. FEBS Lett. 585, 1450–1460.
Sharma, V.K., Chandrashekaran, M.K., 2005. Zeitgebers (*time cues*) for biological clocks. Current Sci. 89, 1136–1146.
Shimizu, I., Matsui, T., Hasegawa, K., 1989. Possible involvement of GABAergic neurons in regulation of diapause hormone secretion in the silkworm, *Bombyx mori*. Zool. Sci. 6, 809–819.
Shine, R., Downes, S.J., 1999. Can pregnant lizards adjust their offspring phenotypes to environmental conditions? Oecologia 119, 1–8.
Shiomi, K., Fujiwara, Y., Yasukochi, Y., Kajiura, Z., Nakagaki, M., Yaginuma, T., 2007. The *Pitx* homeobox gene in *Bombyx mori*: regulation of *DH-PBAN* neuropeptide hormone gene expression. Mol. Cell. Neurosci. 34, 209–218.

Shweiki, D., 1999. The physical imperative in circadian rhythm: a cytoskeleton-related physically resettable clock mechanism hypothesis. Med. Hypotheses 53, 413–420.

Sonoda, J., Pei, L., Evans, R.M., 2008. Nuclear receptors: decoding metabolic disease. FEBS Lett. 582, 2–9.

Sweeney, B.M., Borgese, M.B., 1989. A circadian rhythm in cell division in a prokaryote, the cyanobacterium *Synechococcus* WH7803. J. Phycol. 25, 183–186.

Tanner, C.J., Branstrator, D.K., 2006. Generational and dual-species exposures to invertebrate predators influence relative head size in *Daphnia mendotae*. J. Plankton Res. 28, 793–802.

Teillet, M.-A., Naquet, R., Batini, C., 2005. Transfer of an avian genetic reflex epilepsy by embryonic brain graft: a tissue autonomous process? Int. J. Dev. Biol. 49, 237–241.

Tobias, M.L., Marin, M.L., Kelley, D.B., 1993. The roles of sex, innervation, and androgen in laryngeal muscle of *Xenopus laevis*. J. Neurosci. 13, 324–333.

Toda, H., Hamani, C., Fawcett, A.P., Hutchison, W.D., Lozano, A.M., 2008. The regulation of adult rodent hippocampal neurogenesis by deep brain stimulation. J. Neurosurg. 108, 132–138.

Wang, Z., Aragona, B.J., 2004. Neurochemical regulation of pair bonding in male prairie voles. Physiol. Behav. 83, 319–328.

Webb, M.-L., Denlinger, D.L., 1998. GABA and picrotoxin alter expression of a maternal effect that influences pupal diapause in the flesh fly, *Sarcophaga bullata*. Physiol. Entomol. 23, 184–191.

Weiss, L., Laforsch, C., Tollrian, R., 2012. The taste of predation and the defences of the prey. In: Brönmark, C., Hansson, L.A. (Eds.), Chemical Ecology in Aquatic Systems. Oxford University Press, Oxford/New York.

Xu, W.H., Denlinger, D.L., 2003. Molecular characterization of prothoracicotropic hormone and diapause hormone in *Heliothis virescens* during diapause, and a new role for diapause hormone. Insect Mol. Biol. 12, 509–516.

Young, K.A., Liu, Y., Wang, Z., 2008. The neurobiology of social attachment: a comparative approach to behavioral, neuroanatomical, and neurochemical studies. Comp. Biochem. Physiol. C Toxicol. Pharmacol. 148, 401–410.

Young, K.A., Gobrogge, K.L., Liu, Y., Wang, Z., 2011. The neurobiology of pairs bonding: insights from a socially monogamous rodent. Front. Neuroendocrinol. 32, 53–69.

Young, L.J., Wang, Z., Insel, T.R., 1998. Neuroendocrine bases of monogamy. Trends Neurosci. 21, 71–75.

Young, L.J., Lim, M.M., Gingrich, B., Insel, T.R., 2001. Cellular mechanisms of social attachment. Horm. Behav. 40, 133–138.

Zhang, T.-Y., Sun, J.-S., Zhang, L.-B., Shen, J.-L., Xu, W.-H., 2004. Cloning and expression of the cDNA encoding the FXPRL family of peptides and a functional analysis of their effect on breaking pupal diapause in *Helicoverpa armigera*. J. Insect Physiol. 50, 25–33.

Zhong, L., Yan, C.H., Huang, H., Lu, C.Q., Xu, J., Yu, X.G., et al., 2007. Preweaning exposure to enriched environment induces hippocampal neurogenesis: experiment with rats. Zhonghua Yi Xue Za Zhi 87, 1559–1563. (in Chinese) (English abstract).

Ziegler, S., Eiseler, T., Scholz, R.-P., Beck, A., Link, G., Hausser, A., 2007. A novel PKD phosphorylation site in the tumor suppressor RIN1 is critical for coordination of cell migration. Available from: <http://www.molbiolcell.org/content/early/2011/01/05/mbc.E10-05-0427.full.pdf>

5 Rise of the Animal Kingdom and Epigenetic Mechanisms of Evolution

The Prelude to the Cambrian Burst of Animal Diversity

The mechanisms and driving forces of evolution are central to biological investigation, but a full discussion of these topics is out of the scope of this chapter. Instead, I intend to focus on the major Cambrian diversification in the animal kingdom, where effects of these mechanisms and forces manifest the best.

It is estimated that life emerged on Earth about 3.8 billion years (Holland, 1997) (i.e., about 1 billion years after the Earth formed as a planet of the solar system). The discovery of fossil stromatolites, colonies of extant cyanobacteria, indicates that life started with prokaryote unicellular microorganisms. Transition to eukaryotic unicellular algae occurred about 1.8–1.7 Mya (Bengtson et al., 2009; Conway Morris, 2003). Metazoans appeared not earlier than 650 Mya (Conway Morris, 2003) and complex animal multicellulars appeared only about 635–543 Mya (Knoll et al., 2004).

Ignoring unconfirmed reports, Valentine et al. (1999) believe the earliest indications of metazoan forms are represented by a number of small disks and fossil creeping traces from Canada about 610–590 Mya, which was immediately followed by the sparely diversified Ediacaran fauna (named after a fossil Lagerstätte in Ediacara Hills, southern Australia, discovered in 1946 by Reginald Sprigg) that lasted from about 580 to 542 Mya. These soft-bodied animals are preserved as impressions in rapid sediment depositions or volcanic ash (Narbonne, 2005) (Figure 5.1). The Ediacaran period was dominated by unicellular organisms but, by the end of the period, multicellular forms became more frequent. The Ediacaran forms have been mostly sessile diploblastic-grade organisms of various forms, including bags, tubes, and leaves, ranging from microscopic to a few millimeters/centimeters to 1 m in length. Their position in the animal kingdom is still uncertain.

Including the Ediacaran organisms in the animal kingdom is questionable; some paleontologists consider them related to bilaterian animals, some to prebilaterian animals, and others to plants, fungi, or even prokaryotes (Erwin, 2009). Peterson et al. (2003) believe Ediacaran forms may belong to quite different groups or kingdoms (Peterson et al., 2003). These organisms lacked basic animal organs, such as mouth, digestive, and other internal organs, eyes, and appendages (hence they could only move by peristaltic movements). Most investigators believe they were membranous organisms of two or more unicellular layers. In Seilacher's vivid expression, Ediacaran forms are "air-mattress-like constructions", mostly of discoidal morphology and of considerable size.

Building the Most Complex Structure on Earth. DOI: http://dx.doi.org/10.1016/B978-0-12-401667-5.00005-5

Bilaterian triplobastic, but protostome and deuterostome lineages seem to first appear around 570 Mya (Erwin, 1999). According to Peterson et al. (2007), the last common ancestor (LCA) of protostomes and deuterostomes evolved 580 Mya.

These Ediacaran forms, known as the "vendobiont" phylum, have no confirmed relationship with the Cambrian forms. Together with radiates and bilaterians, they form their own Pre-Cambrian group before gradually becoming extinct at the outset of the Cambrian period. According to Peterson et al.:

> *Ediacaran "survivors" in the Cambrian can be counted on the fingers of one hand.*
>
> *Peterson et al. (2003)*

If this is true, the group may be considered a failed experiment in multicellular life, although there is evidence that a small number of the Ediacaran forms survived into the Cambrian. The extinction of the Ediacaran biota coincided with the onset of the Cambrian fauna about 542 Mya, which marks the beginning of the Phaneorzoic eon. Among the suggested causes for the demise of the Ediacaran fauna are the evolution of the more successful Cambrian forms, which preyed on immobile soft-bodied Ediacaran species (Bengtson, 2002), as well as a global short-term anoxia, and widespread methane release (Seilacher, 1984).

Cambrian Explosion

For five-sixths of its existence, life on Earth evolved slowly, mainly in the form of unicellular organisms and in accordance with the Darwinian principle of gradualism. Then abruptly, about 543 Mya, a new, eumetazoan fauna emerged in a still little understood way.

Metazoan life began with the fossil record of the Cambrian[1] period, dated at 542–485 Mya. Almost all the extant and several extinct metazoan phyla and body plans emerged in geologically rapid succession (phylum *Bryozoa* was the last to evolve by the end of the Cambrian about 490 Mya) within the evolutionarily short span of about 10 million years from about 530–520 Mya (i.e., within less than 2% of the time from the base of the Cambrian to the present) (Conway Morris, 2000; Roth et al., 2007; Valentine et al., 1999).

> *Little high-level morphological innovation occurred during the subsequent 500 million years in that much of animal disparity, as measured by the Linnean taxonomic ranking, was achieved early in the radiation.*
>
> *Erwin et al. (2011)*

In 1948, Preston Cloud (1912–1991) concluded that the discontinuation of the Cambrian fauna from the Pre-Cambrian was not an illusion that derived from the

[1] Cambrian derives from Cambria, the Latinized form of the Cymry, the Celtic name of the Welsh people. The term Cambrian period was introduced in 1836 by paleontologist Adam Sedgwick (1785–1873), one of Darwin's mentors, who studied the fossil fauna of the period in rocks from Wales.

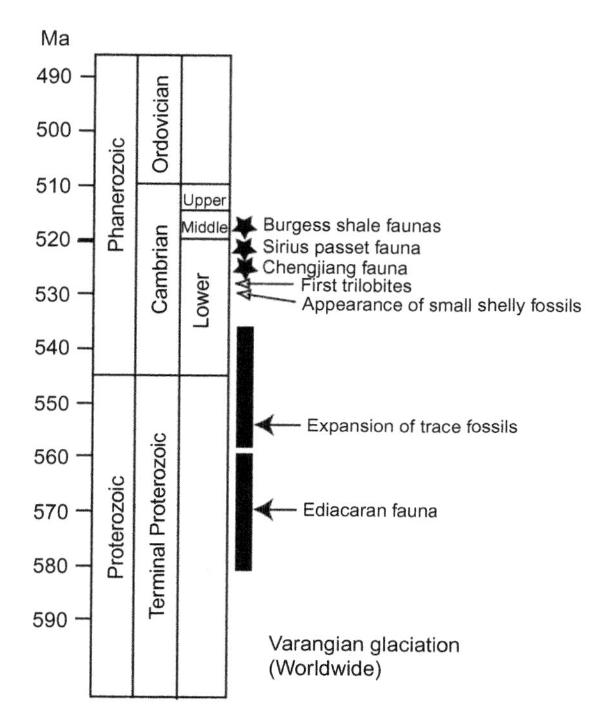

Figure 5.1 The location of the Cambrian period in geological time with some important fossil occurrences and events.
Source: Levinton (2001).

incompleteness of the fossil record but instead was a sudden event that took place more than half a billion years ago (Levinton, 2001, pp. 46–48). For gradualist educated biologists, the exceptionally rapid tempo of evolution during the Cambrian represents an unresolved puzzle that since the early 1970s has been called the *Cambrian explosion*.

Is the Cambrian Explosion a Fact or a Paleontological Artifact?

No consensus exists on whether the Cambrian explosion is a sudden evolutionary or Simpsonian event. Some authors believe that the divergence of metazoan clades occurred much earlier in the Earth's history; hence, the Cambrian explosion may be a normal, gradual Darwinian process of evolution (Ayala et al., 1998; Fortey et al., 1996; Lieberman, 2003). But there are two serious arguments against these conjectural conclusions. Firstly, their estimations are based on molecular clock estimates alone; that is on base substitutions that have taken place in particular genes, which many biologists, including one leading neoDarwinian, generally admit to be often erratic and unreliable because even within clades, rates of change are different for different genes and vary over time (Ayala, 1997; Valentine et al., 1999). It is no wonder that the molecular clock estimates of the time of divergence of metazoan clades by different authors have led to values that vary as widely as about 800 million years, ranging from more than 1500 (Bromham et al., 1998) to 670 Mya (Ayala et al., 1998). So, for example, Runnegar (1986) estimated the time of divergence of the animal phyla (protostome and deuterostomes) at 700 million years, Wray et al. (1996) push it back to mid-Proterozoic, about 1200 Mya, which predates the Cambrian explosion by 400–600 million years, while Wang et al. (1999) estimate that it occurred about 1000 Mya.

The second insurmountable problem is that no fossil record or other empirical evidence corroborates or substantiates their estimates on the Proterozoic origin of metazoans and protostome and deuterostome divergence. It is not easy to answer why the bilaterians did not leave fossils, as they did later, if they existed at the time.

Deniers of the Cambrian explosion have conjectured that the Cambrian explosion may be a paleontological artifact because soft-bodied, Pre-Cambrian metazoans that evolved long before left no fossils and it was only during the Cambrian period that they evolved exoskeletons and skeletons that made their fossilization possible. This assumption is also rejected because most of the known Cambrian fossils are from soft-bodied metazoans (Conway Morris, 1998) and skeletization of metazoans is not a universal feature of the Cambrian explosion. Besides, microscopic, nonmetazoan fossils of soft-bodied Pre-Cambrian unicellulars abound. For the above reasons, the molecular clock faces increasing criticism.

What to do with molecular data then? Keep in mind that Bromham et al. (1998) had this advice:

> *Molecular dates can be misleading.... The molecular clock is unlikely ever to replace the fossil record as the primary source of information on evolution in deep time. But it has a critical role to play as an alternative historical narrative, potentially complementing the biases and gaps of the paleontological record.*
>
> *Bromham (2009)*

The reality of the Cambrian explosion and its impact on the extant animal world in our time is acknowledged by an overwhelming majority of biologists:

> *The Cambrian explosion is real and its consequences set in motion a sea-change in evolutionary history.*
>
> *Conway Morris (2000)*

The Cambrian Explosion Conundrum

Although not fully understood, the "profound discontinuity" between the Cambrian and Pre-Cambrian forms (Conway Morris, 1989) is a well-known fact. The emergence and evolutionary success of the Cambrian fauna stand in stark contrast to the preceding Ediacaran biota. The sudden upsurge in the diversification of animal forms and the rapidly increasing complexity of their structure is still waiting for a scientific explanation.

Darwin was also puzzled:

> *To the question why we do not find rich fossiliferous deposits belonging to these assumed earliest periods prior to the Cambrian system, I can give no satisfactory answer.*
>
> *Darwin (1859)*

To solve the enigma, Darwin assumed that the sudden appearance of animal diversity and the absence of transitional forms in the Cambrian fossil record might be related to the incompleteness of the record. However, the discovery of the Ediacaran biota of the Pre-Cambrian eon failed to provide any transitional forms that would connect them to animal diversification and the appearance of about 40 body plans that evolved within the petite window of the Cambrian:

> *the appearance of the remains and traces of bilaterian animals in the Cambrian remains abrupt.*
>
> *Erwin et al. (2011)*

Attempts are made to explain the scarcity of Ediacaran fossils lacking skeletons or exoskeletons of Ediacaran forms. This is rejected by the relative abundance of fossils of soft-bodied Cambrian cnidarians, ctenophores, echinoderms, and so on.

Even a glimpse at the Cambrian explosion and the blossoming evolution of the animal kingdom to the present leads to two important observations. First, the sudden emergence of all animal body plans at the phylum level, immediately after the extinction of the Ediacaran fauna, unequivocally implies the absence of transitional forms. Second, that during the Post-Cambrian evolution, no new phyla evolved even as the rate of evolution of new species and higher taxa was relatively rapid, implying that diversification of species and other taxa did not follow a "bottom-up" pattern as would be conventionally expected.

There is no question that since it emerged about 3.8 billion years ago, life evolved significantly. Paleontology and other biological disciplines provide a rational outline of that evolution until the Cambrian, vindicating the gradualist view of evolution. The troubling question in the current debate on the Cambrian explosion is what caused the unprecedented and unrivalled burst of diversification of animal forms in only about 10 million years. This is a mandatory question.

The emergence of about 40 phyla, the highest rank of animal groups in the Linnaean system, within the narrow Cambrian time window, is also perplexing for the neoDarwinian prediction that higher taxa result from the evolution of lower ones (species→genus→family→order→class→phylum). The Cambrian fossil evidence clearly shows that the direction of evolution is the other way around: the Bauplans of the phyla preceded those of lower taxonomic ranks (i.e., classes, families, orders, genera, and species), rejecting the basic bottom-up neoDarwinian tenet in favor of the "top-down" hypothesis that predicts the origin of all the extinct and extant metazoan taxa, from the class down to the species levels, from those early Bauplans of the Cambrian phyla (Erwin et al., 1987).

Like Darwin himself, we still wonder:

What might have caused this Big Bang in the diversification of animal forms?

There is no convincing explanation, let alone consensus, on factors that enabled the unprecedented surge of evolutionary rates during the Cambrian explosion. The early trend in evolutionary studies emphasized environmental factors as causes for this eruptive diversification.

Presumed External Factors Involved in the Cambrian Explosion

Several hypotheses have been presented positing that the Cambrian explosion was perhaps triggered by external factors, by specific changes in the environment at the time. The idea of a trigger *per se* suggests the evolution of a mechanism that may be set in motion, but the bone of contention in understanding the evolution is the intrinsic mechanism of the evolutionary change rather than the extrinsic factors that may trigger or facilitate activation of the mechanism. Studying each of these factors on its own, or all of them together, elevates the understanding of the possible selective forces that might have been in effect then, but it contributes little—if anything at all—to an understanding of the mechanisms that generated the phenomenal evolutionary diversification.

Ecological Factors

Some authors think that the expansion of phytoplankton during the transition from Ediacaran to Cambrian may have contributed to the evolutionary innovations of the Cambrian fauna (Butterfield, 1997, 2001). Others believe that predation might have been the main driving force of Cambrian diversification (Conway Morris and Bengtson, 1994). Growing predator pressure and competition for ecospace might have forced shallow marine fauna to explore deeper settings by the end of the Cambrian period (Crimes, 2001).

Glaciation

According to the snowball Earth hypothesis, for a long period of time, extending from 720 to about 630 Mya, the Neoproterozoic Earth experienced a freezing period during which a thick layer of ice covered oceans and separated the Earth's atmosphere from the ocean. This snowball Earth, also known as Cryogenian, resulted from a number of glaciations, the last of which was the Marinoan glaciation that lasted from 660 to 630 Mya. The exhaustion of oxygen in the ocean might have led to the extinction of algae and extremely low temperatures plausibly caused the extinction of the bulk of preglacial microorganisms and plants. During the deglaciation period following the Marinoan glaciation, new conditions and new ocean chemistry and niches might have favored Ediacaran fauna evolution and the Cambrian explosion. This again fails to explain why a surge in animal diversification did not occur after the three earlier glaciations (Erwin, 2005) and why the Cambrian occurred only tens of millions of years after the preceding glaciation (Conway Morris, 2003).

Oxygenation of Oceans and Earth Atmosphere

It is believed that the gradual transition from the anoxic Proterozoic oceans into the oxygenated atmosphere and oceans increased the chances of success of evolutionary novelties and inventions. However, it is not easy to prove that the oxygen enrichment of the oceans and the atmosphere coincided with the advent of the Ediacaran fauna (Erwin, 2005). It has been suggested that the oxygen concentration required by active macrobenthic animals may have reached about 100 million years before the Cambrian radiation occurred. Even if such a coincidence of these ecological and geochemical changes with the Cambrian radiation existed, it does not prove a causal relationship between them.

At best, these external factors might have represented no more than selection pressures that indirectly (by providing selective advantages/disadvantages) may have influenced the course of metazoan evolution, and they do not tell us anything about the mechanisms that produce evolutionary innovations and novelties. These specific environmental changes also fail to explain the long periods of evolutionary stasis prior to the Cambrian explosion.

Before I present my hypothesis and relevant substantiating evidence, I will briefly outline a few of the best known hypotheses for the evolution of the organic world.

Current Hypotheses[2] of Organic Evolution

Natural selection is generally accepted as the driver of biological evolution. Hence, the basic criterion used here to classify the hypotheses will be the mechanism of evolutionary change.

Almost all evolution hypotheses posit that the cause of the evolutionary change or novelty lies within the evolving organism, rather than in its environment. The essential difference between these hypotheses is in the nature of changes; are these changes spontaneous and random or is there any directionality in their occurrence? Based on this criterion, the hypotheses of evolution fall into one of the two groups, genetic or epigenetic hypotheses.

Genetic hypotheses see the origin of evolutionary changes as spontaneous and random changes in the genetic material, in genes, regulatory sequences, or other

[2]Due to the large number of competing scientific proposals for explaining organic evolution, I prefer to use the general term hypotheses in describing each of them.

"noncoding" DNA segments. All of these hypotheses may be considered under the intellectual umbrella of the neoDarwinian paradigm.

All of the theories that consider evolution as a result of mechanisms other than (or in addition to) changes in the genetic information may be included in the epigenetic group of hypotheses. Some of these epigenetic hypotheses attempt to explain only specific aspects of evolution and some accept the role of genetic mutations as a complementary cause of evolutionary change.

Having resolved the theoretical problems of selecting useful evolutionary changes, the crux of the problem that any hypothesis on the evolution of the organic world currently faces is identifying the ultimate source of the information for evolutionary change. This is the sum and substance of the evolution theory and it is the criterion I am applying for the classification of several hypotheses of evolution reviewed briefly below.

The Modern Synthesis—The neoDarwinian Hypothesis of Evolution

Despite its limitations, the modern synthesis that emerged by the early 1930s from the eclectic merger of Darwinian doctrine with classical genetics remains the most widely accepted paradigm. Essentially, the modern synthesis posits that the evolution of living organisms results from the action of natural selection on existing genetic variations, including the occurrence of new gene mutations and genetic recombinations. The genetic variations produce microevolution that is evolution at the molecular and cellular levels. Accumulation of microevolutionary changes, under the action of natural selection and over time, may lead to macroevolutionary changes (i.e., evolution at the level of species and higher taxonomic ranks). Both microevolution and macroevolution depend on the same process of selecting genetic variations. Speciation, the formation of a new species, is primarily related to the geographic isolation of populations (allopatric speciation), which is the cause of genetic (prezygotic and postzygotic) isolation or the reproductive isolation of populations, even if they come into contact again.

The neoDarwinian synthesis focused primarily on developing a mathematical apparatus to study allele frequencies and their selection in populations. Taking the role of changes in allele frequencies for granted, the modern synthesis did not try to show, even theoretically, how these changes in allele frequencies may induce evolutionary changes, especially in morphology, at both subspecific and supraspecific levels. The paradigm posits that natural selection produces order from the randomness of changes in genes. But natural selection first comes to the stage after the emergence of the adaptive change. What we need to know is not how the evolutionary changes in morphology, life history, and behavior are selected but how they arise. What is the mechanism behind the evolutionary change?

It should be pointed out, however, that the identification of the Darwinian variability with spontaneous gene mutations or genetic variability in general (existing alleles) is inaccurate. Darwin repeatedly emphasized, especially in the later editions of *The Origin of Species*, that the natural selection is aided by:

> *the inherited effects of the use and disuse of parts; and...by the direct action of external conditions, and by variations which seem to us in our ignorance to arise spontaneously. It appears that I formerly underrated the frequency and value of these latter forms of variation, as leading to permanent modifications of structure independently of natural selection.*
>
> *Darwin (1872)*

Writing of spontaneous changes, he also pointed out that:

> *Mere chance...alone would never account for so habitual and large a degree of difference as that between the species.*
>
> *Darwin (1872, p. 86)*

Natural selection is a force extrinsic to the organism, while the cause of the evolutionary change is internal to the organisms, which is often a response to changes in environment. Natural selection does not induce changes; it acts on changed individuals. In relation to the evolutionary change, natural selection is a *post factum* action, while we are interested in the cause and the mechanism of the evolutionary change. It is a truism to say that evolutionary change emerges before natural selection comes into play. Indeed, the emergence of the evolutionary change is the *raison d'être* of natural selection; there is no need for selection before evolutionary change emerges. This temporal order shows that natural selection alone cannot be the cause of evolution.

No empirical evidence demonstrates that acquisition of a new gene or a change in a gene *per se*, not involving nongenetic factors, produces a morphological novelty. Or to put it in other words, while it can be demonstrated that new or changed genes are related to, or even necessary for, evolution of inherited phenotypic changes, there is no evidence that evolution of a gene would be sufficient for evolving a new phenotypic change. Paraphrasing an old adage, genes "can but don't know" how to do anything. They do not know when, how much, or where in the organism they have to be expressed. Their expression depends on the flow of epigenetic information. Regulation of the expression of nonhousekeeping genes in multicellular organisms is determined by extracellular signals, which in turn represent the terminal elements of signal cascades or networks. In unicellulars, the control and patterns of gene expressions are determined by extragenomic elements with the cytoskeleton as a central player (see page 22, *The Control System in Unicellulars*).

The paradigm has not made its case on the role of genetic variation (gene mutations and genetic recombination) and changes in allele frequencies in populations as inducers of evolutionary novelties in animal morphology, behavior, and life histories; it has not shown in any concrete case, the specific steps through which the genetic information is embodied in the animal morphology.

Accumulation of a body of challenging evidence from various fields of biological study is exposing the inadequacy of the neoDarwinian paradigm to account for a number of important biological phenomena, such as the sudden emergence of new

Bauplans in the highest Linnaean ranks of phyla during the Cambrian explosion, transgenerational developmental plasticity, "genetic polymorphism," sympatric speciation, induction of inherited changes without changes in genes, and transmission of learned behaviors to the offspring.

The modern synthesis has shown a limited capacity to assimilate new, often groundbreaking, knowledge from other fields of biological investigation. Since its inception as a hypothesis, it sidestepped the entire field and facts of experimental embryology, which Darwin considered to be:

second to none in importance

Darwin (1859)

and that for his theory, these facts were:

by far the strongest single class of facts

Darwin (1860)

Modern synthesis has deemed it impossible to deal with the long known facts on the role of epigenetic inheritance. More than adequate evidence on epigenetic hereditary changes in animals and plants has existed for a long time; Beisson and Sonneborn (1965) and others produced experimental evidence on the transmission of "acquired characters" in Protozoa; abundant evidence on induction and transmission at a cellular level of epigenetic marks (DNA methylation and histone modification) and of epigenetically induced changes is accumulating at an accelerated pace. No attempts are made to deal with the epigenetic control of gene expression and the sudden heritable changes in cases of transgenerational developmental plasticity that affect not one individual but whole populations simultaneously. Ignoring this body of evidence is unexplainable at best.

The neoDarwinian synthesis is hardly compatible with new biological discoveries, especially in the field of evo–devo. Among these discoveries is the conserved common genetic toolkit (responsible for production of transcription factors (TFs) and a number of other genes) determining the development of the Bauplan across animal species, indicating that the eruptive morphological diversity of the Cambrian period was unrelated to genetic variation or diversity of genes and gene products predicted by the modern synthesis. Gerhart and Kirschner pointed out that:

where we most expect to find variation, we find conservation, a lack of change.

Gerhart and Kirschner (2007)

While positing that natural selection drives morphological change, the modern synthesis does not pay the deserved attention to the mechanism and origin of the changes that always precede the selection, although:

selection has no innovative capacity: it eliminates or maintains what exists.

Müller (2003)

It is believed that the so-called master control genes that produce proteins that act as TFs to control expression of other genes and induce development of widely different structures. This also does not support the modern synthesis paradigm. Even if this were the case, it is not easy to understand how these highly conserved and least evolving genes could promote evolution. Moreover, sometimes the evolutionary progress is characterized by the loss of *Hox* genes. For example, the sea anemone, *Nematostella vectensis*, an anthozoan of the phylum *Cnidaria*, has more *Hox* genes than *Drosophila*, which is a far more complex organism with five times as many cell types (Ryan et al., 2006) or the worm *Caenorhabditis elegans*, which has lost half of its ancestral *Hox* genes (Aboobaker and Blaxter, 2010). If it is argued that evolution depends on changes in expression patterns of genes rather than gene evolution, increasing evidence shows that expression of *Hox* genes and cell differentiation are epigenetic processes.

Evo–devo studies created new difficulties for the neoDarwinian explanations of evolution. These studies led to the new biological concept that the nature and development of species depends not on the number or kinds of genes in its genome but instead primarily on the expression patterns of the control genes.

These failures are the reason why more and more biologists are becoming inclined to favor alternative hypotheses on the driving forces and mechanisms of organic evolution. These failures have stimulated the development of numerous alternative explanations of organic evolution. By the same token, the abundance of hypotheses on evolution is an eloquent indication of the serious challenges the theory of evolution is confronting.

Other Hypotheses on Evolution

Updating the Explanations of the neoDarwinian Paradigm

Numerous attempts are made to reconcile the modern synthesis with the new developments in biology, especially in evo–devo, primarily by incorporating into it the idea of a genetic control of gene expression.

The "Regulatory" Hypothesis of Morphological Evolution

As pointed out by Sean B. Carroll, this is a special hypothesis on the evolution of animal morphology rather than a general hypothesis on the evolution of the living world (Carroll, 2008). In a nutshell, the hypothesis posits that the evolution of animal morphology does not depend on gene mutations and evolution of genes (including toolkit genes and *Hox* genes) and proteins, which generally are functionally conserved. Evolution of the morphological diversity in the animal kingdom is determined by changes in the spatiotemporal expression of the same generally conserved gene proteins. Such changes may result from accidental changes in the base sequences of the cis-regulatory elements (CREs) of genes. In turn, changes in the CREs may lead to changes in the elements and wiring of gene regulatory networks (GRNs), which are the primary sources of morphological novelties.

According to Carroll's hypothesis, CREs is ultimately responsible for the evolution and rewiring of GRNs; hence, the evolution of CREs is "the predominant mechanism underlying the evolution of form" (Carroll, 2008). The evolution of CREs is believed to have resulted from four processes:

1. Mutations in existing CREs (i.e., changes in the base sequences, which may create new binding sites for new TFs, thus enabling recruitment of new genes in GRNs).
2. Remodeling of existing CREs. Changes in the number, affinity, and topology of transcription factor binding sites may change expression of proteins.
3. Loss of transcription factor binding sites.
4. Recruitment of transposable elements.

The empirical support for the hypothesis is still inadequate and, generally, still waiting to be confirmed. The evidence on relevant changes in base sequences (i.e., alterations like those that led to changes in the binding function of CREs), is so rare that they would hardly account for the enormous amount of morphological diversification in the animal kingdom. At a fundamental methodological level, Carroll ignores that patterns of expression of TFs are themselves generally controlled ultimately by extracellular epigenetic signals such as hormones, growth factors, and cytokines, which commonly are downstream signal cascades that start in the central nervous system (CNS).

Looking at the accidental random changes in the sequences of CREs as the ultimate source of the evolution of animal morphology, the hypothesis sticks to the neo-Darwinian mechanism of evolution via *random* changes. It does, however, displace the focus from changes in genes to changes in their regulatory elements as the main venue of the evolutionary process. In doing so, the hypothesis is vulnerable to the same criticism as the basic neoDarwinian hypothesis.

The GRN Hypothesis of Evolution

Central to this hypothesis developed by Eric H. Davidson and Douglas H. Erwin (Davidson, 2005; Davidson and Erwin, 2006, 2007; Erwin and Davidson, 2009) are the GRNs, groups of functionally linked regulatory, and signaling genes responsible for production of TFs and receptors.

GRNs are hierarchical structures composed of modular circuits where those at the apex of the hierarchy, the stable "kernels" of GRNs, control the early development and formation of the Bauplan. The middle portion is involved in morphogenesis and is less stable. The inferior portion, which controls differentiation gene batteries' activities, is the most labile portion. Any mutational change in the middle and peripheral portions may bring about either unfavorable or favorable changes in the GRN and in the animal development, but changes in the "kernels" are generally disastrous because they affect early development. The lethality of changes in "kernels" eliminated carriers of changes during early development. The hypothesis assumes that the calamitous effects of changes in kernels are the reason why kernels are so stable and why no new Bauplan at the phylum level evolved during the last 400–500 million years. Mutations in the middle portion and the periphery of the GRN may change the developmental GRN architecture and "[affect] the expression of multiple genes downstream" (Erwin and Davidson, 2009).

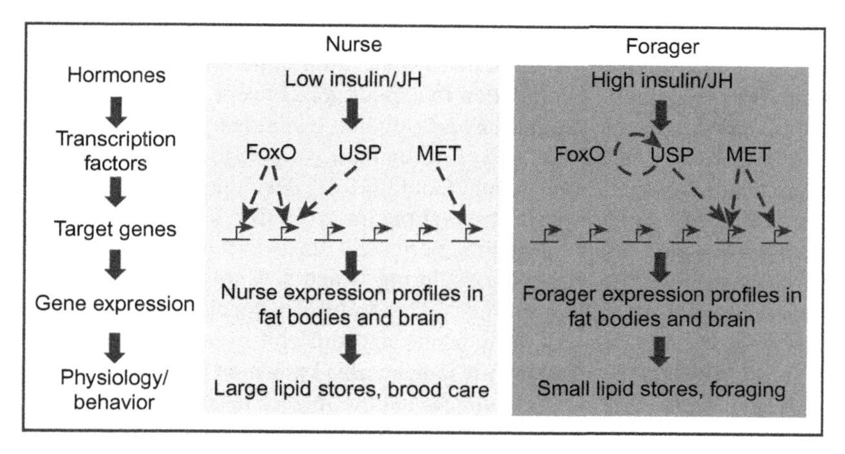

Figure 5.2 Theoretical model for the role of GRNs in the regulation of worker division of labor. Signaling through Ilps and JH is low in nurses and higher in foragers. These hormones regulate gene expression through interactions with TFs, some of which have already been identified in other insect species. Known transcriptional regulators include FoxO, which is involved in insulin action, as well as ultraspiracle (USP) and methoprene-tolerant (MET), both of which are associated with JH. Increased insulin signaling in foragers is likely to repress *FoxO* target genes by preventing the FoxO protein from binding to their promoters. Increased JH signaling causes increased USP expression in honeybees, as well as other hypothetical changes in target gene activation by USP and MET. According to this framework, interactions among these and other TFs lead to distinct gene expression profiles of nurses and foragers in the brain and fat bodies. These hormonally controlled GRNs are hypothesized to be causal for behavioral maturation and stable lipid loss.
Source: Ament et al. (2010).

The hypothesis further assumes that both the internal (middle portion) subcircuits and the peripheral (inferior) subcircuits can be redeployed in various parts of the organism where they produce phenotypic changes, but they fail to show how this can happen or which mechanism is responsible for the redeployment.

The hypothesis holds GRNs as self-regulated entities, but from a systems biology view they, and most TFs, are just downstream elements of systemic signal cascades/networks starting in the CNS (hormones, growth factors, secreted proteins, etc. are their proximate activators). This is illustrated in Figure 5.2, which shows how hormones (insulin and juvenile hormone (JH)), secreted under ultimate CNS control in ants, control the activity of GRNs that determine phenotypic changes (e.g., physiology, morphology, and behavior).

The Hypothesis of Punctuated Equilibria—Challenging the neoDarwinian Hypothesis "from Within"

In 1971, Niles Eldredge observed that the gradualism of evolution propounded by the neoDarwinian theory was incompatible with paleontological evidence of the sudden appearance of changes in the fossil record, without intermediate fossil forms. He used Ernst Mayr's concept of allopatric speciation in peripheral populations isolated from the main stock of population to explain the concept and proposed that the result could be explained by migration and isolation (Eldredge, 1971).

Based on earlier observations as well as on their own observations of the gaps in paleontological evidence, Eldredge and Gould (1972) developed the theory of punctuated equilibria to explain these observations as reflecting a nongradual mode of evolution that was characterized by long periods of stasis in the evolution of species, which were interrupted by short periods during which new species suddenly sprung, hence the term "punctuated equilibria": "The norm of a species or, by extension, a community, is stability. Speciation is a rare and difficult event that punctuates the system in homeostatic equilibrium." It is generally known as the theory of punctuated equilibria. The theory handles evolution at the species level:

> *Punctuated equilibria is a model for discontinuous tempos of change at one biological level only: the process of speciation and the deployment of species in geological time.*
>
> *Gould and Eldredge (1977)*

There is abundant evidence to support the "punctuated" model of evolution: the existence of microorganisms such as cyanobacteria that remained almost unchanged (Schopf, 1994) since the dawn of the evolution of life on Earth more than 3 billion years ago, or the existence of "living fossils" such as the gingko tree, *Ginkgo biloba*, believed to have remained virtually unchanged for 270 million years, and the coelacanth, which has remained unchanged for 409 million years (Figure 5.3). These are paradigmatic examples of the extremely long periods of stasis in the evolution of living forms, although gradual evolution can be illustrated with equivalent examples.

The Hypothesis of Facilitated Variation

Marc W. Kirschner and John C. Gerhart presented their hypothesis for the first time in *The Plausibility of Life* (2005). It is intended to explain the evolution of metazoans alone. Admitting the role mutations play in producing genetic variability, the hypothesis posits that phenotypic variability and animal evolution depend primarily on internal developmental processes. Besides changes in regulatory circuits, the hypothesis acknowledges the possibility of involvement of rare neoDarwinian gene mutations and of the Baldwin effect in the emergence of evolutionary changes.

The hypothesis posits that there is a group of core components, consisting of the genetic toolkit and processes that generate and operate the animal's phenotype. Although these core components and processes are highly conserved, the physiological adaptability of animals provides them with a capacity to form "weak linkages" between the elements of the causal chains of regulatory circuits. The capacity facilitates phenotypic variation. Central to the authors' illustration of the capacity to form weak linkages is the phenomenon of the inherent developmental plasticity, *sensu*,

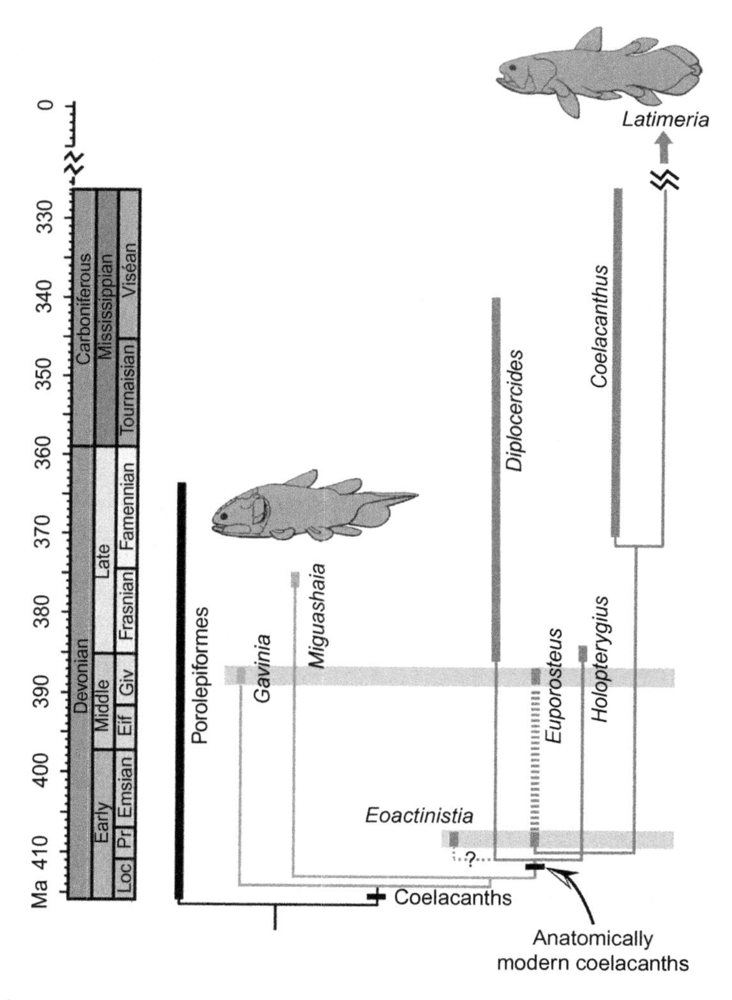

Figure 5.3 Primitive coelacanths are in light blue and the anatomically modern coelacanths including the extant *Latimeria* are in red. *Euporosteus* is positioned either crownward of *Diplocercides* or as its sister taxon, indicating that the distinctive body plan of anatomically modern coelacanths must have been established no later than 409 Mya. *Euporosteus* also lends support to the possibility that *Eoactinistia* may represent an early member of the anatomically modern coelacanths with the dentary sensory pore. *Abbreviations*: Eif, Eifelian; Giv, Givetian; Loc, Lochkovian; Pr, Pragian. (For interpretation of the references to color in this figure legend, the reader is referred to the web version of this book.)
Source: Zhu et al. (2012).

M.J. West-Eberhard, an inherent potential of animals to switch to alternative developmental states in response to external or internal factors. Weak regulatory linkages enable the animal to adaptively respond to various external or internal stimuli, by switching between two or more alternative states at different levels from the cellular to organismic levels. In this developmental plasticity, the authors see the "cryptic source" of evolutionary change (Kirschner and Gerhart, 2005).

Such phenotypic changes may be stabilized by the emergence of genetic accommodation. So, facilitated variation posits that evolutionary novelties are generated primarily by developmental processes, while natural selection plays no role in the emergence of evolutionary novelties.

They believe the facilitation evolved as a "by-product of properties of core processes such as robustness, adaptability, modularity, exploratory behavior, and capacity for regulatory weak linkage" (Gerhart and Kirschner, 2007). According to the hypothesis, the robustness and adaptability are essential for using core processes in various combinations and quantities, thus reducing the need for regulatory changes in the process of the evolution of novelties.

The supporting evidence for the hypothesis is hardly adequate or plausible. For instance, the illustration of the exploratory behavior of the core components with the example of neural crest cells is hardly convincing because these cells do not have "many different options"; before leaving the neural tube, they are provided with information on where to go and what to do. The stabilization of microtubules within the cell could also hardly be considered a "futile cycle of outgrowth and shrinkage" but is based on a processing of information on the state of the cytoplasm.

Although the hypothesis of facilitated variation considers the phenotypic variability rather than random mutational changes as the main source of evolutionary change, it does not provide the source of the information that is used to select among the developmental "options," the one option that is adaptive under the specific external or internal conditions.

The Developmental Plasticity Hypothesis

M.J. West-Eberhard's hypothesis marks a considerable progress in understanding the origin of diversity and mechanisms of evolution (West-Eberhard, 1989, 2003, 2005). Her studies on developmental plasticity culminated in the publication of the influential *Developmental Plasticity and Evolution* (2003). West-Eberhard believes evolutionary changes and speciation result from the reorganization of the phenotype in response to changes in the environment. The phenotypic reorganization may be followed by selection and genetic accommodation of the new phenotype via the occurrence of relevant mutations or the existing genetic variation. Ultimately, this may lead to reproductive isolation and the evolution of new species. While genetic changes occur spontaneously and randomly, phenotypic reorganization is an adaptive response to external/internal stimuli.

New phenotypes can arise in response to genetic or environmental inputs in living organisms which developmentally are very plastic. These inputs may lead to the reorganization of the phenotype, which West-Eberhard terms "developmental recombination," resulting in a new phenotype.

If the new phenotype increases the fitness of the organism in the environment and offers the carrier some advantage over other individuals, natural selection then comes into play, favoring the spread of the phenotype in the population. From this point onward, a process of "genetic accommodation" may begin. This process consists in incorporating into the relevant developmental pathway a newly evolved gene or an allele already present in the population. Because genetic accommodation follows the phenotypic change, West-Eberhard concludes that genes are followers rather than leaders in evolution.

She supports her hypothesis with empirical evidence. For example, the recurrence of two morphologically distinct species, the limnetic and benthic species pairs of the three-spined stickleback (*Gasterosteus aculeatus*) complex of closely related fish species, illustrates the developmental recombination; early on, some of these species display limnetic morphology and feeding behavior (living in the water column and feeding on plankton), but later in the life they appear benthic in both morphology and feeding behavior, suggesting that the recurrent species pairs of the fish evolved via a developmental switching (heterochrony) of an ancestral species that entertained both limnetic and benthic morphology and behavior during its lifetime.

Empirical evidence on the role of hormone levels in expressing of parental care by male birds with low testosterone levels and the incubation of eggs by males (like sandpipers) also support her hypothesis of developmental recombination.

According to the hypothesis, evolution through developmental recombination and its adaptive advantages counteract changes in genes and explain why the same genetic toolkit is used recurrently in living organisms.

The hypothesis also articulates that speciation precedes reproductive isolation, suggesting that speciation may occur in sympatry, whereas the neoDarwinian hypothesis argues that allopatric speciation is almost the exclusive mode of speciation.

Developmental plasticity's great achievement is the first hypothesis to represent an empirically supported mechanism of production of evolutionary changes in which epigenetic factors play the leading role. While changes in the developmental pathways (e.g., in hormonal pathways) certainly produce phenotypic changes, the hypothesis fails to demonstrate the ultimate source of information necessary to produce these adaptive, nonrandom, changes in developmental pathways.

The Genetic Assimilation Hypothesis

Essentially, this hypothesis dates back to the American psychologist James Mark Baldwin (1861–1934), who argued theoretically that in a changed environment, a new phenotype such as a learned behavior may help an organism survive or increase its reproductive success. After a number of generations, inheritance mechanisms may evolve making the behavior inherited even in the absence of a stimulus.

In the 1950s, Conrad Hal Waddington (1905–1975) demonstrated experimentally in *Drosophila* his hypothesis of genetic assimilation where *Drosophila* acquired traits developed in response to stressful stimuli (high temperature or ether vapor, both of which are not mutagenic) and produced phenotypic changes (ultrabithorax and crossveinlessness) in a small number of individuals. By selecting and crossing these changed individuals for many generations, he obtained individuals that expressed the new traits even in the absence of the inducing environmental stimuli. To explain the phenomenon, it is assumed that the evolution of the innate trait from an acquired trait was enabled through the repetitive crosses between the individuals that showed the new trait with individuals that simply had the genes responsible for the trait. Hence, the environmental stimuli merely exposed the existing genetic variation. The fact that neither of the new traits seems to be adaptive makes it impossible to qualify them as an evolutionary change that would survive under natural conditions.

In modern biological literature, Pigliucci and Murren (2003), and Pigliucci et al. (2006) not only revitalized Waddington's hypothesis, but also enriched it in one essential respect: they argued that the emergence of the developmentally induced new phenotype may not be related with/depend on the existing genetic variation in the population. But it may also precede the evolution of the genetic determinants of the new phenotype. For them, this renders the hypothesis to "be a particular case of the broader possibility envisioned by West-Eberhard, specifically when the origin of the trait (step 1) is due to an environmental, rather than a genetic, change" (Pigliucci et al., 2006). The evidence to support the hypothesis is scarce at best. As for the possible role of genetic assimilation in evolution, the proponents of the hypothesis believe it "is of course a matter for empirical investigation" (Pigliucci et al., 2006).

The Hypothesis of the Genotype Network

Andreas Wagner's hypothesis of the origins of evolutionary innovations relies more on a computational approach to the origins of evolutionary innovations than the preceding theories. The theory envisages populations as collections of genotypes in genotypic spaces within which genotype networks of individuals of the same phenotype exist. Mutations can change the position of individuals within the genotype space. A change in the environment may make the phenotype of an individual better adapted to the changed environment. The population can find this existing to superior genotype within the genotype network via several small mutations with little phenotypic change. Within the genotype space, genotype *neighborhoods* also exist, which allow that genotype to be reached by any other genotype in the genotype space via one or a number of mutations. The genotype networks thus allow small step changes in the genotype that can be reached by any genotype via small step mutations and genotype neighborhoods allow different genotypes to search for different new phenotypes in the neighborhood.

According to the theory, there are three classes of systems that bring about novelties in the living world, all of which exist in respective genotype spaces of possible metabolic networks. The first class comprises hundreds to thousands of chemical reactions catalyzed by specific enzymes in metabolic networks. According to the hypothesis, a metabolic novelty may arise more often via the emergence of new combinations of enzymes in metabolic networks rather than via the evolution of new enzymes through gene mutations. The problem with this system class is that it is investigated by computational methods alone, because it is currently unfeasible through experiment. The validation of the mechanism has to wait until empirical evidence becomes available.

The second system class involves regulatory circuits comprising primarily transcriptional regulatory circuits, in which a transcriptional factor regulates the expression of another. The hypothesis posits that mutations in regulatory proteins/sequences can produce inherited changes in relationships between the transcriptional regulators in a circuit that can induce changes in the gene expressing phenotype with all the ensuing consequences in development and evolution. The evidence supporting this idea is ambiguous and it does not explain how changes in relationships between regulators (not in genes or gene sequences) may be inherited.

The third system class comprises the established neoDarwinian concept of gene mutations, discussed earlier.

When the hypothesis posits that external stimuli may induce an adaptive change in the phenotype, it is not clear if it implies random changes or adaptive nonrandom changes. If the first is the case, then it faces the same statistical unlikelihood. If the latter is the case, then the source of information is beyond the three system classes. To the extent that the hypothesis relies on random mutational events, it faces all the difficulties that modern synthesis does.

The Hypothesis of Epigenetic Variation

About two decades ago (1995), Eva Jablonka and Marion Lamb concluded that the modern synthesis was unable to account for a number of nongenetic phenomena such as DNA methylation, X-chromosome inactivation, genome imprinting, and chromatin modifications that influence patterns of gene expression. They postulated the existence of an epigenetic inheritance system (EIS), which would expand the concept of biological inheritance and acknowledge a Lamarckian role in evolution (Jablonka and Lamb, 1995). They and others (Avital et al.) further developed the hypothesis in its present form (Jablonka and Lamb, 2008; Jablonka and Raz, 2009).

Living organisms are endowed with varying degrees of developmental plasticity that they express in response to environmental stimuli. Developmental plasticity is the primary source of heritable non-DNA epigenetic variations arising in response to induced stimuli that are transmitted to later generations of individual cells or organisms. These variations fall into one of three main categories, epigenetic, behavioral, and symbolic, with the latter relevant for human evolution only.

Epigenetic variations are related to an EIS that is responsible for the cellular epigenetic inheritance. It comprises *self-sustaining feedback loops* or transcriptional circuits, *structural inheritance* of the type of the ciliate cortical inheritance, chromatin markings, which include both the DNA methylation and histone modifications and, finally, the *RNA-mediated inheritance*.

Epigenetic mechanisms may also induce nonrandom changes in the genetic structures and genetic information, which is adaptive rather than random as assumed by the modern synthesis:

> *The origin of many genetic variations, especially under conditions of stress, is not random, is often predictable.*
>
> *Jablonka (2008)*

This clearly emphasizes the role of stress conditions in the induction and transmission of epigenetic variation.

The supporting empirical evidence on epigenetic variation is adequate (Jablonka and Raz, 2009). One of the most impressive illustrations of the role of the epigenetic factors in evolution is the common toadflax, *Linaria vulgaris*, which sometimes forms a variant with a very different floral system. It is believed that "epigenetic variants in every locus in the eukaryotic genome can be inherited, but in what manner, for how long, and under what conditions has yet to be qualified" (Jablonka and Raz, 2009). The involvement of hormones in the induction of heritable epigenetic variations is no longer mere speculation: several of the mammalian examples presented suggest that changes in hormonal stimuli induce heritable epigenetic changes.

Despite the unquestionable contribution of the hypothesis in advancing our understanding of the epigenetic factors in evolution, the hypothesis fails to show the material basis, the mechanism, and the source of information necessary to produce epigenetic marks in strictly determined DNA sites and chromosomes. The main concern with the evidence, especially surrounding multicellular plants and animals, is that epigenetic changes they present seem to have primarily deleterious effects or little bearing on evolution.

The Hypothesis of Generic Factors

Stuart Newman and Gerd Müller developed a hypothesis on the role of physical "generic" factors especially in the early development and in the early metazoan evolution (Müller and Newman, 2005; Newman, 2006). They believe that at the early stages of evolution, about 600 Mya, a period which they termed "physical" or "pre-Mendelian," living systems used several physical mechanisms for embryonic patterning. These physical factors might have been responsible for determining body plans during the Cambrian explosion and explain the embryonic hourglass puzzle (Newman, 2011).

In their hypothesis, the adhesivity of unicellular organisms was the initial cause for the forming of primitive multicellular aggregates and the emergence of such cell aggregates made possible the activity of a number of physicochemical factors (viscoelasticity, differential cell adhesivity, biochemical diffusion, etc.) that are active in nonliving systems, but not in unicellulars. They believe spontaneously arising cell aggregates, in which cells fail to separate after cell division, can form primitive multicellular organisms. They posit that the formation and morphology of these cell aggregations was determined by physicochemical factors and processes rather than by gradual changes in genes and natural selection. These factors were the major sources of morphological innovation in evolution at the early stages of evolution of multicellular systems.

Individual cells in these early cell aggregates tend to associate with cells of equal or comparable adhesivity, thus creating islands within "lakes of their less cohesive neighbors" (Newman and Müller, 2000) separated by nonmixing interfaces between them. Physical factors played a determining role in the formation of cell layers, spherical and tubular structures, sheaths, and so on. The differential adhesivity and other physical processes may have been behind the formations of embryonic layers, gastrulation, lumina, and various tubular structures, rods, spheres, segmentation, compartmentalization, and three-dimensional patterning (Müller, 2007; Newman and Müller, 2000).

The existence of thresholds of activity of various inducers might have promoted the emergence of morphological novelties up to the subphylum level. At this stage of the evolution of multicellularity, genes were playing the role of "suppliers of building blocks and catalysis with little direct influence on the architectural outcome in these pre-Mendelian systems" (Newman and Müller, 2000).

Over time, the physically determined morphogenesis was captured by the genetic system. The advent of the pre-Mendelian mechanism of heredity made possible the evolutionary stabilization of the physically determined biological forms. The pre-Mendelian physical factors continue to retain, to various extents, their role in development and are "of decisive importance even in contemporary ontogenies" (Newman and Müller, 2000). Initially, Nature did not have to bother coding "for what happens naturally in the physicochemical universe" (Noble, 2011).

The Control System Hypothesis of Evolution

The Control System in Animals

To function, animals require the creation and maintenance of a relatively constant internal environment in which their cells can perform their normal functions. The unavoidable and continued degradation of the animal structure and creation of a "constant" internal environment requires the establishment of a dynamic equilibrium, in which the organism continually replaces the structural elements, molecules, and cells it loses every moment.

To do this, the animal organism must be able to do the following:

- Assess the amount of lost structures and the regions where they occur.
- Generate and send to relevant regions, organs, or cells the information to compensate for the lost structures and restore normal function.

These are basic functions of a control system. Human intelligence fails to imagine how a living system, which is beyond compare more complex than any nonliving system, would maintain its normal functioning state without a control system that continually restores the system's normal state. Multicellular organisms could not emerge until a control system for regulating and maintaining the multicellular structure in a state of dynamic equilibrium would be "invented."

Common sense tells us that a living system, with its innumerable variables and unmatched complexity, be it a unicellular ameba or a human being, must have a control system.

Eumetazoans are hierarchical systems with several distinct but structurally integrated and functionally coordinated levels of organization at the:

- molecular level, with genes and all other molecules that make up a cell;
- cell level, with organelles, including chromosomes, centrosomes, cytoskeleton, and cell membrane;
- tissue level;
- organ level;
- systemic level.

Each level of organization has its own control, but the hierarchical character of the control system requires the coordination of control at different levels. In this interaction and coordination, a relationship is established between the control levels, in which lower levels of control, including genomic control, are subordinate to higher levels of control. At the top of the hierarchy is the systemic control that integrates and coordinates the function of all the lower level controls. Anticipating the modern concept of systems biology, in the context of the emergence of phenotype, as early as 1939, Austrian biologist Paul A. Weiss (1898–1989) pointed out that "phenotypes, and the mechanisms that underlie them, depend on, and subordinate to, the law which rules the complex as a unit" (Crews et al., 2012).

The control system in animals (sponges excluded) is an ICS, which epitomizes the holistic nature of animals.

There is no general theory of control systems in animals, plants, or unicellulars. Yet the discipline of physiology tells us that all the vital functions of animals (blood circulation, excretion, digestion, reproduction, endocrine functions, behavior, etc.) are under neural control with the brain as the controller of the control system. The study of the functions of living organisms is, in fact, the science of control systems for living organisms. Animal behavior too is centrally determined by the CNS and empirical evidence shows that life histories and their evolution in animals are determined by the CNS. New in this ICS hypothesis is that it additionally attributes the CNS a controlling role in the development of animal morphology. In Chapter 3, more than adequate evidence demonstrating that the CNS controls the development of animal morphology up to adulthood is presented (Cabej, 2005, 2008, 2012).

That the ICS works to restore the normal state of the system implies that it "knows" the norm of the immense number of variables at the molecular, cellular, and supracellular levels. In many cases, it is empirically demonstrated that this "knowledge" is coded in the form of established set points in the CNS (see page 16, *Homeostasis*, and page 79, *Epigenetic Information and Signal Cascades*).

The nervous system's omnipresence and its immense computational capabilities allow it to monitor the homeostasis (in the broadest meaning of the word) via interoceptors and exteroceptors, which afferently transmit the data on the various parameters of the system to the CNS. By processing these data, the CNS detects deviations from the norm by comparing the actual data with the norm and then sending instructions for restoring the normal state to relevant regions of the body. Thus, the CNS serves as the controller of the ICS in animals.

Herein I outline the model of the ICS in eumetazoans, a top-down model, in which the information for restoring the homeostatic variables flows from the brain, the controller of the ICS, sequentially down through the lower levels of organization ending at the molecular level of gene expression.

The animal control system consists of the following:

- The sensory division of the peripheral nervous system (PNS), a dense and pervasive presence of sensory receptors, which are located at the surface of the body to receive external stimuli (exteroceptors), sensory receptors that receive stimuli arising within the body (interoceptors), and those that are located primarily in muscles, tendons, and joints and receive information on the length and stretch of these parts (proprioceptors). All types of receptors send their information to the CNS where they are perceived.
- The incoming information is processed in the CNS, the controller of the ICS, which monitors the state of the system and continually provides its output in the form of signal cascades.

The integration and processing of the interoceptive input on the state of the system, as well as decision-making for restoring the normal state, are functions of the *controller* of the ICS, which is the CNS (see Figure 1.12). By comparing the information on the actual state with the "normal" state, the CNS identifies deviations from the norm and sends instructions for activating signal cascades to restore the normal state.

Being hierarchical, the ICS model is an epigenetic top-down model where controls at all levels are coordinated. In this hierarchy, the lower levels of control are subordinate to, although they interact with each other, the control of the higher levels. The genome is also subordinate to rules of the whole, that is, of the cell, the organism, and the ICS, not the other way around. The neoDarwinian bottom-up model positing that the genome controls the functions and the structure of animals contradicts the experimental evidence to date and, in doing so, puts the cart before the horse.

The epigenetic top-down model is supported by two fundamental biological phenomena:

1. Expression of housekeeping genes in animals is regulated by epigenetic signals from the cytoskeleton (see page 128–131).
2. Expression of systemic (nonhousekeeping) genes is controlled and regulated by extracellular signals, which ultimately originate in the ICS (nervous system).

Cell differentiation is generally considered an epigenetic process (Maruyama et al., 2011; Reik, 2007) and serves as a major and universal illustration of the top-down flow of information in the process of development. All cell types in an organism have the same genotype or DNA, but differ from each other due to the different epigenetic information (extracellular signals) they are given through signal cascades starting in the CNS.

Is it possible that the CNS might control events at a cellular level?

This question may be less troubling than it seems when we bear in mind that our brain contains up to 1 trillion neurons, each connected to thousands of others, which makes the brain capable of executing as many as 100 billion operations per second.

But a glance at the general scheme (Figure 5.4) of neuroendocrine regulation is our best empirical answer; the brain, via the pituitary, neurohormonally controls and regulates the expression of specific genes in a dozen distant organs.

The epigenetic mechanisms of the cell are not reflected in this figure.

In mammal cells, the cytoskeleton is in contact with both the extracellular matrix (ECM) and the nuclear matrix, acting as a transducer of extracellular signals to the cell nucleus. Intermediary filaments (IFs) are intimately connected with the

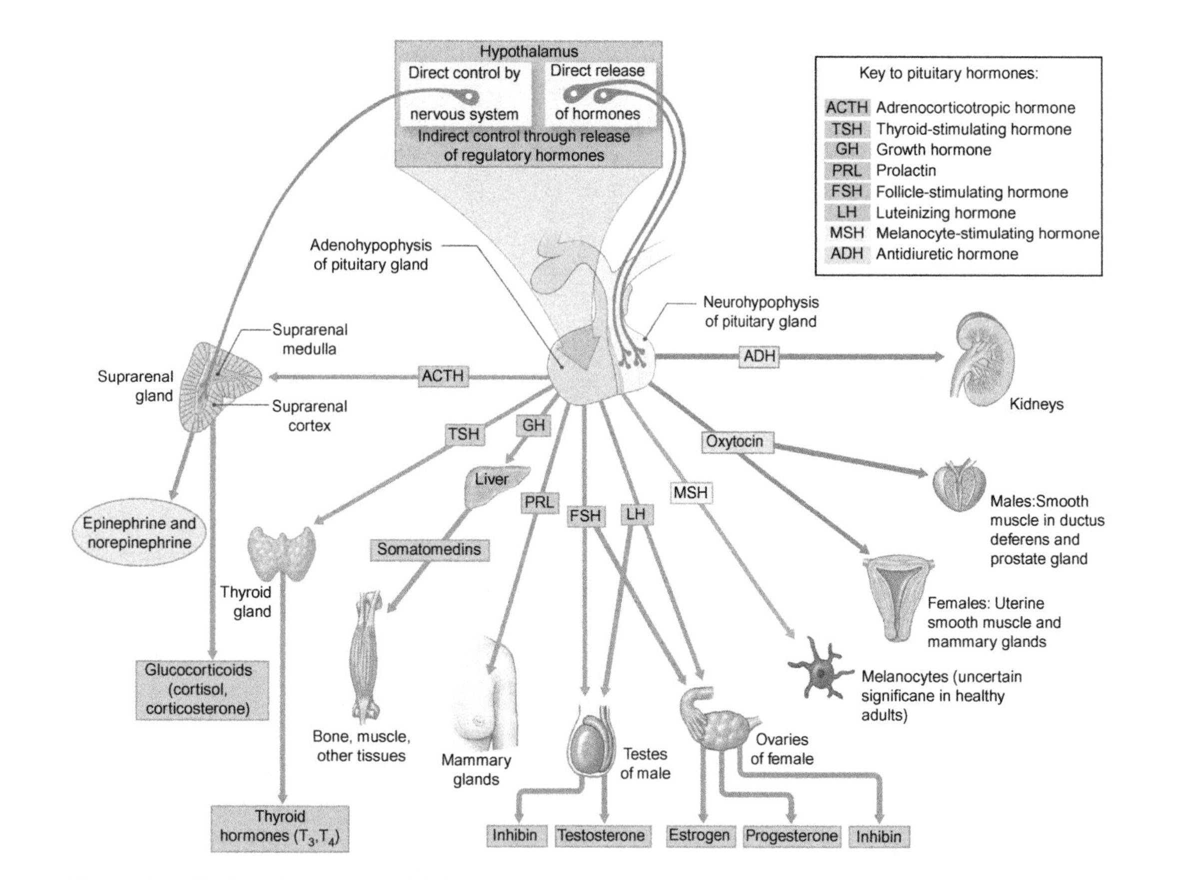

Figure 5.4 Pituitary hormones and their targets.
Source: Copyright © 2009 Pearson Education, Inc., publishing as Pearson Benjamin Cummings.

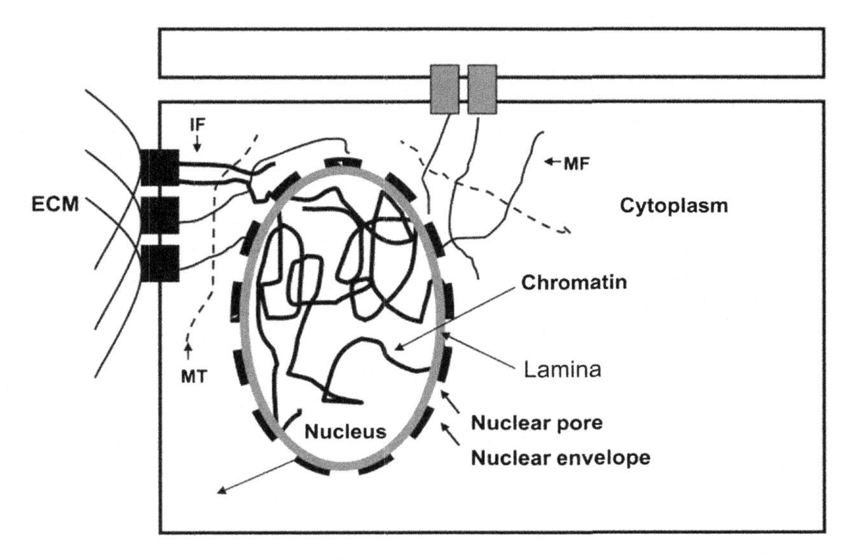

Figure 5.5 The structural continuum of the cell. A wide variety of reports has demonstrated that ECM fibers make contact with cell membrane receptors (black boxes) that are themselves in contact with the intracellular cytoskeleton networks . Some of the cytoskeleton elements make contact with proteins located in the nuclear envelope, which also bind to elements of the nuclear shell (lamina). Finally, the lamina interacts with chromatin via proteins that bind both lamins, the structural components of the lamina, and chromatin-associated proteins. A similar continuum also exists starting from cell–cell contacts (gray boxes). *Abbreviations*: MF, actin microfilaments; MT, microtubules; IF, intermediate filaments. (For interpretation of the references to color in this figure legend, the reader is referred to the web version of this book.) *Source*: Lelièvre (2009).

structural elements of the nuclear matrix (Figure 5.5) and are an integral part of the nuclear skeleton (Tolstonog et al., 2002).

The ECM is a noncellular network of various components secreted by cells, which fills the intercellular spaces. The cell responds to changes in the ECM composition and thickness with a global reorganization of chromosomes and changes in the expression of 990 genes. The mediator and inducer of these changes is the cytoskeleton (disruption of actin filaments exposes chromatin, whereas disruption of microtubules and intermediate filaments sequesters it) (Lelièvre, 2009; Maniotis et al., 2005). Via membrane receptors that have access to the ECM (Figure 5.5), the cytoskeleton senses the changes in the ECM and responds by sending signals for adaptive expression of genes.

The Evolution of the Integrated Control System in Metazoans

An Obscure Control System in Sponges

Sponges are the simplest of extant metazoans that evolved probably more than 600 Mya. Multicellular structures, believed to have been the ancestors of modern sponges, appear in the fossil record about 580 Mya (Li et al., 1998). They have a system of canals through which water circulates bringing food and oxygen to cells in the interior of the body. Most cells maintain a relatively high independence and the number of cell types is low, about 7–8. Sponges have no precise form and symmetry as eumetazoans do. They also lack true tissues and organs (Ereskovsky and Dondua, 2006). A sponge's structure and functions at the supracellular level are modest and its control system is barely discernible.

Sponges are a "dead end of evolution"; there is no evidence that they, or other simple creatures, such as placozoans, evolved into any higher groups of extant or extinct metazoans (Figure 5.6).

Why did sponges fail to evolve into other metazoan groups of higher structural and functional complexity as the eumetazoans did?

Any comparison of the differences between sponges and their possible sister group, cnidarians, the simplest of known eumetazoans, reveals that sponges are genetically not inferior; the sponge, *Amphimedon queenslandica*, genome encodes

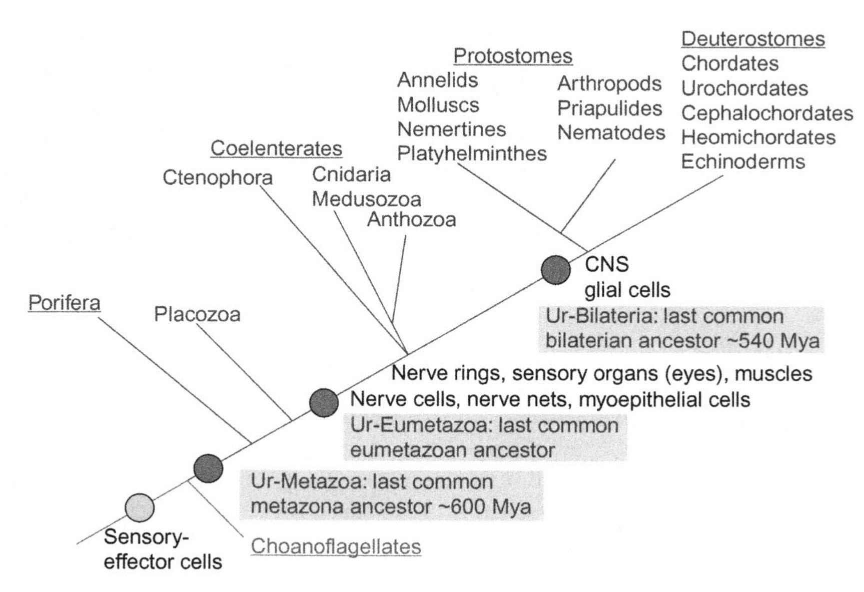

Figure 5.6 Tree representing the main steps in the formation of first-evolved nervous systems along the animal phylogeny (blue branches). This tree was deduced from the cellular and physiological analyses performed in choanoflagellates that behave as sensory-effector cells, in Porifera (sponges) that lack synaptic transmission but exhibit contractile behavior in response to their environment, and in coelenterates (grouping Ctenophora—comb jellies—and Cnidaria). Coelenterates show coordinated behaviors thanks to their nervous systems that already include sensory organs, nerve rings, and neuromuscular transmission. Bilaterians (protostomes and deuterostomes) that originated later share a putative common ancestor (Urbilateria) equipped with a CNS and differentiating glial cells. (For interpretation of the references to color in this figure legend, the reader is referred to the web version of this book.)
Source: Galliot and Quiquand (2011).

about 30,000 genes (Srivastava et al., 2010), while a cnidarian, *Hydra littoralis*, is believed to have only about 13,000 genes. Sponges have more genes than humans do and share about 70% of their genes with humans (Stout et al., 2007). They also have genes for the cell types and tissues they lack, such as muscle cells and neurons. Mocking conventional biological thought the evolution of sponges and eumetazoans from the Urmetazoa "required" a loss in the number of genes (Harcet et al., 2010).

Sponges use developmentally important cell signaling and adhesion genes are involved in the morphogenetic processes in higher invertebrates and even vertebrates, from worms to humans (Nichols et al., 2006). For example, *Oscarela carmela* expresses many of the signaling genes and has all but one (the nuclear hormone receptor pathway) of the seven major bilaterian signaling pathways: Wnt, TGF-β, Hedgehog, receptor tyrosine kinase, Jak/STAT, nuclear hormone receptor, and Notch, which are responsible for the development of limbs, eyes, vertebrate segmentation, and for the assembly of neural circuits in bilaterians (Nichols et al., 2006). Sponges also have hormones that may be involved in their growth and reproduction.

Such data speak unambiguously against any correlation or causal relationship between genome's size or gene number and the structural complexity or evolutionary progress in *Animalia*. What are sponges missing that would have changed their fate as a dead end of evolution?

That eumetazoans, such as cnidarians, also evolved other types of cells such as muscle cells, epithelial, dermal, and other cells is a water cooler topic to the main problem of the evolution of a centralized ICS that is necessary to develop and maintain the complex biological structures of eumetazoans. The absence of a centralized control system in sponges can explain why they lack precise morphology, symmetry, or size as eumetazoans do. Their morphology and size are determined in part by physical factors (viscoelasticity, differential cell adhesivity, biochemical diffusion, etc.) (Newman and Müller, 2000) and in part by their "obscure" control system at the supracellular level (Cabej, 2008, 2012), a type of "paracrine prenervous system" that is responsible for their contractility in response to various stimuli (Nickel, 2010).

It is very likely that the most salient difference between them and eumetazoans is that the last succeeded in evolving a different cell type, the neuron, and the resulting nervous system (Cabej, 1999, 2004; Stanley, 1992).[3]

[3]Recently, I found that Steven M. Stanley was the first to propose the possible role of the neuron as an intrinsic driver of the animal evolution, in a short but insightful 1992 article in the little accessible *Geological Society of America Abstracts with Programs*. Deviating from the mainstream biological thought that looked into extrinsic environmental factors as the cause of the delay of adaptive radiation, Stanley suggested that the "boring" delay of evolutionary progress and innovations for hundreds of millions of years was caused by an intrinsic barrier:

> *A likely barrier was the difficulty of evolving the neuron. Prior to the origin of the neuron, effective use of muscles for feeding and locomotion was impossible; preneural animals could not have evolved in grade far beyond the extant Placozoa. The neuron is highly complex, employing a positive feedback system, and is quite similar among all animal taxa, from jellyfishes to humans. Thus, the neuron was almost certainly of monophyletic origin and should be viewed as a defining trait for the Metazoa.... Sponges illustrate the limitations of evolution sans (without - N.C.) neurons, having existed in a state of adaptive stagnation for more than half a billion years as simple, sessile creatures that feed via single cells.*
>
> *Stanley (1992)*

Rise of the Eumetazoan Control System: Evolution of the Synapse, Neuron, and the Diffuse Neural Net in Cnidarians

It is believed that cnidarians and ctenophores are "living fossils" that evolved between 634 and 604 Mya (Peterson and Butterfield, 2005). They are the lowest group of animals that differentiate unpolarized neurons and have a diffuse nervous system of interconnected neurons throughout the animal body. Their nervous system represents a primitive ICS.

Evolution of the radial symmetry in cnidarians is related to the differentiation of the neuron and the resulting diffuse neural net; their body may be divided into two similar parts by passing a plane along the central axis. This is not only the simplest form of symmetry, but it also represents a progress in their morphology when compared to the lack of symmetry in sponges. In distinction from sponges, their form is clearly determined. They have evolved diploblasty—the development of two embryonic layers—endoderm and ectoderm. Cnidarian have 7–8 cell types, among which are muscle cells, nerve cells, interstitial cells, gland cells, and cnidocytes. Tentacles provide the food to the digestive cavity where it is digested under the influence of the secretions of gland cells of the endoderm (gastroderm). Each cell in the body gets its nutrients, exchanges gases, and eliminates waste via water diffusion.

Cnidarians have the most primitive nervous system known in extant metazoans, a minimally evolved diffuse neural net with neurons that transmit electrical impulses to all the neurons of the net throughout the animal body. The diffuse cnidarian net has some homeostatic functions, especially in relation to morphogenesis and apoptosis (e.g., for eliminating the damaged cells). For example, the homeostasis of the neuronal population in cnidarians is maintained by a balance between the neuropeptide Hym 33H, which inhibits the differentiation of neurons, and the epitheliopeptide Hym 355, which stimulates it (Watanabe et al., 2009). This evolutionary progress is related to the evolution of their primitive ICS rather than evolution of genes or genetic mechanisms.

Cnidarians have no head or a homologous organ and no locomotor organs. They did not evolve any true organs or organ systems for digestion, respiration, circulation, waste excretion, body support structure (exoskeleton or endoskeleton) like other eumetazoans did. In this regard, they represent another evolutionary "dead end."

The crucial and, with a high likelihood, first step in the advent of the neuron is the evolution of the synapse. A synapse precursor may have been the intermediate step in the transition from the poriferan biochemical/paracrine system to the primarily electrochemical nervous system of eumetazoans (Nickel, 2010).

Let us keep in mind that all the essential genes of simple eumetazoans with a nervous system, such as the anthozoan cnidarian *N. vectensis*, were present in the unicellular choanoflagellates, such as *Monosiga brevicollis* (Erwin, 2009) and almost all the neurogenic genes and proteins involved in the differentiation of the neuron and in the development of the nervous system are found in sponges.

Because of their role in communication between neurons, it is assumed that synapses, small gaps between neurons through which electrical or chemical signals are transmitted from one neuron to another, evolved before axons and dendrites. Being a defining feature and integral part of the neuron, the synapse may be considered to have evolved at the same time and along with neurons. This is the "synapse first" model of Ryan and Grant (2009).

Crucial in transmitting signals through synapses is the group of postsynaptic proteins known as postsynaptic density (PSD) proteins, embedded in the postsynaptic membrane. PSD is a discoidal structure composed of several hundred proteins (scaffold proteins linked to the actin cytoskeleton as well as receptor molecules, protein kinases, phosphatases, etc.) and protein complexes. Genes coding for PSD proteins are found across the animal kingdom (Alie and Manuel (2010); Galliot and Quiquand, 2011; Kosik, 2009).

Once evolved in cnidarian-like animals, the neuron seems to have conserved not only its synaptic composition and structure but also its functions and morphology to an exceptional degree; it secretes almost all essential neurotransmitters and their receptors throughout the metazoan species (Figure 5.7).

A close examination of the differences between cnidarians and ctenophores on one side, and the bilaterians on the other, shows that the only relevant difference is that the latter evolved a higher organized CNS with a greater number of neurons. Centralization of the nervous system in bilaterians increased its computational capabilities and its ability to generate the epigenetic information required for erecting specialized supracellular structures of different types of cells and tissues into organs, and organs into organ systems.

The diffuse neural net in cnidarians represents a not yet full-fledged ICS. The animals of this eumetazoan group have not yet acquired the function of a central regulator of homeostasis like the one we saw in bilaterians. This is why cnidarians and ctenophores did not evolve in their respective specialized organs and larger bodies, but still rely on diffusion that is efficient only in these thin-walled diploblastic animals.

The primitive diffuse nervous system that led to the evolution of these primitive eumetazoans soon became the cause of their evolutionary stagnation. Further progress into the evolution of the animal world waited for a major evolutionary innovation, in the form of the centralized nervous system, which brought expanded computing capabilities, that performs the functions of a more advanced ICS.

The advent of the neuron in organisms comparable to extant cnidarians and ctenophores about 600 Mya marks the evolution of an early offshoot that is higher in the tree of life than sponges, yet not evolved enough to join the Cambrian club.

Evolution of the Integrated Control System and the Cambrian Explosion

Hard fossil evidence on the evolution of bilateria exists only for the lower Cambrian with only one case, that of Kimberella, dated in the transition period between the Ediacaran and Cambrian, about 555 Mya. Hence:

> *the most parsimonious interpretation of the Cambrian fossil record is that it represents a broadly accurate temporal picture of the origins of the bilaterian phyla.*
>
> *Budd (2003); Budd and Telford (2009)*

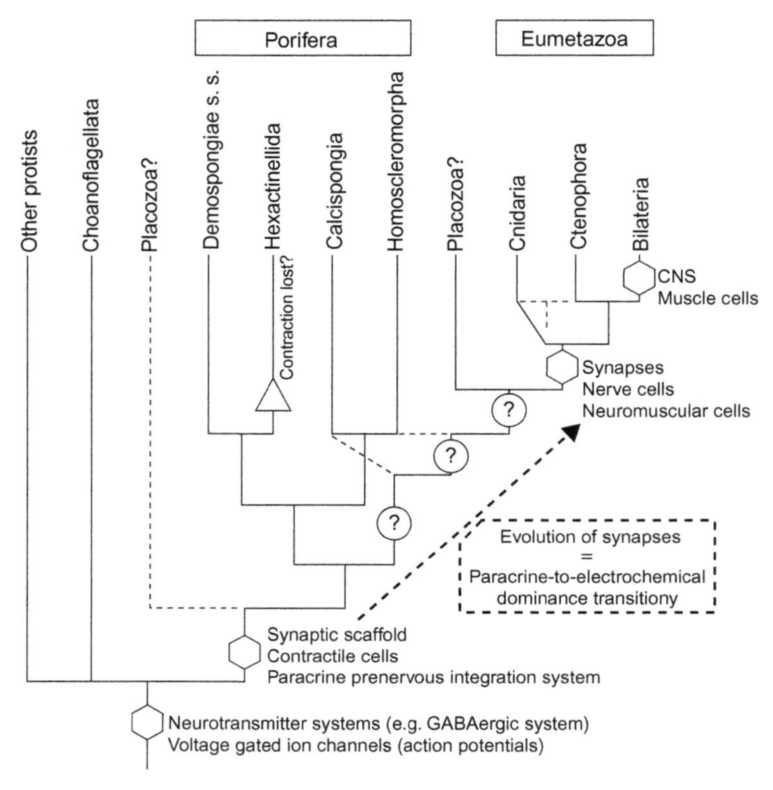

Figure 5.7 Summary of the coevolution of integration and effector systems in Metazoa. The phylogenetic tree follows, and others, with alternative scenarios presented as dashed lines. The position of Placozoa remains a matter of debate but will be of special interest for our understanding as soon as more details on the placozoan integration system are known. The evolution of the ultrastructure of synapses remains unresolved, but the synapse gene repertoire seems to be plesiomorphic to all Metazoa. *Abbreviations*: CNS, central nervous system; GABA, γ-amino butyric acid.
Source: Nickel (2010).

Bilateria gave rise to the explosive radiation of millions of species and higher taxa that dominate the animal kingdom as we know it: 99% of extant animals belong to Bilateria (Finnerty et al., 2004). Why does Bilateria enjoy such a huge taxonomic superiority over its sister groups?

From the neoDarwinian view, the Cambrian explosion is a great paradox. It is an event that cannot be predicted by the conventional biological wisdom. The most widely accepted neoDarwinian explanations involve changes in genes, especially those in the genetic toolkit (which ironically is highly conserved), in GRNs, and regulatory sequences (see *The "Regulatory" Hypothesis of Morphological Evolution, p. 247, and The GRN Hypothesis of Evolution, p. 248*). Here it is necessary to reiterate that no correlation is shown to exist between *Hox* genes, considered as major players in animal development, and the evolution of animals.

Hox genes may be traced back to the LCA of bilaterians and cnidaria (Schierwater and Kamm, 2010). The cnidarian *N. vectensis*, for example, has more homeodomains than *Drosophila* (Ryan et al., 2006) and the worm *C. elegans* has half the *Hox* genes of its ancestor (Aboobaker and Blaxter, 2010). There is no evidence to prove gene duplications as "one of the primary driving forces in the evolution of genes and genomes" (Roth et al., 2007). To the contrary, the evidence shows that the Cambrian falls in the "silent periods" of gene duplications. There is no link between these bursts and the Cambrian explosion (Miyata and Suga, 2001; Suga et al., 1999).

A bare glimpse on the Cambrian explosion would not have missed the temporal coincidence of five major transitions at that point in time: the emergence of the centralization of the nervous system, triploblasty, bilateral symmetry, evolution of organs and organ systems, and cephalization. Let us remember that this evolutionary leap occurred in a sister group of cnidarians that showed none of the above characters. If one excludes miracles, the simultaneous emergence of four major transitions requires a causal explanation.

A closer examination into the causal net between the four coinciding evolutionary developments may help us recognize the possible causal agent(s).

My hypothesis is that the centralization of the nervous system, which represents evolution of the full-fledged ICS in animals, was the driving force behind this momentous landmark of animal evolution.

Centralization of the nervous system is associated with the evolution of the bilateral symmetry. This is in clear distinction to the cnidarians, where the diffuse nervous system is generally associated with radial symmetry (Figure 5.8). Bilateral symmetry and cephalization enabled directed forward movement and increased the efficiency of movement, which is essential for a predatory lifestyle (both for preying on and escaping from predators) of the Cambrian biota. Directed movement implies the existence of a centralized nervous system. In individual development, the bilaterian symmetry during development first appears at the neurula stage.

> *In the ontogeny of vertebrates, bilateral symmetry appears at gastrulation in the stage of the primitive streak (Hensen node) and primitive streak, at the time of the notochord process and the neural plate. Not only is bilateral symmetry established at this time but the three fundamental axes of the body and of the future neural tube also become evident: a longitudinal axis with rostral and caudal ends (cephalization), a vertical axis with dorsal and ventral surfaces and a horizontal axis with medial (proximal) and lateral (distal) positions along the axis.*
>
> *Flores-Sarnat and Sarnat (2008)*

This is clearly before the expression of the A–P (anterior–posterior)-related *Hox* genes; hence, if ontogeny tells us anything about evolutionary history, it is that the above suggests that the evolution of body axes in bilaterians is not related to *Hox* genes and their colinear expression in extant animals.

Triploblasty seems related to (not that it is a cause of!) the development of organs and organ systems, since most organs develop from the mesoderm, which is absent

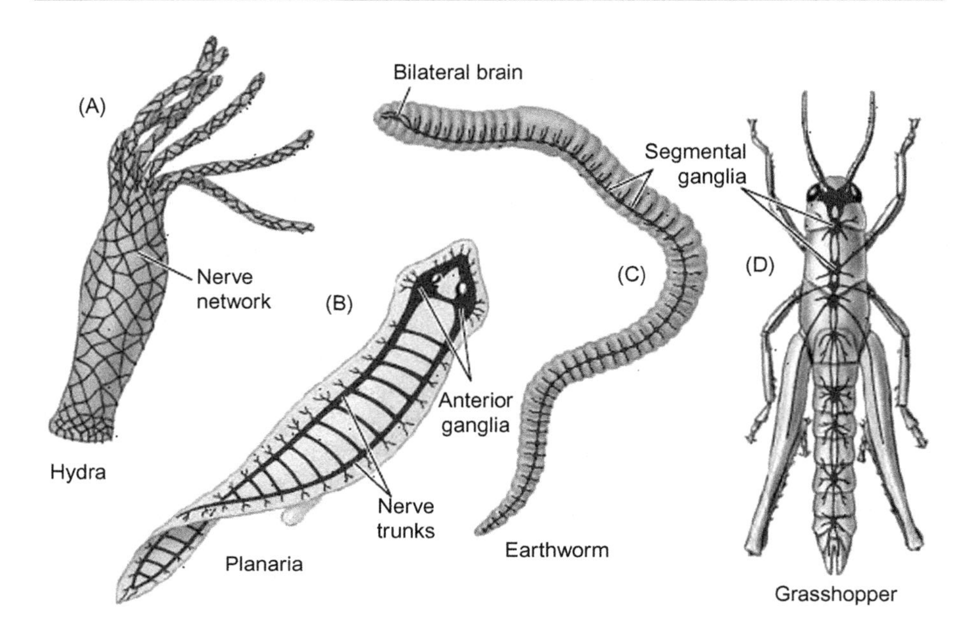

Figure 5.8 Invertebrate nervous systems. (A) Nerve net of radiates, the simplest neural organization. (B) Flatworm system, the simplest linear-type nervous system of two nerves connected to a complex neuronal network. (C) Annelid nervous system, organized into a bilobed brain and ventral cord with segmental ganglia. (D) Arthropod nervous system with large ganglia and more elaborate sense organs.
Source: Biocyclopedia. Available from: http://www.biocyclopedia.com/index/general_zoology/invertebrates_development_of_centralized_nervous_systems.php.

in diploblastic animals (cnidarian and ctenophores). There is no hint that triploblasty is related to bilaterality neither to the evolution of the nervous system, which develops from the ectoderm, nor to cephalization.

Bilateral symmetry is a property or geometric concept, not a functioning structure; it is an effect of development rather than a causal agent. Hence, there is no way for it to induce the formation of organs, a nervous system, or cephalization.

Evolution of cephalization is related to the evolution of the CNS, but it is causally not related to the evolution of organs and organ systems or to bilaterality. When it comes to the evolution of the CNS, the temporal coincidence with all the aforementioned structures and qualities displays a clear causal relationship. Signals and cellular elements from the CNS determine or participate in the formation of all organs, cephalization, and the development of bilateral symmetry (see Chapter 3, *Epigenetic Control of Animal Development*).

None of the aforementioned, individually or collectively, proves that the centralized nervous system determines the bilateral symmetry in bilaterians. Now, half a billion years after the Cambrian, we possess a wealth of fossil evidence on the astounding radiation of animal phyla, but we are left with little means of determining with any degree of certainty the mechanisms that drove the unprecedented rapid evolution of the Cambrian biota. Paleobiological and comparative biological techniques have created a general picture of the dynamics of the Cambrian diversification, but it can only indirectly, if at all, help us determine the driving forces of this unique event in the evolution of animals.

However, it seems that some general biological principles may be a valuable means of acquiring ample scientific insights into the mechanisms and processes that drove the Cambrian explosion.

Other Insights into the Possible Role of the Centralized Nervous System in the Cambrian Diversification

The history of the evolution of bilaterians is buried deep in the Cambrian, but extant eumetazoans are "historical" structures. For two centuries, or about 150 years since Haeckel, most biologists have continued to hold that the evolutionary history of animals may be retrieved from individual development. Indeed, the evolution of animals, essentially, is the visible result of the evolution of animal developmental processes. Most evolutionary changes or novelties reflect specific changes in the developmental pathways. Developmental changes and evolutionary changes are two sides of the same coin. Commenting on Haeckel's biogenetic law, almost a century ago, English zoologist Walter Garstang (1868–1949) pointed out:

> *Ontogeny does not recapitulate phylogeny, it creates it.*
>
> *Garstang (1922)*

In this sense, development runs the show of evolution. What controls development controls evolution. So what controls development, itself?

I postulate that mechanisms controlling the development of morphological traits during individual development are responsible for the evolution of these traits in their Cambrian ancestors. This principle of phylogenetic actualism stems from Haeckel's biogenetic law.

In my previous work (Cabej, 2004, 2008, 2012) I developed and supported with adequate empirical evidence, the hypothesis that the individual development in eumetazoans takes place in two stages and is under a bigenerational control. During the first stage, from the unicellular state (egg or zygote) until the phylotypic stage, development takes place under epigenetic control of the parental cytoplasmic factors provided with the gamete(s). At the onset of the phylotypic stage, the exhaustion of the parental epigenetic information coincides with formation of the operative CNS, which takes over the continuing development until adulthood. It is the source of the epigenetic information provided in the form of biochemical inducers that trigger signal cascades for cell differentiation and tissue and organ growth. The evidence in support of the hypothesis shows that "the nervous system has a central role in animal evolution" (Jablonka, 2009) (see page 149, *Epigenetic Control of Postphylotypic Development in Animals* for an extended discussion of the crucial role of the nervous system in development).

In this context, the postulate of the phylogenetic actualism that the present mechanisms of development of phenotypic traits in general are similar to those that brought about the evolution for the first time around, syllogistically leads to the conclusion that the CNS played a leading role in the Cambrian morphological diversification.

A second source of corroborative evidence on the role of the nervous system in the development and evolution of Cambrian biota derives from studies in biological regeneration. The phenomenon is observed in lower invertebrates, such as coelenterates (cnidarians and ctenophores) and echinoderms and up through higher vertebrates such as amphibians, reptiles, and even mammals. In all, the studied species regeneration depends on innervation, suggesting that the nervous system plays a morphogenetic role in regeneration (Kumar and Brockes, 2012).

Within a week after the experimental removal of its eyes, the flatworm *Polycelis nigra* regenerates new eyes. But it does not do so if its brain is removed. Eyes regenerate if the head lysates or brains of individuals from other flatworm species are implanted (Kumar and Brockes, 2012).

Almost two centuries ago, in 1823, Tweedy John Todd (1789–1840), an English physician, observed that salamanders could regrow amputated limbs, providing for the first time experimental evidence that:

> *Nerves…may also serve to maintain the morphological and developmental integrity.*
>
> *Singer and Géraudie (1991)*

Regeneration of amputated limbs by amphibians also requires local innervation. After limb amputation, the apical ectodermal cap covered by an ectodermal layer develops on the stump. The remaining cell types in the neighborhood dedifferentiate to form a cluster of pluripotent cells, the regeneration blastema, from which the new limb develops. In the absence of local innervation in the early larvae of *Xenopus*, the blastema does not form and the limb does not regenerate. Experimental deviation of the sciatic nerve to wound sites in certain regions around the hind leg induces the formation of supernumerary limbs.

An interesting example of the morphogenetic role of the innervation during regeneration in mammals comes from experiments on the healing of the punched ear lobe in Murphy Roths Large (MRL) mice. When a normally innervated ear is punched, it heals completely by developing normal cartilage and epithelial structures, including hairs, whereas in the denervated ear the regeneration is defectuous (Figure 5.9).

Centralization of the Nervous System and Evolution of the Animal Complexity in Bilaterians

We have shown that the centralization of the nervous system coincided with the advent of bilaterality, triploblasty, and the formation of organs and cephalization in the Cambrian biota. Now, we will briefly review the concomitant evolution of the

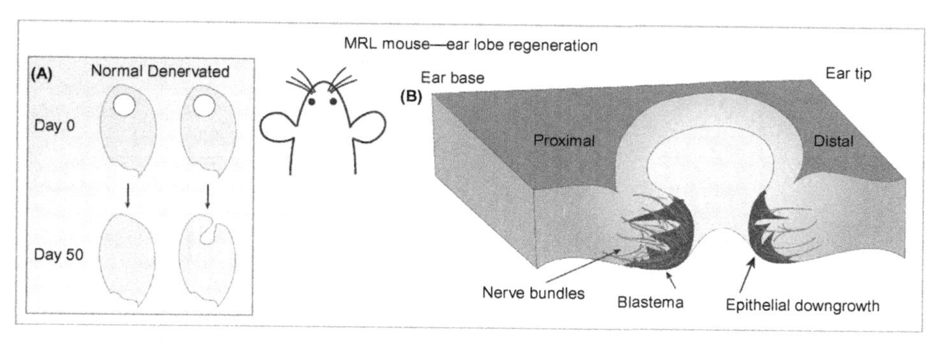

Figure 5.9 Innervation and mammalian regeneration. Regeneration of the pinna in the MRL mouse. (A) Injury induced by an ear punch is completely healed by day 50 in the presence of normal innervation. Transection of the nerves to the ear lobe prior to ear punch results in abnormal wound healing and necrosis of the tissue. (B) Diagrammatic representation of a regenerating ear-punch hole. Quantitative image analysis has shown that the density of innervation is higher in the proximal ear base of the pinna compared to the distal tip. *Source*: Kumar and Brockes (2012).

CNS and the structural complexity, as it manifests in the evolution of specialized animal organs and structures, through the course of the evolution of bilaterians.

Acoela (Acoelomates)

Acoela are the most primitive group of extant bilaterians with a rudimentary centralized nervous system. The group consists of 340 species and 21 families known so far (Hooge and Tyler, 2006). Their nervous system shows only a minute anterior–posterior gradient of neuronal concentration (Semmler et al., 2010), which is far from a "compact brain." The nervous system consists of one or more commissural rings in the anterior part that are connected with longitudinal nerve cords along the length of the body; hence, they are termed "orthogon" for their nervous system and the designation "commissural brain" for the anterior concentration of neurons (Raikova et al., 2001) (Figure 5.10).

Acoela belong to the phylum *Acoelomorpha* and are believed to be the earliest offshoot of bilaterians (Bery et al., 2010; Jondelius et al., 2002; Peterson et al., 2008). They tend to be free-living unsegmented marine animals that lack a coelom (body cavity), have no anus and no appendages, and move via the ciliary epithelium on their ventral surface.

Substantial evolutionary progress is observed in flatworms (*Platyhelminthes*). The progress in the organization and centralization of the nervous system is reflected in the evolution of the excretory and digestive systems. Flatworms lack circulatory and pulmonary systems, but because of their small body sizes their cells are close enough to nutrients and oxygen.

Recent studies in the planarian flatworm, *Schmidtea mediterranea*, have identified more than 140 neuropeptides and peptide hormones revealing the complexity of the nervous system of *S. meditteranea*, one of the most primitive centralized nervous systems in invertebrates. Many of these are neurohormones secreted only by neurons

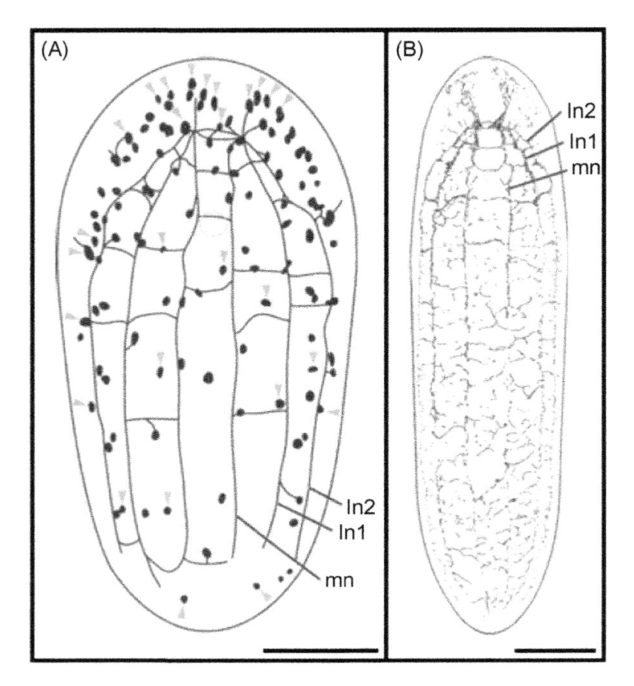

Figure 5.10 Semi-schematic sketch drawing of the juvenile (A) and the adult (B) serotonergic nervous system of the acoelomate *Symsagittifera roscoffensis*. Six longitudinal neurite bundles extend along the anterior–posterior body axis: two median (mn) and four lateral (ln1 and ln2) ones.
Source: Semmler et al. (2010).

in the CNS. Chemical signals, especially members of the NPY-8 family of neurohormones released by the cephalic ganglia, are essential for the development and maintenance of reproductive organs. In *S. mediterranea*:

> *peptides (e.g., NPY-8 peptides) from the nervous system promote events associated with reproductive maturation (i.e., the production of differentiated germ cells)*
>
> *Collins et al. (2010)*

Evolution of the "True" Brain

The true compact brain in the anterior of the animal body is believed to have emerged four times, thrice in protostomes (i.e., arthropods, mollusks, and annelids) and once in the latest common ancestor of cephalochordates, urochordates, and vertebrates (Figure 5.11).

Annelids

The earliest annelid fossil comes from the lower to middle Adtabanian, about 520 Mya (Conway Morris and Peel, 2008).

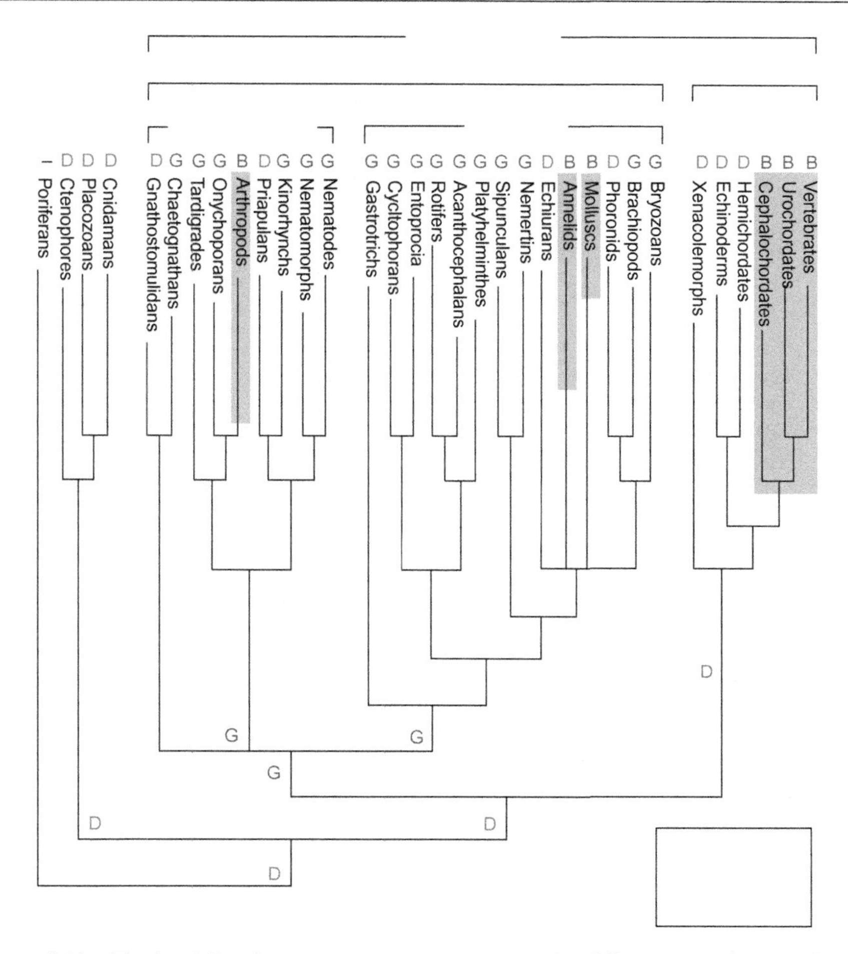

Figure 5.11 The last bilaterian common ancestor possessed a diffuse nerve plexus and brains independently evolved at least four times among bilaterians.
Source: Northcutt (2012).

The annelid nervous system consists of a primitive compact brain in the anterior of the body connected with two ventral nerve cords that connect with ganglia in each segment. Annelids have evolved specialized sense organs (eyes, taste buds, statocysts, etc.) and a relatively complex neuroendocrine system that comprises secretory neurons with projections throughout the worm body. They secrete more than 40 neuropeptides and neurohormones that circulate in hemolymph. Most are mammalian-like neuropeptides, such as angiotensin-like peptides, oxytocin/vasopressin, and enkephalins (Salzet, 2007; Veenstra, 2011).

Morphologically, annelids display a great progress in comparison to flatworms; their digestive system, besides mouth and guts, also has an anus. The body is segmented with a cavity, although it is not a true coelom. They have evolved a closed

Figure 5.12 Fossilized specimen of *Fuxianhuia protensa*, representative of an arthropod group found in Chengjiang Lagerstätte. *Abbreviations*: A1, antenna; Ab, abdomen; Es, eye stalk; Ey, eye; Hs, head shield; Oc, optic capsule; Th, thorax. Scale bar, 1 cm.
Source: Ma et al. (2012).

blood circulation system with a number of vessels and annelid blood contains hemoglobin for oxygen transport. The excretory system in lower annelids has a metanephridium in each segment and a duct that opens at the body surface. In other annelids, the excretory system consists of a single duct. Many aquatic annelids evolved gills for exchanging gases with the environment. Most of them have separate sexes and produce free-swimming larvae. A number of species reproduce asexually.

Arthropods

Arthropod-like animals evolved during the Cambrian period. A fossilized specimen of *Fuxianhuia protensa* is 520 Mya old (Figure 5.12). Today they represent the most successful phylum. With more than 1 million species, arthropods encompass 90% of animal species on Earth. Insects are the largest arthropod group.

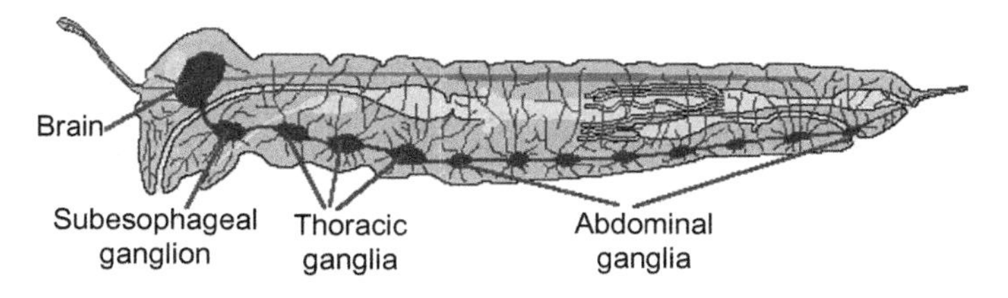

Figure 5.13 Generalized diagrammatic representation of the insect nervous system.
Source: Introduction to Insect Anatomy. Available from: http://www.earthlife.net/insects/anatomy.html.

Insects may be the first group to display clear cephalization. Evolution of a rigid chitin exoskeleton protects them from predators and drought. The body consists of the head, thorax, and abdomen, composed of a varying number of segments. They have simple eyes with a single lens and compound eyes of multiple lenses. As arthropods, they have jointed appendages. Most insects develop via metamorphosis and undergo ecdysis.

The insect nervous system (Figure 5.13) is composed of a brain that develops from the fusion of three pairs of anterior ganglia (group of neurons) and the ventral cord, which includes the subesophageal ganglion formed from fusion of three pairs of ganglia, and pairs of thoracic and abdominal ganglia for each segment linked together by connective cords that run the length of the body. The ganglia innervate respective segments. In insects, the ICS displays the hierarchic character. The brain and the subesophageal ganglion secrete neurohormones that stimulate specialized endocrine glands such as the prothoracic gland and the corpora allata to respectively secrete hormones ecdysone and JH. Thus the CNS controls the growth, development, and reproduction in insects.

Vertebrates

Vertebrate brain consists of CNS and PNS. The hypertrophied CNS consists of the brain (i.e., forebrain, midbrain, and hindbrain) (Figure 5.14) and spinal cord. The PNS consists of the visceral (autonomic) and somatic systems consisting of nerves (axon bundles) innervating respectively the internal organs and muscles, skin, and joints. In vertebrates, an accelerated increase in the size and the proportion of the CNS and the brain-to-body weight with a strong trend toward increasing cortex size is observed. The CNS evolved many specialized centers for new functions and to perfect older functions. The CNS grows and its structure and function becomes more complex ascending the higher vertebrate taxa (Figure 5.14). Ever increasing elaboration of the structure and function of the brain in higher vertebrates, especially in humans, made the human brain the most complex of the known structures.

In vertebrates, the hierarchical character of the ICS becomes more conspicuous as it is seen in the schematic representation of the flow of epigenetic information from the brain structures down to particular genes (Figure 5.15).

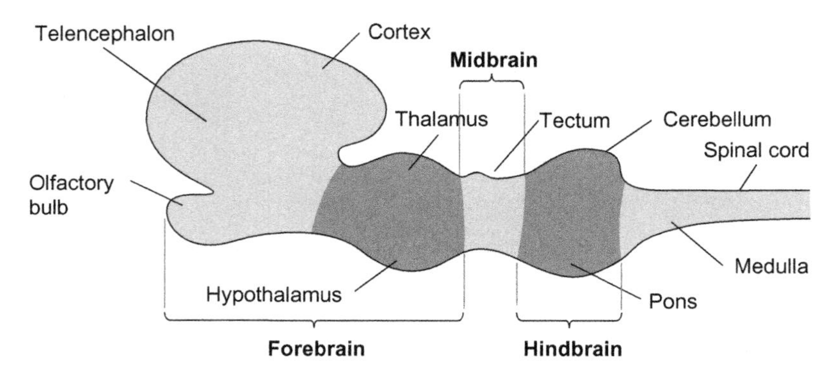

Figure 5.14 Major divisions of the vertebrate brain. The hindbrain comprises the medulla (which contains centers that regulate several autonomic visceral functions such as breathing, heart and blood vessel activity, swallowing, vomiting, and digestion), the pons (which also participates in some of these activities), and the cerebellum.
Source: Bownds (2001).

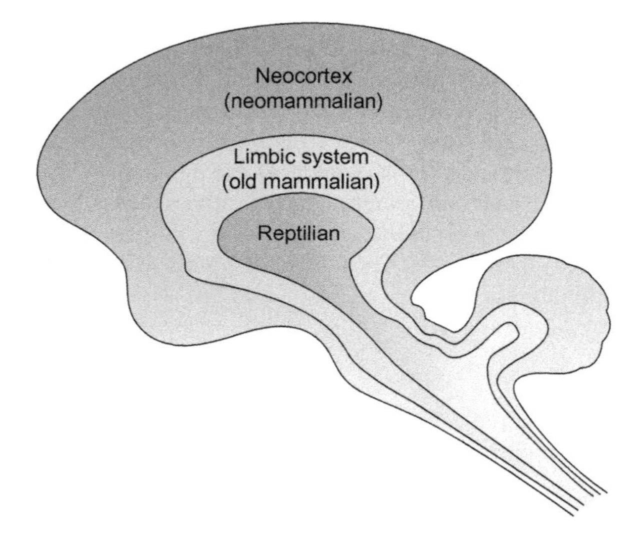

Figure 5.15 The model of the triune brain proposed by Paul MacLean, indicating brain areas that are added during vertebrate evolution. The reptilian brain is the main seat of innate or instinctive behaviors regulating primitive survival issues. The old mammalian brain, or the limbic system, expresses innate motivational value systems that interact with the newer cortex, or neocortex, which manages propositional information and declarative knowledge about the world.
Source: Bownds (2001).

The increased structural sophistication of the CNS in vertebrates proceeded in parallel to the increased complexity in morphology, physiology, and behavior of the group. Among the novelties of the vertebrate brain, the neural crest, a transient embryonic structure, plays a special role in the development and evolution of almost all vertebrate organs. The vertebrate head is considered an evolutionary addition to the body of protochordates (Gans and Northcutt, 1983). In the head's case, the cranial crest took over most of the functions of the mesoderm.

On the Generation of Epigenetic Information in the CNS

We have seen that the animal organism responds to external and internal stimuli adaptively, by activating signal cascades and GRNs that tend to avoid possible harmful effects of the stimuli. We have also seen that the stimuli are received, perceived, and "classified" as such in the CNS where the relevant signal cascades start. Adaptive responses are neither random nor default; hence the animal uses information to select and perform them. The fact that perception of stimuli and the activation of cascades starts in the CNS suggests that there may also be the source of information for adaptive responses.

The data/information on the internal or external environment are received and processed in the synapses of the sensory cells. This induces the movement of ions across neuronal membranes, leading to potential differences. Thus sensory cells convert external stimuli into electronic messages presented as electrical spikes. The messages are transmitted to the CNS via afferent pathways. They may be encoded in the form of the spike frequencies, spike duration, number, or patterns (Stafford, 2010). In the CNS, the sensory information is processed further in various centers until an output is produced in the form of a chemical signal (neurotransmitter, neuromodulator, etc.).

Not all data on external or internal agents are taken for a stimulus requiring a response. The nervous system is bombarded with a myriad of data on the external and internal environment and it must distinguish between the normal innocuous variables and variables that pose problems. It is in the CNS where the distinction is made and the chaff is separated from the wheat. What is classified as innocuous is neglected as "noise." The rest are classified as stimuli requiring response.

The CNS makes these decisions by comparing the data on environmental variables with respective set points: when the variable is outside the limits of the set point, it is taken as a stimulus or as a problem requiring a solution. The processing of the stimulus then follows. So, for example, a small drop of 1°C in the environmental temperature is received by skin thermoreceptors and transmitted to a specific center in the hypothalamic preoptic area (POA), where it is ignored as "noise: unworthy of a response." But when the temperature drop is greater, say 10°C, it falls outside the set point and is classified as a stimulus to which the CNS responds adaptively: the POA stimulates the thermogenesis (other CNS centers such as the brainstem and spinal cord are also involved in thermogenesis) by activating the production of thyroid hormones (TH) by the thyroid gland via the following signal cascade:

> Secretion by the hypothalamic neurons of the thyroid-releasing hormone (TRH)→Secretion of thyroid-stimulating hormone (TSH) by the pituitary→Secretion of TH by the thyroid glands.

Neural mechanisms involved in the increased thermogenesis are constrictions of skin vessels for reducing heat loss, shivering, and migration toward warmer places.

When the environmental temperature is elevated, the CNS activates mechanisms that intend to maintain the organism's temperature within a species-specific set point of the CNS.

Certainly, in determining the activation and the level of activation of the thermogenesis, the hypothalamus processes information about the brain and core temperature and it will continue to stimulate or inhibit thermogenesis until the body temperature comes within the normal range. The hypothalamic temperature set point is found experimentally (Hammel et al. (1963)). It is neither arbitrary nor inalterable. Set points are adjustable not only in evolution but also during individual development. Modification of the temperature set points is induced experimentally in chickens and rabbits. Rearing rabbits in cold temperatures lowers their temperature set points (Tzschenke and Nichelmann, 1997). Changes in the incubation temperatures also alter the set point in chickens (*Gallus domesticus*). Warm-blooded animals modify their temperature set points to adapt their physiology to various stress conditions. They also modify their temperature set points in response to cyclic changes in estrogen levels.

Neural responses are normally adaptive, as they tend to avert possible negative effects of stimuli or utilize their possible positive effects. In a generalized case, the processing output in the CNS starts a signal cascade, which ends with the activation of one or more genes in a specific cell, tissue, or organ (Figure 5.16). For example, the output of the processing in a bird's brain of the changes in the photoperiod (lengthening of the day) or in social stimuli (e.g. song of conspecifics) leads to the activation of gonadotropin-releasing hormone (GnRH)-secreting hypothalamic neurons. In turn, the secreted GnRH starts the well-known hypothalamic–pituitary–gonadal cascade, which induces its reproductive activity.

While neural electrical signals are encoded in the form of variations in intensity, frequency, duration, and patterns of electrical spikes, chemical signals released by the CNS for starting signal cascades are often encoded in amplitudes or pulses of the release of neurohormones.

An interesting example of encoded hypothalamic epigenetic information is offered by the pulsatile patterns of GnRH secretion. GnRH is secreted by a few hundred hypothalamic neurons in humans. It is secreted in brief pulses that vary from about 1 h during the follicular phase to about 6 h during the luteal phase. The pulse frequency is the form in which the hypothalamus instructs the pituitary cells on which of the two alternative forms of gonadotropic hormones, luteinizing hormone (LH) or follicle-stimulating hormone (FSH), to be produced (Walker et al., 2010). Slower pulses of the GnRH secretion stimulate the pituitary cells to produce LH, whereas higher pulses stimulate the secretion of FSH, with all the consequences in the reproductive organs and functions (Krakauer et al., 2002).

Now we know of a potential mechanism used by the pituitary cells to decode the epigenetic information about which of the two alternative pituitary gonadotropins to secrete. A slower pulse frequency activates the secretion of the immediate *Egr1* and

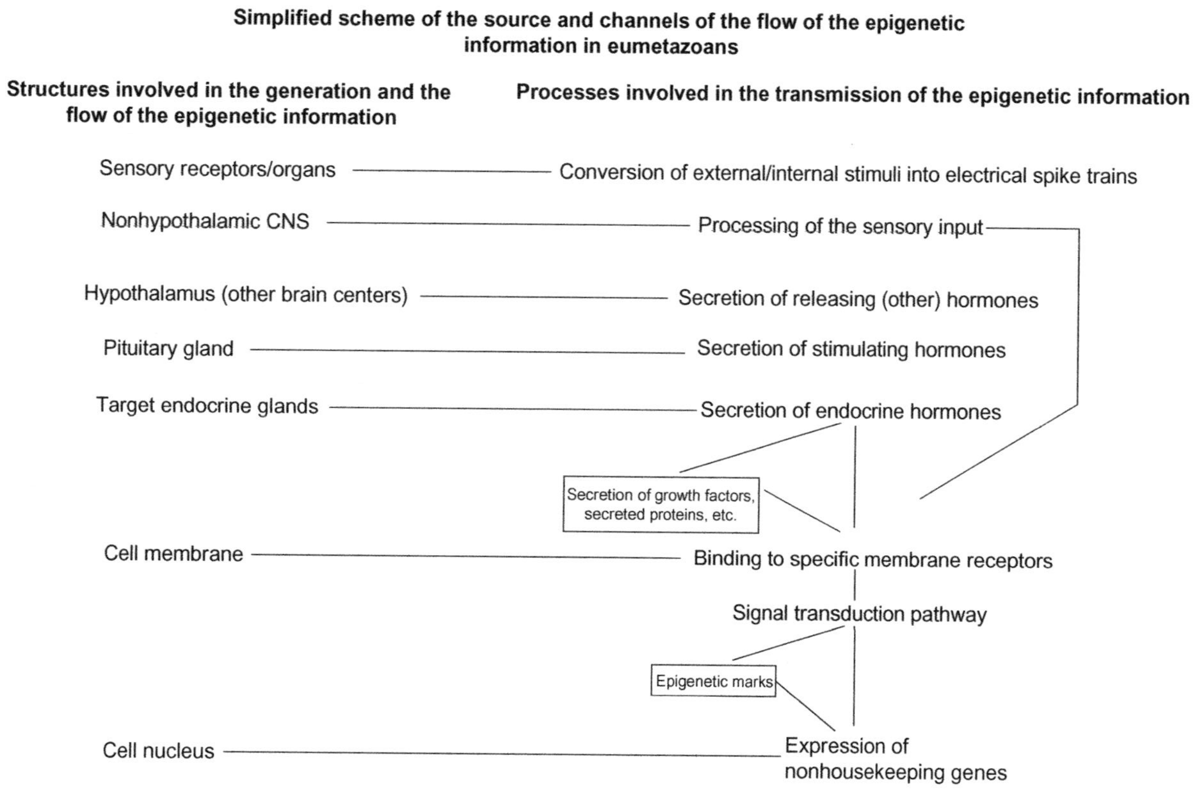

Simplified scheme of the source and channels of the flow of the epigenetic information in eumetazoans

Structures involved in the generation and the flow of the epigenetic information	Processes involved in the transmission of the epigenetic information
Sensory receptors/organs	Conversion of external/internal stimuli into electrical spike trains
Nonhypothalamic CNS	Processing of the sensory input
Hypothalamus (other brain centers)	Secretion of releasing (other) hormones
Pituitary gland	Secretion of stimulating hormones
Target endocrine glands	Secretion of endocrine hormones
	Secretion of growth factors, secreted proteins, etc.
Cell membrane	Binding to specific membrane receptors
	Signal transduction pathway
	Epigenetic marks
Cell nucleus	Expression of nonhousekeeping genes

Figure 5.16 Simplified model of generation and flow of epigenetic information for expression of systemic genes in response to external or internal stimuli in vertebrates.

Egr2 genes, activating β-promoter and LH secretion, whereas high pulses activate the Nab repressor genes that suppress LH synthesis (Lawson et al., 2007).

The processing of the stimuli in the CNS generates the epigenetic information for signal cascades that control and regulate the expression of nonhousekeeping (systemic) genes and individual development in animals. The evolution of the neural processing in the CNS was a prerequisite for the increased morphological complexity that characterized it.

The role of neural processing as a source of epigenetic information implies that it is intended to figure out an adaptive response to an environmental challenge. It tends to prevent, resist, or manage the anticipated harmful effects of environmental actions/agents. This assertion may raise concerns on the teleological implications of neural processing, because the words "intention" and "purpose" in everyday speech invoke consciousness. However, despite whatever philosophical implication, it is commonplace in scientific literature to speak of the purpose of animals, especially when describing animal behavior. A sea gull's dive has as clear a purpose as that of an eagle.

It may be argued that one cannot attribute intentions and purposes to the neural processing taking place at electrical and chemical levels. But bear in mind, human consciousness also results from the processes taking place at these "low" levels (see page 79, *Epigenetic Information and Signal Cascades*).

The Antientropic Demon and the Advent of Cambrian Biota—Fine-Tuning Gene Expression

Multicellular animals are believed to have appeared between 110 and 60 million years before the Cambrian period. By any account, this is a long period of evolutionary stasis that needs an explanation. Common sense says that a barrier existed at the time and time was needed to break through it. In 1992, Steven M. Stanley came up with an idea that the neuron evolution might have triggered the Cambrian explosion. Insightful as it is, in hindsight, the idea was not accurate. Ediacaran eumetazoans of the cnidarian sort existed at least 30 million years before the Cambrian phenomenon. As mentioned earlier, the Cambrian diversification coincides not with the neuron but with the transition to the centralization of the nervous system. Cnidarians that did not venture out were evolutionary losers.

Why might the centralization of the nervous system have been such an indispensable innovation for the further evolution of the animal world? The centralization amplified the computational capabilities of the nervous system required for evolving, erecting, and maintaining more complex structures in the Cambrian biota. The acquisition by the centralized control system of the craftsmanship to monitor, work out instructions, and transmit them for restoring the normal state down to the cell level, including the ability to reproduce the organism, enabled the major transition to the higher stage of structural and functional complexity of the Cambrian biota. The centralization of the nervous system provided animals with an unprecedented antientropic demon (*sensu* Maxwell) that also made possible four indispensable functions of the control system in the development and reproduction of animals:

- Spatial restriction of gene expression in the animal body.
- Manipulative expression of genes (see page 197, *Making Environmental Signals Intelligible to Genes*).
- Selective recruitment of genes.
- Epigenetic marking of genes (see page 71, *Neural Origin of Epigenetic Information in Epigenetic Structures*; page 206, *Where Does the Epigenetic Information for DNA Methylation Come from?*; and page 202, *The Source of Information for Selecting Sites of Histone Modification*).

The Biological Demon 1—Precise Spatial Restriction of Gene Expression

During development, the developing human embryo produces billions to trillions of cells of more than 400 different types. The type to which a cell is differentiated at this stage depends on the genes it expresses and the patterns of gene expression. Since the different cell types are intertwined in complex patterns in tissues, the organism must possess a highly precise mechanism, at the microscopic level, capable of providing individual cells the right morphogen. Needless to say, the textbook mechanism of gradients of concentrations of morphogens does not come into account. No plausible genetic mechanism or hypothesis exists to explain how the organism can determine the intricate arrangement of cell type patterns of cells in animal structures. Indeed, even simpler biological tasks, such as the localized action of circulating hormones on specific cells, tissues, and organs, while preventing their action throughout the rest of the body, is beyond the explanatory power of modern genetics.

The spatial restriction of the expression of genes has been a crucial requirement, a *sine qua non*, for cell differentiation, histogenesis, and organogenesis, for the evolution of the animal world from the beginning of metazoan life.

How can we learn about this mechanism of restricted gene expression employed by Cambrian biota more than a half a billion years ago? The principle of phylogenetic actualism again tells us that we may extrapolate mechanisms of the restricted spatial expression of genes in extant animals to the Cambrian biota. In view of their indispensability, these mechanisms have been operative from the onset of the Cambrian period.

We know the neuroendocrine system, via various hormonal cascades and GRNs, controls expression of numerous genes. But we do not know how the organism restricts the action of thousands of factors that circulate throughout the animal body to target cells, tissues, or organs only.

A considerable number of examples of local innervation are involved both in the local expression of specific genes and in restricting the action of circulating hormones to specific regions of the body that are known as targets of hormonal action (Cabej, 2004). Local innervation restricts the expression of hormonally induced genes to particular regions, where needed, by inhibiting the expression of specific receptors, which act as mediators to these hormonal actions. This is a novel mechanism and the only demonstrated mechanism of spatial restriction of expression of hormonally induced genes (see page 200, *Restricting Gene Expression to Relevant Cells Alone: Binary Neural Control of Gene Expression*).

A clear example of the global neural control gene expression is the regulation of muscle growth in insects. Muscle growth in insects results from the action of two antagonistically acting neural hormones, the muscle growth-inducing insulin-like peptides (Ilps) produced by a group of 14 neurons in the *Drosophila* brain, and muscle growth-inhibiting neurohormone prothoracicotropic hormone (PTTH) secreted by several other neurons (Géminard et al., 2006).

A signal cascade originating in the hypothalamus also controls and regulates the muscle growth in vertebrates. Based on the processing of afferent signals on the state of muscle growth and other internal stimuli, such as nutritional status and exercise, one of the two alternative hypothalamic neurotransmitter pathways (and other extrahypothalamic pathways) (Jorgensen et al., 1993; Ruaud and Thummel, 2008) is activated:

- Secretion of growth hormone-releasing hormone (GHRH) by a group of hypothalamic neurons.
- Secretion of its antagonist growth hormone releasing–inhibiting hormone (GHRIH) or somatostatin.

Each of the neurohormones binds to its specific receptors in pituitary cells. The first (GHRH) stimulates the secretion of growth hormone (GH), while the second (GHRIH) inhibits GH secretion. GH stimulates muscle formation or growth and secretion of insulin-like growth factor-1 (IGF-1), which is synergistic with GH.

Since the early 1980s it was demonstrated that the distribution pattern of myoblasts in sites of future muscles is determined by local innervation patterning of segmental muscles in *Drosophila* (Currie and Bate, 1991; Lawrence and Johnston 1984). The denervation of muscles in *Manduca sexta* prevents proliferation and migration of myoblasts to the proper sites causing formation of muscleless legs (Consoulas and Levin, 1997).

Experiments of denervation of indirect flight muscles in *M. sexta* have shown that:

> *The motoneuron influences both the number of cells available for fusion, as well as potentially regulates the fusion events themselves. This in our view is an elegant mechanism for controlling muscle fiber differentiation during myogenesis, and may have evolved as a way to ensure that muscle primordial develops into muscles that meet the diverse demands placed on them by the nervous system.*
>
> *Fernandes and Keshishian (2005)*

The dorsal external oblique (DEO1) muscle in *M. sexta* larvae consists of five muscle fibers, but out of them one is lost and the surviving fibers develop into the adult DEO1. Experiments of Hegstrom and Truman revealed that before ecdysis, the terminal arbor of the local motoneuron withdraws from all the fibers but the one that develops into adult DEO1. It was also found that in the absence of the motoneuron, muscle cells express the ecdysone receptor EcRA, which causes muscle fibers' apoptosis. The muscle fiber that is innervated expresses the EcR-B1, which allows the growth of the muscle. These experiments demonstrated that the choice by the motoneuron of the type of ecdysone receptor that will be expressed determines whether the ecdysone will play its muscle growth-inhibiting action (when EcRA is expressed) or not (when the local innervation induces expression of EcR-B1).

[B]oth exposure to ecdysteroids and local influences from the motoneuron are required for the upregulation and the maintenance of this high level of EcR-Bl expression that is associated with muscle regrowth.

Hegstrom and Truman (1996)

Innervation determines which of EcR isoforms expresses the growing muscle.

Hegstrom et al. (1998)

Results of these experiments were corroborated by other *in vitro* and *in vivo* experiments, which showed that local innervation performs its myogenetic function by releasing diffusible substances or by transmitting electrical signals in tissues (Launay, 2001).

As mentioned earlier, muscle growth in insects is cerebrally regulated by the types of neurohormones, the growth-promoting Ilps, and the growth-inhibiting PTTH, with ecdysone serving as the mediator of the muscle-inhibiting action of the PTTH. The above experiment shows that muscle growth is under a dual neural control, a central neural control via the circulating hormones and neurohormones, and an adjacent control via local innervation. This is a binary neural mechanism of control of gene expression. It is indispensable for the development of animal structure and must have evolved no later than the evolution of the centralized nervous system in planarian-like animals during the Cambrian period.

The role of the ICS in regulating gene expression in animals, both globally and locally, may be illustrated through the experimental evidence on brain surveillance and the regulation of bone homeostasis.

The brain integrates and processes internal stimuli to generate and provide instructions for bone remodeling to the bone skeleton (Elefteriou, 2008).

Neurons in the central nervous system integrate clues from the internal and external milieux, such as energy homeostasis, glycemia, or reproductive signals, with the regulation of bone remodeling.

Elefteriou (2008)

It sends these instructions in the form of chemical signals to increase/decrease production of osteoblasts and osteoclasts via two complementary pathways.

The hypothalamus regulates bone homeostasis remotely, via four signal cascades, and locally, via sympathetic innervation. The integrated regulation of bone homeostasis is depicted in Figure 5.17.

The local regulation of bone homeostasis is related to the action of leptin, a hormone that acts surprisingly in a nonhormonal way (i.e., not directly in bones, but in the brain by stimulating the sympathetic innervation), which induces noradrenaline release on osteoblasts, thus inhibiting osteoblast proliferation and bone loss (Figure 5.18).

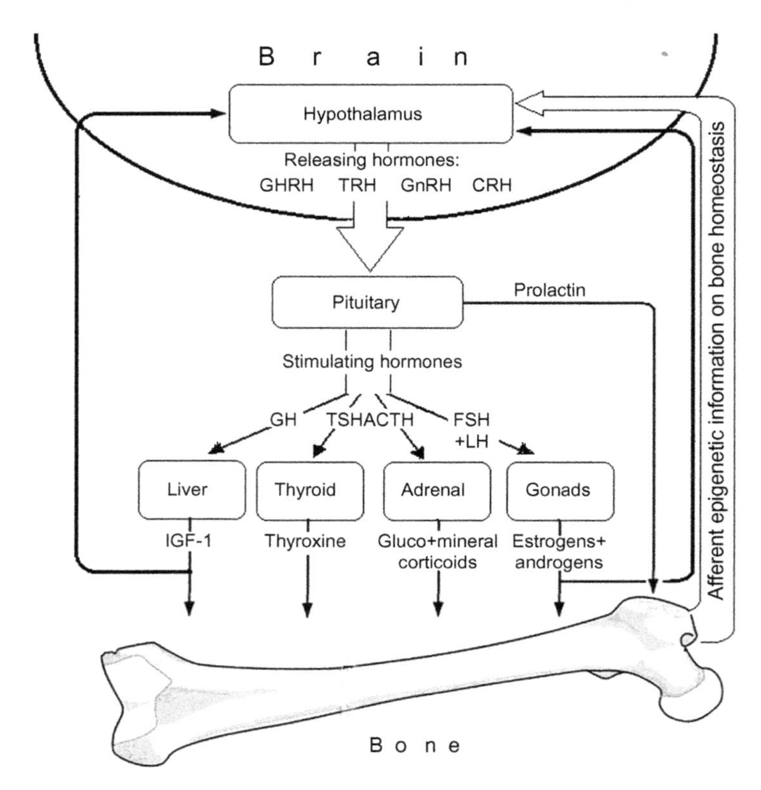

Figure 5.17 Simplified diagram of the central neural control and regulation of osteogenesis in vertebrates via the hypothalamic–pituitary–terminal endocrine gland (i.e., thyroid, gonads, and adrenals) axes and via the hypothalamic–pituitary axis (i.e., prolactin). Note that the orders for activation/inactivation of the five-signal cascades or central regulatory axes for bone homeostasis ultimately originate in the brain. *Abbreviations*: ACTH, adrenocorticotropic hormone; CRH, corticotropin-releasing hormone; FSH, follicle-stimulating hormone; GH, growth hormone; GHRH, growth hormone-releasing hormone; GnRH, gonadotropin-releasing hormone; IGF-1, insulin-like growth factor-1; TRH, thyrotropin-releasing hormone; TSH, thyroid-stimulating hormone (thyrotropin); LH, luteinizing hormone (lutropin).

One of the most important neuropeptides involved in the regulation of bone homeostasis is the hypothalamic neuropeptide Y (NPY) and one of its receptors, Y2 (Baldock et al., 2002; Reid et al., 2005).

Let us see several other examples of the spatial restriction and fine-tuning by the binary neural control system in animals.

In insects, production of ecdysteroids in the prothoracic gland is regulated by the combined action of two groups of neuropeptides: ecdysteroid-inducing PTTH (prothoracicotropic hormone), released by brain neurons, and three ecdysteroid-inhibiting neuropeptides released by central and peripheral secretory neurons (Yamanaka et al., 2006, 2010).

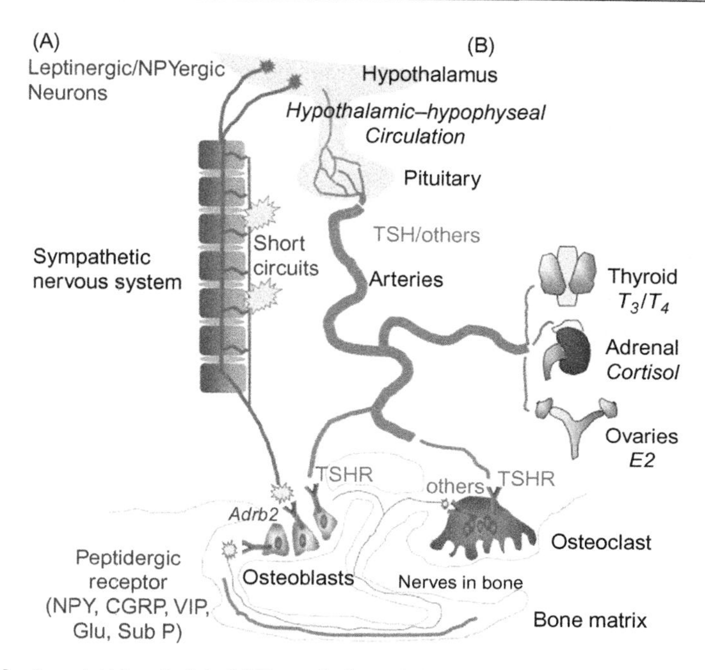

Figure 5.18 Local (A) and global (B) regulation of skeletal homeostasis by the hypothalamus. Short circuits refer to nonleptinergic, non-NPYergic release of adrenergic neurotransmitters, as well as epinephrine release from the adrenals. *Abbreviations*: NPY, neuropeptide Y; CGRP, calcitonin gene-related peptide; VIP, vasoactive intestinal peptide; Sub P, substance P; Glu, glutamate; TSH, thyroid-stimulating hormone; TSHR, TSH receptor; *Adrb2*, β2-adrenergic receptor; T_3 and T_4, thyroid hormones; E2, estradiol.
Source: (slightly simplified) Zaidi (2005).

Production of the JH by corpora allata in insects is stimulated by the brain hormones allatotropins and inhibited by the JH-inhibiting action of the local nerves nervi corporis allati I and II (NCA I and II) (Kou and Chen, 2000; Stay et al., 1996). The global-inhibiting action of brain neurohormones, allatostatins, is counteracted by the JH-stimulating action of nerves innervating the corpora allata releasing dopamine in these glands (Granger et al., 1996).

The global neural control of male reproductive activity in vertebrates via the hypothalamic–pituitary–testicular axis is complemented by another neural pathway from the spinal cord that plays a critical role in the function of Leydig cells in testicles (Lee et al., 2002). Besides a central hypothalamic pathway, regulation of the ovarian function is critically involved in the ovarian innervation (Morales-Ledesma et al., 2004), which only "targets a circumscribed subpopulation of ovarian cells" (Mayerhof et al., 1997).

Many other examples can be presented to demonstrate the existence of an animal's binary (central and adjacent) neural control of spatial restriction of gene expression. Indispensable as it is to develop and maintain animal structures, the binary control must have been ready to go from the dawn of the Cambrian biota.

The Biological Demon 2—Selective Recruitment of Genes

The existence of the genetic toolkit and its high conservancy across the animal taxa indicates that evolution employs basically the same genes for producing most different animal structures. Modern biology has demonstrated that the evolution of the animal phenotypes depends on gene expression patterns and in the incorporation of genes in new GRNs rather than on the evolution of genes themselves. This mechanism of gene recruitment liberated metazoans from the need to evolve new genes for new structures and functions. From this view, gene recruitment restrained the evolution of animal genes in the animal kingdom and can partially explain the discrepancies or absence of a clear correlation between gene number and animal phenotypic complexity.

Two questions arise in relation to the nature of gene recruitment. The first considers the mechanism of recruitment and the second relates to the mechanism of maintaining the gene "recruited" in the next generation.

Modern biology has no clear-cut answer for any of these questions. However, attempts to answer the first question on the mechanism of gene recruitment have been made. Among these attempts, the most intriguing is Sean Carroll and John True's hypothesis (Carroll, 2005; True and Carroll, 2002).

They believe that gene recruitment may be a result of gene mutations, gene duplication, and mutations in the regulatory sequences of genes, especially in the *Hox* genes that are important in the development of body plans. Not only are observations on a relationship between mutations in *Hox* genes and changes in their functions scarce but also, more importantly, they do not show whether the change in the function or recruitment for a new function occurred as a result of mutations or preceded the occurrence of mutations. Later on, Carroll admits that these events are too rare to account for the observed morphological diversification in the animal world (Carroll, 2005). Relating to the possible role of gene duplication in gene recruitment, he also acknowledges that:

> *the frequency of duplication events is not at all sufficient to account for the continuous diversification of lineages.*
>
> *Carroll (2005)*

Hence he believes that the primary sources of morphological evolution are changes in regulatory sequences. His argument is based on the example of the human *FOXP2* gene, which differs from that of the chimp in only two coding sequences and that mutations in this gene can lead to speech disorders in humans (Lai et al., 2001). But these facts in themselves do not hint, let alone prove, that changes in the regulatory sequences of the *FOXP2* gene may be related to the evolution of speech in humans. Even if such changes in regulatory sequences are found down the road, this will not prove that the change caused the evolution of speech in humans.

Human speech is one of the most complex neural and motoric behaviors that we are aware of.

While it is not impossible for the *FOXP2* gene (or other genes) to be involved in the evolution of speech, it would be a gross simplification to consider speech a genetic phenomenon related to the presence or absence of a gene or changes in a gene or its regulatory sequences. Human speech is a primarily epigenetic phenomenon involving complex and still little understood neural processing, cognitive, and motor functions; only the last element in the speech circuitry, the regulation of air flow through the larynx to produce the sound is an intricate process involving the actions and harmonizations of actions of numerous nerves, muscles, and ligaments of the larynx and neighboring regions and organs. This process cannot be considered a genetic phenomenon, by any stretch of imagination.

The mechanism of gene recruitment was another *sine qua non* for evolution of animals in the Cambrian period and even earlier. Understanding that mechanism is a critical requirement for the development of the theory of evolution. Again, the only available approach for understanding how animals that lived over half a billion years ago recruited their genes is to study the phenomenon in their extant descendants.

We know of animals where the gene recruitment may be subject to experimental induction. For example, when all the female species of *Daphnia magna* are reared under unfavorable stressful conditions (e.g., scarce food and crowding), they produce sexually reproducing (male and female) offspring. The developmental mechanism of this major phenotype is known: the processing of environmental data generates the information necessary for activating a new signal cascade starting in the brain (X-organ/sinus gland complex) that stimulates/inhibits production of the methyl farnesoate by the mandibular organ and inhibition/expression of the methyl farnesoate receptor, leading to the expression/suppression of genes for producing mictic (male and female) offspring (Figure 5.19).

In the above and a number of other cases, the recruitment of genes in new pathways, signal cascades, or GRNs starts with chemical signals released by the mother's brain as a result of processing of the various environmental stimuli. But if mechanisms of gene recruitment, which are indispensable for evolution of the animal world, operate in these extant lower invertebrates, why doubt that they were operative at the Cambrian (Cabej, 2011). We do not know how old these neural mechanisms of gene recruitment are, but there is no reason to believe that they evolved only in our time. If the processes of development really do reflect the evolutionary history of species, as the postulate of the phylogenetic actualism posits, the neural mechanisms of similar gene recruitment evolved since the dawn of animal evolution, at least at the beginning of the Cambrian explosion.

Leaving evo-devo knowledge and theoretical arguments aside, we delve into the evolutionary history of the recruitment of specific genes. Salamanders of the *Plethodontid* family have two main pheromone genes, sodefrin precursor-like factor (SDF) and plethodontid receptivity factor (PRF). Between 50 and 100 million years ago, they recruited SPF for regulating male sexual behavior and female receptivity. The red-legged salamander, *Plethodon shermani*, has both of those genes, but about 27 million years ago it recruited PRF to replace SPF as regulator of the male sexual

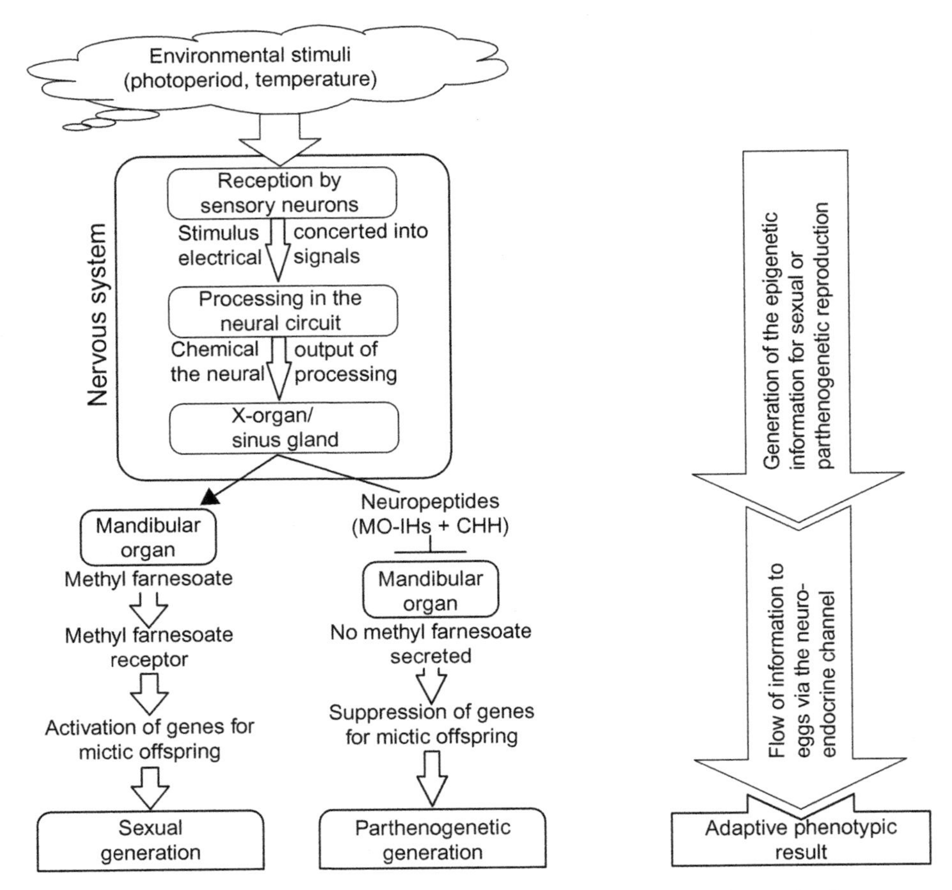

Figure 5.19 Illustration of the recruitment of genes in two different pathways leading to birth of asexual (parthenogenetic) and sexual generations in *D. magna*. The point of divergence between the two pathways starts in the crustacean's brain with the failure of the X-organ/sinus gland to synthesize and secrete neuropeptides MO-IH1, MO-IH2, and CHH (left cascade) and recruitment of the methyl farnesoate in the case of sexual generation-producing Daphnias and recruitment of the above genes and prevention of the synthesis of methyl farnesoate in the case of production of the asexual generation. Most insects have photoperiodic clocks and calendars that allow them to measure the length of the day and to count days. Based on this information, they decide whether to produce diapausing eggs or nondiapausing eggs, in order to rear their offspring under favorable climatic conditions. When these insects perceive the shortening of the day and the drop in environmental temperatures, they produce diapausing eggs, in which the embryo holds off the development until the spring. In a simplified form, the mechanism of diapausing-egg production is this: when the mother's photoperiodic clock and calendar has counted a specific number of short days, the mother stimulates a specific firing pattern of specific neurons, which inactivates in the GABAergic inhibitory mechanism brain, thus enabling the recruitment of the diapause hormone (DH) in the signal cascade that induces the formation of diapausing eggs.
Source: Cabej (2011).

behavior (Palmer et al., 2007). If the nervous system recruits the gene in development, why cannot it have recruited the same gene in *P. shermani* 27 million years ago?

Butterflies of a number of families determine their pupal body color based on the color of the pupation background for the purpose of avoiding predation during this sessile stage of life. The pupae of *Papilonidae* on dark background secrete the neurohormone pupal melanization-reducing factor (PMRF) that determines their dark body color. On light or green vegetation backgrounds, they do not produce the neurohormone and display green-yellowish color (Starnecker and Hazel, 1999). In both cases, the coloration is adaptive.

Pupae of the *Pieridae* and *Nymphalidae* families on a dark background do not release the neurohormone and display dark color. The reverse occurs on a light colored background, where pupae secrete PMRF but develop a light body color. The mechanism of PMRF secretion is a neural mechanism: the visual stimuli on the background are transmitted to the brain, where the color is perceived and the neurohormone is released by secretory neurons.

In all the representative cases presented above, the recruitment of the gene/gene product in the relevant pathways is under neural control and this neural mechanism may have been operative since these insects evolved this diphenism (Cabej, 2011).

References

Aboobaker, A., Blaxter, M., 2010. The nematode story: *Hox* gene loss and rapid evolution. Adv. Exp. Med. Biol. 689, 101–110.

Alie, A., Manuel, M., 2010. The backbone of the post-synaptic density originated in a unicellular ancestor of choanoflagellates and metazoans. BMC Evol. Biol. 10, 34.

Ament, S.A., Wang, Y., Robinson, G.E., 2010. Nutritional regulation of division of labor in honey bees: toward a systems biology perspective. Syst. Biol. Med. 2, 1–11.

Ayala, F.J., 1997. Vagaries of the molecular clock. Proc. Natl. Acad. Sci. U.S.A. 94, 7776–7783.

Ayala, F.J., Rzhetsky, A., Ayala, F.J., 1998. Origin of the metazoan phyla: molecular clocks confirm paleontological estimates. Proc. Natl. Acad. Sci. U.S.A. 95, 606–611.

Baldock, P.A., Sainsbury, A., Couzens, M., Enriquez, R.F., Thomas, G.P., Gardiner, E.M., et al., 2002. Hypothalamic Y2 receptors regulate bone formation. J. Clin. Invest. 109, 915–921.

Beisson, J., Sonneborn, T.M., 1965. Cytoplasmic inheritance of the organization of the cell cortex in *Paramecium aurelia*. Proc. Natl. Acad. Sci. U.S.A. 53, 275–282.

Bengtson, S., 2002. Origins and early evolution of predation. Paleont. Soc. Papers 8, 289–317.

Bengtson, S., Belivanova, V., Rasmussen, B., Whitehouse, M., 2009. The controversial "Cambrian" fossils of the Vindhyan are real but more than a billion years older. Proc. Natl. Acad. Sci. U.S.A. 106, 7729–7734.

Bery, A., Cardona, A., Martinez, P., Hartenstein, V., 2010. Structure of the central nervous system of a juvenile acoel, *Symsagittifera roscoffensis*. Dev. Genes Evol. 220, 61–76.

Bownds, D., 2001. Layers of brain. Available from: <http://dericbownds.net/bom99/Ch03/Ch03.html>

Bromham, L., 2009. Molecular dates for the Cambrian explosion: is the light at the end of tunnel an oncoming train? Palaeontologia Electron. 9 (2E), 3. Available from: <http://palaeo-electronica.org/2006_1/editor/focus.htm>

Bromham, L., Rambaut, A., Fortey, R., Cooper, A., Penny, D., 1998. Testing the Cambrian explosion hypothesis by using a molecular dating technique. Proc. Natl. Acad. Sci. U.S.A. 95, 12386–12389.
Budd, G.E., 2003. The Cambrian fossil record and the origin of the phyla. Integr. Comp. Biol. 43, 157–165.
Budd, G.E., Telford, M.J., 2009. The origin and evolution of arthropods. Nature 457, 812–817.
Butterfield, N.J., 1997. Plankton ecology and the Proterozoic–Phanerozoic transition. Paleobiology 23, 247–262.
Butterfield, N.J., 2001. Ecology and evolution of Cambrian plankton. In: Zhuravlev, A.Y., Riding, R. (Eds.), The Ecology of the Cambrian Radiation Columbia University Press, New York, NY, pp. 200–216.
Cabej, N.R., 2004. Neural Control of Development. Albanet, Dumont, NJ, pp. 121–177.
Cabej, N.R., 2008. Epigenetic Principles of Evolution. Albanet, Dumont, NJ, pp. 123–216.
Cabej, N.R., 2011. Neural control of gene recruitment in metazoans. Dev. Dyn. 240, 1–8.
Cabej, N.R., 2012. Epigenetic Principles of Evolution, first ed. Elsevier Inc., Amsterdam/Boston, pp. 127–254.
Carroll, S.B., 2005. Evolution at two levels: on genes and form. PLoS Biol. 3, 1159–1166.
Carroll, S.B., 2008. Evo–devo and an expanding evolutionary synthesis: a genetic theory of morphological evolution. Cell 134, 25–36.
Collins III, J.J., Hou, X., Romanova, E.V., Lambrus, B.G., Miller, C.M., Saberi, A., et al., 2010. Genome-wide analyses reveal a role for peptide hormones in planarian germline development. PLoS Biol. 8 (10), e1000509. doi: 10.1371/journal.pbio.1000509.
Consoulas, C., Levin, R.B., 1997. Accumulation and proliferation of adult leg muscle precursors in *Manduca* are dependent on innervation. J. Neurobiol. 32, 531–553.
Conway Morris, S., 1989. Burgess shale faunas and the Cambrian explosion. Science 246, 339–346.
Conway Morris, S., 1998. Early metazoan evolution: reconciling paleontology and molecular biology. Amer. Zool. 38, 867–877.
Conway Morris, S., 2000. The Cambrian "explosion": slow-fuse or megatonnage? Proc. Natl. Acad. Sci. U.S.A. 97, 4426–4429.
Conway Morris, S., 2003. The Cambrian "Explosion" of metazoans. In: Müller, G.B., Newman, S.A. (Eds.), Origination of Organismal Form: Beyond the Gene in Developmental and Evolutionary Biology A Bradford Book, MIT Press, Cambridge, MA/London, England, pp. 13–32.
Conway Morris, S., Bengtson, S., 1994. Cambrian predators: possible evidence from boreholes. J. Paleontol. 68, 1–23.
Conway Morris, S., Peel, J.S., 2008. The earliest annelids: Lower cambrian polychaetes from the Sirius Passet Lagerstätte, Peary Land, North Greenland. Acta Palaeontol. Pol. 53, 137–148.
Crews, D., Gillette, R., Scarpino, S.V., Manikkam, M., Savenkova, M.I., Skinner, M.K., 2012. Epigenetic transgenerational inheritance of altered stress responses. Proc. Natl. Acad. Sci. 109, 9143–9148.
Crimes, T.P., 2001. Evolution of the deep-water benthic community. In: Zhuravlev, A.Y., Riding, R. (Eds.), The Ecology of the Cambrian Radiation Columbia University Press, New York, NY, pp. 275–297.
Currie, D., Bate, M., 1991. The development of adult abdominal muscles in *Drosophila*: myoblasts express *twist* and are associated with nerves. Development 113, 91–102.
Darwin, C.R., 1859. On the Origin of Species by Means of Natural Selection, or the Preservation of Favoured Races in the Struggle for Life. John Murray, London, p. 450.

Darwin, C.R., 1860. Letter to Asa Gray, 1860 In: Darwin, F. (Ed.), The Life and Letters of Charles Darwin, vol. 2 D. Appleton and Company, New York, NY, pp. 131. 1896.
Darwin, C.R., 1872. The Origin of Species by Means of Natural Selection or the Preservation of Favored Races in the Struggle for Life, sixth ed. John Murray, London, p. 421, with additions and corrections.
Davidson, E., 2005. Gene regulatory networks. Proc. Natl. Acad. Sci. U.S.A. 102 (14), 4935.
Davidson, E.H., Erwin, D.H., 2006. Gene regulatory networks and the evolution of animal body plans. Science 311, 796–800.
Davidson, E.H., Erwin, D.H., 2007. Gene regulatory networks and the evolution of animal body plans. Science 311, 796–800.
Eldredge, N., 1971. The allopatric model and phylogeny in Paleozoic invertebrates. Evolution 25, 156–167.
Eldredge, N., Gould, S.J., 1972. Punctuated equilibria: an alternative to phyletic gradualism. In: Schopf, T.J.M. (Ed.), Models in Paleobiology Freeman, San Francisco, CA, pp. 82–115.
Elefteriou, F., 2008. Regulation of bone remodeling by the central and peripheral nervous system. Arch. Biochem. Biophys. 473, 231–236.
Ereskovsky, A.V., Dondua, A.K., 2006. The problem of germ layers in sponges (Porifera) and some issues concerning early metazoan evolution. Zool. Anz. 245, 65–76.
Erwin, D., Valentine, J., Sepkoski, J.J., 1987. A comparative study of diversification events: the early Paleozoic versus the Mesozoic. Evolution 41, 1177–1186.
Erwin, D.H., 1999. The origin of body plans. Amer. Zool. 139, 617–629.
Erwin, D.H., 2005. Development, ecology, and environment in the Cambrian metazoan radiation. Proc. Calif. Acad. Sci. 56 (Suppl I), 24–31.
Erwin, D.H., 2009. Early origin of the bilaterian developmental toolkit. Philos. Trans. R. Soc. Lond. B Biol. Sci. 364, 2253–2261.
Erwin, D.H., Davidson, E.H., 2009. The evolution of hierarchical gene regulatory networks. Nat. Rev. Genet. 10, 141–148.
Erwin, D.H., Valentine, J.W., Sepkoski, J.J.J., 1987. A comparative study of diversification events: the early Paleozoic versus the Mesozoic. Evolution 41, 1177–1186.
Erwin, D.H., Laflamme, M., Tweedt, S.M., Sperling, E.A., Pisani, D., Peterson, K.J., 2011. The Cambrian conundrum: early divergence and later ecological success in the early history of animals. Science 334, 1091–1097.
Fernandes, J., Keshishian, H., 2005. Motoneurons regulate myoblast proliferation during adult myogenesis in *Drosophila*. Dev. Biol. 277, 493–505.
Finnerty, J.R., Pang, K., Burton, P., Paulson, D., Martindale, M.Q., 2004. Origins of bilateral symmetry: Hox and dpp expression in a sea anemone. Science 304, 1335–1337.
Flores-Sarnat, L., Sarnat, H.B., 2008. Axes and gradients of the neural tube for a molecular/morphological genetic classification of nervous system malformations, 3rd series In: Sarnat, H.B. Curatolo, P. (Eds.), Handbook of Clinical Neurology, 87 Elsevier, pp. 5. (3–11).
Fortey, R.A., Briggs, D.E.G., Wills, M.A., 1996. The Cambrian evolutionary "explosion": decoupling cladogenesis from morphological disparity. Biol. J. Linn. Soc. 57, 13–33.
Galliot, B., Quiquand, M., 2011. A two-step process in the emergence of neurogenesis. Eur. J. Neurosci. 34, 847–862.
Gans, C., Northcutt, R.G., 1983. Neural crest and the origin of vertebrates: a new head. Science 220, 268–274.
Garstang, W., 1922. The theory of recapitulation: a critical restatement of the biogenetic law. Proc. Linn. Soc. Lond. Zool. 35, 81–101.

Géminard, C., Arquier, N., Layalle, S., Bourouis, M., Slaidina, M., Delanoue, R., et al., 2006. Control of metabolism and growth through insulin-like peptides in *Drosophila*. Diabetes 55 (Suppl. 2), S5–S8.

Gerhart, J., Kirschner, M., 2007. The theory of facilitated variation. Proc. Natl. Acad. Sci. U.S.A. 104 (Suppl. 1), 8582–8589.

Gould, S.J., Eldredge, N., 1977. Punctuated equilibria: the tempo and mode of evolution reconsidered. Paleobiology 3, 115–151.

Granger, N.A., Sturgis, S.L., Ebersohl, R., Geng, C., Sparks, C., 1996. Dopaminergic control of corpora allata activity in the larval tobacco hornworm, *Manduca sexta*. Arch. Insect. Biochem. Physiol. 32, 449–466.

Hammel, H.T., Jackson, D.T., Stolwijk, J.A.J., Hardy, J.D., Stroeme, S.D., 1963. Temperature regulation hypothalamic proportional control with an adjustable set point. J. Appl. Physiol. 18, 1146–1154.

Harcet, M., Roller, M., Ćetković, H., Perina, D., Wiens, M., Müller, W.E.G., et al., 2010. Demosponge EST sequencing reveals a complex genetic toolkit of the simplest metazoans. Mol. Biol. Evol. 27, 2747–2756.

Hegstrom, C.D., Truman, J.W., 1996. Synapse loss and axon retraction in response to local muscle degeneration. J. Neurobiol. 31, 175–188.

Hegstrom, C.D., Riddiford, L.M., Truman, J.W., 1998. Steroid and neuronal regulation of ecdysone receptor expression during metamorphosis of muscle in the moth, *Manduca sexta*. J. Neurosci. 18, 1786–1794.

Holland, H.D., 1997. Evidence for life on earth more than 3850 million years ago. Science 275, 38–39.

Hooge, M.D., Tyler, S., 2006. Concordance of molecular and morphological data: the example of the Acoela. Integr. Comp. Biol. 46, 118–124.

Jablonka, E., 2009. The nervous system in development and evolution. BioEssays 31, 687–689.

Jablonka, E., Lamb, M.J., 1995. Epigenetic Inheritance and Evolution: The Lamarckian Dimension. Oxford University Press, Oxford, pp. 79–110.

Jablonka, E., Lamb, M.J., 2008. Soft inheritance: challenging the modern synthesis. Genet. Mol. Biol. 31, 389–395.

Jablonka, E., Raz, G., 2009. Transgenerational epigenetic inheritance: prevalence, mechanisms and implications for the study of heredity and evolution. Q. Rev. Biol. 84, 131–176.

Jondelius, U., Ruiz-Trillo, I., Baguñà, J., Riutort, M., 2002. The Nemertodermatida are basal bilaterians and not members of the Platyhelminthes. Zool. Scr. 31, 201–215.

Jorgensen, E.V., Schwartz, I.D., Hvizdala, E., Barbosa, J., Phuphanich, S., Shulman, D.I., et al., 1993. Neurotransmitter control of growth hormone secretion in children after cranial radiation therapy. J. Pediatr. Endocrinol. 6, 131–142.

Kirschner, M.W., Gerhart, J.C., 2005. The Plausibility of Life. Yale University Press, New Haven/ London.

Knoll, A.H., Walter, M.R., Narbonne, G.M., Christie-Blick, N., 2004. A new period for the geologic time scale. Science 305, 621–622.

Kosik, K.S., 2009. Exploring the early origins of the synapse by comparative genomics. Biol. Lett. 5, 108–111.

Kou, R., Chen, S.J., 2000. Allatotropic and nervous control of corpora allata in the adult male loreyi leafworm, *Mythimna loreyi* (Lepidoptera: Noctuidae). Physiol. Entomol. 25, 273–278.

Krakauer, D.C., Page, K.M., Sealfon, S., 2002. Module dynamics of the GnRH signal transduction network. J. Theor. Biol. 218, 457–470.

Kumar, A., Brockes, J.P., 2012. Nerve dependence in tissue, organ, and appendage regeneration. Trends Neurosci. 35, 691–699.

Lai, C.S.L., Fisher, S.E., Hurst, J.A., Vargha-Khadem, F., Monaco, A.P., 2001. A forkhead-domain gene is mutated in a severe speech and language disorder. Nature 413, 519–523.
Launay, T., 2001. Expression and neural control of myogenic regulatory factor genes during regeneration of mouse soleus. J. Histochem. Cytochem. 49, 887–899.
Lawrence, P.A., Johnston, P., 1984. The genetic specification of pattem in a *Drosophila* muscle. Cell 36, 775–782.
Lawson, M.A., Tsutsumi, R., Zhang, H., Talukdar, I., Butler, B.K., Santos, S.J., et al., 2007. Pulse sensitivity of the luteinizing hormone β promoter is determined by a negative feedback loop involving early growth response-1 and Ngfi-A binding protein 1 and 2. Mol. Endocrinol. 21, 1175–1191.
Lee, S., Miselis, R., River, C., 2002. Anatomical and functional evidence for a neural hypothalamic-testicular pathway that is independent of the pituitary. Endocrinology 143, 4447–4454.
Lelièvre, S.A., 2009. Contributions of extracellular matrix signaling and tissue architecture to nuclear mechanisms and spatial organization of gene expression control. Biochim. Biophys. Acta 1790, 925–935.
Levinton, J.S., 2001. Genetics, Paleontology, and Macroevolution, second ed. Cambridge University Press, Cambridge/New York, p. 449.
Li, C.W., Chen, J.Y., Hua, T.E., 1998. Precambrian sponges with cellular structures. Science 279, 879–882.
Lieberman, B.S., 2003. Taking the pulse of the Cambrian radiation. Integr. Comp. Biol. 43, 229–237.
Ma, X., Hou, X., Edgecombe, G.D., Strausfeld, N.J., 2012. Complex brain and optic lobes in an early Cambrian arthropod. Nature 490, 258–261.
Maniotis, A.J., Valyi-Nagy, K., Karavitis, J., Moses, J., Boddipali, V., Wang, Y., et al., 2005. Chromatin organization measured by AluI restriction enzyme changes with malignancy and is regulated by the extracellular matrix and the cytoskeleton. Am. J. Pathol. 166, 1187–1203.
Maruyama, R., Choudhury, S., Kowalczyk, A., Bessarabova, M., Beresford-Smith, B., Conway, T., et al., 2011. Epigenetic regulation of cell type—specific expression patterns in the human mammary epithelium. PLoS Genet. 7 (4), e1001369. doi: 10.1371/journal.pgen.1001369.
Mayerhof, A., Dissen, G.A., Costa, M.F., Ojeda, S.R., 1997. A role for neurotransmitters in early follicular development: induction of functional follicle-stimulating hormone receptors in newly formed follicles of the rat ovary. Endocrinology 138, 3320–3329.
Miyata, T., Suga, H., 2001. Divergence pattern of animal gene families and relationship with the Cambrian explosion. BioEssays 23, 1018–1027.
Morales-Ledesma, L., Betanzos-Garcia, R., Dominguez-Casala, R., 2004. Unilateral vagotomy performed on prepubertal rats at puberty onset of female rat deregulates ovarian function. Arch. Med. Res. 35, 279–283.
Müller, G.B., 2007. Evo–devo: extending the evolutionary synthesis. Nat. Rev. Genet. 8, 943–950.
Müller, G.B., Newman, S.A., 2005. The innovation triad: an EvoDevo agenda. J. Exp. Zool. B Mol. Dev. Evol. 304, 487–503.
Narbonne, G.M., 2005. The Ediacara biota: neoproterozoic origin of animals and their ecosystems. Available from: <http://pqdweb?RQT=318&pmid=38590&TS=1249654597&clientId=10763&VInst=PROD&VName=PQD&VType=PQD>
Newman, S., 2006. The developmental genetic toolkit and the molecular homology–analogy paradox. Biol. Theory 1, 12–16.
Newman, S.A., 2011. Animal egg as evolutionary innovation: a solution to the "embryonic hourglass" puzzle. J. Exp. Zool. B Mol. Dev. Evol. 316, 467–483.

Newman, S.A., Müller, G.B., 2000. Epigenetic mechanisms of character origination. J. Exp. Zool. B Mol. Dev. Evol. 288, 304–317.

Nichols, S.A., Dirks, W., Pearse, J.S., King, N., 2006. Early evolution of animal cell signaling and adhesion genes. Proc. Natl. Acad. Sci. U.S.A. 103, 12451–12456.

Nickel, M., 2010. Evolutionary emergence of synaptic nervous systems: what can we learn from the non-synaptic, nerverless Porifera? Invertebr. Biol. 129, 1–16.

Noble, D., 2011. Differential and integral views of genetics in computational systems biology. Interface Focus 1 (1), 7–15.

Northcutt, R.G., 2012. Evolution of centralized nervous systems: two schools of evolutionary thought. Proc. Natl. Acad. Sci. U.S.A. 109 (Suppl. 1), 10626–10633.

Palmer, C.A., Watts, R.A., Houck, L.D., Picard, A.L., Arnold, S.J., 2007. Evolutionary replacement of components in a salamander pheromone signaling complex: more evidence for phenotypic molecular decoupling. Evolution 61, 202–215.

Peterson, K.J., Butterfield, N.J., 2005. Origin of the Eumetazoa: testing ecological predictions of molecular clocks against the Proterozoic fossil record. Proc. Natl. Acad. Sci. U.S.A. 102, 9547–9552.

Peterson, K.J., Waggoner, B., Hagadorn, J.W., 2003. A fungal analog for newfoundland Ediacaran fossils? Integr. Comp. Biol. 43, 127–136.

Peterson, K.J., Summons, R.E., Donoghue, P.C.J., 2007. Molecular palaeobiology. Palaeontology 50, 775–809.

Peterson, K.J., Cotton, J.A., Gehling, J.G., Pisani, D., 2008. The Ediacaran emergence of bilaterians: congruence between the genetic and geologic fossil records. Philos. Trans. R. Soc. Lond. B Biol. Sci. 363, 1435–1443.

Pigliucci, M., Murren, C.J., 2003. Perspective: Genetic assimilation and a possible evolutionary paradox: can macroevolution sometimes be so fast as to pass us by? Evolution 57, 1455–1464.

Pigliucci, M., Murren, C.J., Schlichting, C.D., 2006. Phenotypic plasticity and evolution by genetic assimilation. J. Exp. Biol. 209, 2362–2367.

Raikova, O.I., Reuter, M., Justine, J.-L., 2001. Contribution to the phylogeny and systematics of the *Acoelomorpha*. In: Littlewood, D.T.J., Bray, R.A. (Eds.), *Interrelationships of the Platyhelminthes*, Taylor & Francis, NewYork, pp. 13–22.

Reid, I.R., Lucas, J., Wattie,, D., Horne, A., Bolland, M., Gamble, G.D., et al., 2005. Effects of a β-blocker on bone turnover in normal postmenopausal women: a randomized controlled trial. J. Clin. Endocrinol. Metab. 90, 5212–5216.

Reik, W., 2007. Stability and flexibility of epigenetic gene regulation in mammalian development. Nature 447, 425–432.

Roth, C., Rastogi, S., Arvestad, L., Dittmar, K., Light, S., Ekman, D., et al., 2007. Evolution after gene duplication: models, mechanisms, sequences, systems, and organisms. J. Exp. Zool. B Mol. Dev. Evol. 308, 58–73.

Ruaud, A-F., Thummel, C.S., 2008. Serotonin and insulin signaling team up to control growth in *Drosophila*. Genes Dev. 22, 1851–1855.

Runnegar, B., 1986. Molecular palaeontology. Palaeontology 29, 1–24.

Ryan, J.F., Burton, P.M., Mazza, M.E., Kwong, G.K., Mullikin, J.C., Finnerty, J.R., 2006. The cnidarian–bilaterian ancestor possessed at least 56 homeoboxes: evidence from the starlet sea anemone, *Nematostella vectensis*. Genome Biol. 7, R64.

Ryan, T.J., Grant, S.G.N., 2009. The origin and evolution of synapses. Nat. Rev. Neurosci. 10, 701–712.

Salzet, M., 2007. Molecular aspect of annelid neuroendocrine system. In: Satake, H. (Ed.), Invertebrate Neuropeptides and Hormones: Basic Knowledge and Recent Advances Transworld Research Network, Trivandrum, India, pp. 17–36.

Schopf, J.W., 1994. Disparate rates, differing fates: tempo and mode of evolution changed from the Precambrian to the phanerozoic. Proc. Natl. Acad. Sci. U.S.A. 91, 6735–6742.

Schierwater, B., Kamm, K., 2010. The early evolution of *Hox* genes: a battle of belief? Adv. Exp. Med. Biol. 689 (81), 90.

Seilacher, A., 1984. Late Precambrian and early Cambrian metazoa: preservational or real extinctions?. In: Holland, H.D., Trendall, A.F. (Eds.), Patterns of Change in Earth Evolution Springer Verlag, Berlin, Heidelberg, New York, Tokyo, pp. 159–168.

Semmler, H., Chiodin, M., Bailly, X., Martinez, P., 2010. Steps towards a centralized nervous system in basal bilaterians: insights from neurogenesis of the acoel *Symsagittifera roscoffensis*. Dev. Growth Differ. 52, 701–713.

Singer, M., Géraudie, J., 1991. The neurotrophic phenomenon: its history during limb regeneration in the newt. In: Donsmore, C.E. (Ed.), A History of Regeneration Research Cambridge University Press, Cambridge/New York, pp. 101–112.

Srivastava, M., Simakov, O., Chapman, J., Fahey, B., Gauthier, M.E.A., Mitros, T., et al., 2010. The *Amphimedon queenslandica* genome and the evolution of animal complexity. Nature 466, 720–726.

Stafford, R., 2010. Constraints of biological neural networks and their consideration in AI applications. Adv Artif. Intell. 2010 Article ID 845723, 6 pp. Available from: <http://www.hindawi.com/journals/aai/2010/845723/>

Stanley, S.M., 1992. Can neurons explain the Cambrian explosion? Geol. Soc. Am. Abstr. Programs 24, A45.

Starnecker, G., Hazel, W.N., 1999. Convergent evolution of neuroendocrine control of phenotypic plasticity in pupal colour in butterflies. Proc. R. Soc. Series B 266, 2409–2412.

Stay, B., Fairbairn, S., Yu, C.G., 1996. Role of allatostatins in the regulation of juvenile hormone synthesis. Arch. Insect Biochem. Physiol. 32, 287–297.

Stout, T., McFarland, T., Appukuttan, B., 2007. Suppression subtractive hybridization identifies novel transcripts in regenerating *Hydra littoralis*. J. Biochem. Mol. Biol. 40, 286–289.

Suga, H., Koyanagi, M., Hoshiyama, D., Ono, K., Iwabe, N., Kuma, K., et al., 1999. Extensive gene duplication in the early evolution of animals before the parazoan–eumetazoan split demonstrated by G proteins and protein tyrosine kinases from sponge and hydra. J. Mol. Evol. 48, 646–653.

Tolstonog, G.V., Sabasch, M., Traub, P., 2002. Cytoplasmic intermediate filaments are stably associated with nuclear matrices and potentially modulate their DNA-binding function. DNA Cell Biol. 21, 213–239.

True, J.R., Carroll, S.B., 2002. Gene co-option in physiological and morphological evolution. Annu. Rev. Cell Dev. Biol. 18, 53–80.

Tzschentke, B., Nichelmann, M., 1997. Influence of parental and postnatal acclimation on nervous and peripheral thermoregulation. Ann. NY Acad. Sci. 813, 87–94. (Abstract).

Valentine, J.W., Jablonski, D., Erwin, D.H., 1999. Fossils, molecules and embryos: new perspectives on the Cambrian explosion. Development 126, 851–859.

Veenstra, J.A., 2011. Neuropeptide evolution: neurohormones and neuropeptides predicted from the genomes of *Capitella teleta* and *Helobdella robusta*. Gen. Comp. Endocrinol. 171, 160–175.

Walker, J.J., Terry, J.R., Tsaneva-Atanasova, K., Armstrong, S.P., McArdle, C.A., Lightman, S.L., 2010. Encoding and decoding mechanisms of pulsatile hormone secretion. J. Neuroendocrinol. 22, 1226–1238.

Wang, D.Y.-C., Kumar, S., Hedges, S.B., 1999. Divergence time estimates for the early history of animal phyla and the origin of plants, animals and fungi. Proc. R. Soc. London B 266, 163–171.

Watanabe, H., Hoang, V.T., Mättner, R., Holstein, T.W., 2009. Immortality and the base of multicellular life: lessons from cnidarian stem cells. Semin. Cell Dev. Biol. 20, 1114–1125.

West-Eberhard, M.J., 1989. Phenotypic plasticity and the origins of diversity. Annu. Rev. Ecol. Syst. 20, 249–278.

West-Eberhard, M.J., 2003. Developmental Plasticity and Evolution. Oxford University Press, New York, NY, pp. 377–417.

West-Eberhard, M.J., 2005. Developmental plasticity and the origin of species differences. Proc. Natl. Acad. Sci. U.S.A. 102, 6543–6549.

Wray, G.A., Levinton, J.S., Shapiro, L.H., 1996. Molecular evidence for deep Precambrian divergences among metazoan phyla. Science 274, 568–573.

Yamanaka, N., Žitnan, D., Kim, Y-J., Adams, M.E., Hua, Y-J., Suzuki, Y., et al., 2006. Regulation of insect steroid hormone biosynthesis by innervating peptidergic neurons. Proc. Natl. Acad. Sci. U.S.A. 103, 8622–8627.

Yamanaka, N., Hua, Y.J., Roller, L., Spalovská-Valachová, I., Mizoguchi, A., Kataoka, H., et al., 2010. Bombyx prothoracicostatic peptides activate the sex peptide receptor to regulate ecdysteroid biosynthesis. Proc. Natl. Acad. Sci. U.S.A. 107, 2060–2065.

Zaidi, M., 2005. Neural surveillance of skeletal homeostasis. Cell Metab. 1, 219–221.

Zhu, M., Yu, X., Lu, J., Qiao, T., Zhao, W., Jia, L., 2012. The earliest known coelacanth skull extends the range of anatomically modern coelacanths to the Early Devonian. Nat. Commun. 3, 772. doi: 10.1038/ncomms1764.

Printed in the United States
By Bookmasters